PROBABILISTIC METRIC SPACES

B. SCHWEIZER
Professor Emeritus of Mathematics
University of Massachusetts

A. SKLAR
Professor Emeritus of Mathematics
Illinois Institute of Technology

DOVER PUBLICATIONS, INC.
Mineola, New York

Bibliographical Note

This Dover edition, first published in 2005, is an unabridged republication of the work first published by Elsevier Science Publishing Co., Inc., in 1983. A new Preface to the Dover edition, list of Errata, Notes, and Supplementary References have been added.

International Standard Book Number: 0-486-44514-3

Manufactured in the United States by Courier Corporation
44514303
www.doverpublications.com

Contents

Preface to the Dover Edition

The republication of our book has given us the opportunity to provide, not only the customary list of errata (fortunately, nearly all minor), but also an extensive set of notes and a supplementary bibliography which together follow the unaltered original text and serve to update it.

The scope of the notes (particularly, those referring to Chapters 5, 6, 7, 11, and 15) and the length of the supplementary bibliography (which is by no means complete) testify to the remarkable growth of interest in and applications of several parts of the subject. Many persons have contributed to this growth: Besides the persons named in the original preface and whose work is still continuing, we thank the following ones, with whom we have personally interacted: J. Aczél, F. Balibrea, G. Dall'Aglio, W. F. Darsow, E. Diday, G.-L. Forti, C. Genest, Tie-Xin Guo, U. Höhle, M. F. Janowitz, E. P. Klement, G. M. Krause, B. Lafuerza-Guillén, R. B. Nelsen, L. Paganoni, J. J. Quesada-Molina, T. Riedel, J. A. Rodriguez-Lallena, C. Sempi, J. Smítal, M. D. Taylor, and E. Trillas.

The original edition of our book has been long out of print, yet has been the subject of increasing demand. We therefore owe special thanks to Dover Publications, and to its editor John Grafton in particular, for making it again available in an expanded edition. And for her professional, insightful, and cheerful assistance in preparing this expanded edition, we thank Kathy Richards.

Sadly, since 1983, we have lost several friends and colleagues who had participated in this enterprise. So we pay tribute here to the memory of our mentor Karl Menger, and also to the memory of Pietro Benvenuti, Alan T. Bharucha-Reid, Violet Fiatte, Bruno Forte, Joseph Kampé de Fériet, Marek Kuczma, and György Targonski.

Preface

The story of this book begins in 1951 when the first author was reading the book *Albert Einstein, Philosopher-Scientist*. There, in a paper by Karl Menger entitled "Modern Geometry and the Theory of Relativity," he came across the following passage:

> Poincaré, in several of his famous essays on the philosophy of science, characterized the difference between mathematics and physics as follows: In mathematics, if the quantity A is equal to the quantity B, and B is equal to C, then A is equal to C; that is, in modern terminology: mathematical equality is a transitive relation. But in the observable physical continuum "equal" means indistinguishable; and in this continuum, if A is equal to B, and B is equal to C, it by no means follows that A is equal to C. In the terminology of the psychologists Weber and Fechner, A may lie within the threshold of B, and B within the threshold of C, even though A does not lie within the threshold of C. "The raw result of experience," says Poincaré, "may be expressed by the relation
>
> $$A = B, \qquad B = C, \qquad A < C,$$
>
> which may be regarded as the formula for the physical continuum." That is to say, physical equality is not a transitive relation.
>
> Is this reasoning cogent? It is indeed easy to devise experiments which prove that the question whether two physical quantities are distinguishable cannot always be answered by a simple Yes or No. The same observer may regard the same two objects sometimes as identical and sometimes as distinguishable. A blindfolded man may consider the simultaneous irritation of the same two spots on his skin sometimes as one, and sometimes as two tactile sensations. Of two constant lights, he may regard the first sometimes as weaker than,

> sometimes as equal to, and sometimes as stronger than the second. All that can be done in this situation is to count the percentage number of instances in which he makes any one of these two or three observations. In the observation of physical continua, situations like the one just described seem to be the rule rather than the exception.
>
> Instead of distinguishing between a transitive mathematical and an intransitive physical relation of equality, it thus seems much more hopeful to retain the transitive relation in mathematics and to introduce for the distinction of physical and physiological quantities a probability, that is, a number lying between 0 and 1.
>
> Elaboration of this idea leads to the concept of a space in which a distribution function rather than a definite number is associated with every pair of elements. The number associated with two points of a metric space is called the distance between the two points. The distribution function associated with two elements of a statistical metric space might be said to give, for every x, the probability that the distance between the two points in question does not exceed x. Such a statistical generalization of metric spaces appears to be well adapted for the investigation of physical quantities and physiological thresholds. The idealization of the local behavior of rods and boards, implied by this statistical approach, differs radically from that of Euclid. In spite of this fact, or perhaps just because of it, the statistical approach may provide a useful means for geometrizing the physics of the microcosm.

Reading this passage crystallized this author's own thoughts on this topic and, in short order, led to his coming to the Illinois Institute of Technology to pursue his graduate studies under Menger's direction. The second author joined the faculty of I.I.T. in 1956, and soon thereafter our collaboration on probabilistic metric spaces began. In the intervening years the subject has grown into a coherent whole which is the theme of this book.

In our development we follow Menger and work directly with probability distribution functions, rather than with random variables. This means that we take seriously the fact that the outcome of any series of measurements of the values of a nondeterministic quantity is a distribution function; and that a probability space, although an exceedingly useful mathematical construct, is in fact and in principle unobservable. This point of view gives our subject a distinct nonclassical flavor. It has also forced us to develop a considerable body of new mathematical machinery. Our book thus divides naturally into two major parts. After the introductory Chapter 1, in which we give an overview and discuss the historical development of the theory, we devote Chapters 2–7 to the development of this machinery. Our presentation here is complete in the sense that all the needed concepts are defined, all the needed results are stated, and virtually all the proofs not available in the literature are given.

The study of probabilistic metric spaces proper begins with Chapter 8, where we give the basic definitions and establish some simple properties. Chapters 9, 10, and 11 are devoted to special classes of probabilistic metric spaces; Chapters 12 and 13 to topologies and generalized topologies; Chapter 14 to betweenness; and the final Chapter 15 to several related structures, such as probabilistic normed and inner-product spaces. Throughout, we are primarily interested in developing those aspects of the theory that differ from the theory of ordinary metric spaces, and not simply in transferring known metric space results to a more general setting.

The book contains no exercises per se (the reader will find work enough in supplying missing proofs!). However, the subject is comparatively new and thus abounds with open problems and unsettled conjectures. The more specific ones among these are listed in the concluding sections of the appropriate chapters.

Readers of this book are advised to proceed as follows: Read Chapter 1. Skim Chapters 2–7 to become acquainted with their contents. Start reading in earnest again with Chapter 8. The necessary portions of the mathematical machinery presented in Chapters 2–7 can be assimilated as and when they are applied in the subsequent text. This is not to say that the material in these chapters is without intrinsic interest—quite the contrary: the mathematical problems here have often seduced us away from the main topic. In its division, our book may be compared to the standard texts in probability theory which find it necessary to devote a considerable amount of space to measure theory before coming to grips with their central theme.

We owe thanks to many: To our teacher K. Menger, who stimulated us and encouraged us in our efforts; to T. Erber, for infecting us with his enthusiasm and for collaborating with us on several applications of our theory to physics; to A. T. Bharucha-Reid, for asking us to write this book and for his cooperation and patience while the work was in progress; to C. Alsina, M. J. Frank, R. M. Moynihan, E. T. Olsen, M. Senechal, A. N. Šerstnev, H. Sherwood, and R. M. Tardiff, for their critical appraisal and constructive criticism of portions of the manuscript; to all our other students and colleagues whose work helped to advance the subject; to our families and friends who gave us their moral support; and to V. Fiatte for years of invaluable superb technical assistance.

Special Symbols

Symbol		Page
$\circ$	composition of functions	19
$*$	convolution	4, 106
$B(p,x)$	open ball with center p and radius x	12
C	copula	15, 83
$\overline{C}$	dual copula	88
$c(A)$	closure of A	39
$c_T F$	T-conjugate transform of F	117
$c_T^* f$	inverse T-conjugate transform of f	119
d_L	modified Lévy metric	45
$d(p,q)$	distance between p and q	2, 30
$df(X)$	distribution function of the r.v. X	44
D_A	probabilistic diameter of A	200
Dom	domain	18
$\mathcal{D}$	set of distribution functions	43
$\mathcal{D}^+$	set of distribution functions F in $\mathcal{D}$ with $F(0) = 0$	48
Δ	set of distribution functions	43
Δ^+	distribution functions F with $F(0) = 0$	48
Δ_T^+	set of T-log-concave distribution functions in Δ^+	119

∇^+	dual space to Δ^+	116
$\mathcal{E}(I)$	PM space of measurable functions on I	142
ε_a	unit step at a (a may be finite or infinite)	7, 44
f^{-1}	inverse of f	20
f^n	nth iterate of f	19
$f^\wedge$	left-continuous quasi-inverse of f	50
$f^\vee$	right-continuous quasi-inverse of f	51
$f(c-)$	left-hand limit of f at c	22
$f(c+)$	right-hand limit of f at c	22
$f(A)$	direct image under f of the set A	21
$f^{-1}(B)$	inverse image under f of the set B	21
F_{pq}	probabilistic distance between p and q	2, 124
F_{pA}	probabilistic distance from p to A	193
F_{AB}	probabilistic Hausdorff distance between A and B	214
$(F, G; h)$	condition relating F and G in Δ	45
$[F, G; h]$	condition relating F and G in Δ^+	48
$\mathcal{F}$	probabilistic metric	124
$g[Q]f$	g is a quasi-inverse of f	19
I	closed unit interval $[0, 1]$	18
(I, λ)	I with Lebesgue measure	26
j_S	identity function on S (may lack subscript)	18
$k(A)$	strong closure of A	193
K_α	$K_\alpha(x, y) = (x^\alpha + y^\alpha)^{1/\alpha}$ for x, y in R^+	76
l^-f	left-hand limit of f	22
l^+f	right-hand limit of f	22
L	function in $\mathfrak{L}$	98, 127
$L_1^+(\Omega)$	set of nonnegative random variables on Ω	26
$\mathfrak{L}$	a set of binary operations on R^+	98
$m^{(k)}(F)$	kth moment of F	52
M	$M(x, y) = \min(x, y)$ for x, y in I	5, 20
M^*	$M^*(x, y) = \max(x, y)$ for x, y in I	5, 74
$\mathbf{M}$	maximal triangle function	97
$N_p(t)$	strong t-neighborhood of p	191

$N_p(\varphi, h)$	(φ, h)-neighborhood of p	219
$\mathcal{O}[a,b]$	a set of associative binary operations on $[a,b]$	59
$(\Omega, \mathcal{A}, P)$	probability space	25
$\mathcal{P}(S)$	power set of S	19
Π	$\Pi(x, y) = xy$ for x, y in I	5, 67
Π^*	$\Pi^*(x, y) = x + y - xy$ for x, y in I	5, 74
φ	profile function	14, 219
$\emptyset$	empty set	25
$Q(f)$	quasi-inverse set of f	49
R	extended real line $[-\infty, \infty]$	18
R^+	extended positive half-line $[0, \infty]$	18
Ran	range	18
$\rho_C, \rho_{C,L}$	binary operation on Δ^+	109
$\rho_{C,L}^{\wedge}$	operation dual to $\rho_{C,L}$	116
S^n	nth Cartesian power of S	19
(S, d)	metric space	2, 30
$(S, \mathcal{F})$	PM space; pre-PM space	124
$(S, \mathcal{X})$	RM space; pre-RM space	149
(S, d, G)	simple space	132
$(S, \mathcal{F}, \tau)$	PM space	124
$(S, d, G; \alpha)$	α-simple space	138
$\sigma_C, \sigma_{C,L}$	binary operation on Δ^+	106
T	binary operation, particularly a t-norm	27, 65, 98, 127
T^n	nth serial iterate of T	87
T^*	binary operation, particularly a t-conorm	73
$\mathbf{T}$	binary operation on Δ^+	97
$\mathcal{T}, \mathcal{T}^*, \mathcal{T}_C$	sets of binary operations on I	97
τ	general triangle function	9, 96
τ_T	triangle function	10, 99
$\tau_{T,L}$	triangle function	99
$\tau_{T^*}, \tau_{T^*,L}$	triangle function	104
$\tau_{T,L}^{\wedge}$	operation dual to $\tau_{T,L}$	116
U_b	uniform distribution on $[0, b]$	44

U_{ab}	uniform distribution on $[a, b]$	44
$V_H(B)$	H-volume of B	79
W	$W(x, y) = \max(x + y - 1, 0)$ for x, y in I	5, 71
W^*	$W^*(x, y) = \min(x + y, 1)$ for x, y in I	5, 74
X_{pq}	random distance between p and q	149
$\mathcal{X}$	function from $S \times S$ to $L_1^+(\Omega)$	149
χ_A	indicator function of A	179
Z	minimal t-norm	71
Z^*	maximal t-conorm	74

PROBABILISTIC METRIC SPACES

1
Introduction and Historical Survey

1.0. Introduction

The 19th century, which marks the beginning of the modern age of science, was an era of great advances in the art of measurement. These advances stimulated a corresponding concern with the accompanying errors. Until the early part of this century, however, it was still believed that through careful design and ample data the error in any measurement could be made arbitrarily small. The advent of quantum mechanics shattered this belief, for here the uncertainties of the measurements are inherent in the measurement process itself and in principle cannot be removed.

Today, in the fourth quarter of the 20th century, the existence of such inherent uncertainties and thresholds is a generally accepted fact. This is true not only in physics but also in areas such as psychometrics [S. Stevens 1959], communication theory [Shannon 1948; Brillouin 1956], and pattern recognition [Duda and Hart 1973; Yakimovsky 1976; Prager 1979]. It is also central to various mathematical disciplines, such as cluster analysis [Janowitz 1978; Shepard 1980] and interval analysis [Moore 1979]. However, in virtually all the mathematical models built to describe these various situations it is assumed that the measurements in question are made with respect to a rigid reference frame. Remarks to the effect that this assumption may be unsatisfactory and that some of the uncertainties should be built into the geometry are scattered here and there in the literature [Poincaré 1905, 1913; Hjelmslev 1923; de Broglie 1935; Black 1937; Weyl 1952; Born 1955; Oppenheimer 1962], along with suggestions on the proper way of doing this [Penrose and MacCullum 1973; Penrose 1975]. There are also some serious attempts in this direction [Eddington 1953; Rosen 1947, 1962; Blokhintsev 1971, 1973; Frenkel 1977]. This book is a direct outgrowth of one of these attempts, namely, the theory of probabilistic metric spaces as initiated by K. Menger in 1942.

1.1. Beginnings

The first abstract formulation of the notion of distance is due to M. Fréchet [1906]. This notion, which was later given the name "metric space" ("metrischer Raum") by F. Hausdorff [1914], is based on the introduction of a function d that assigns a nonnegative real number $d(p,q)$ (the distance between p and q) to every pair (p,q) of elements (points) of a nonempty set S. This function is assumed to satisfy the following conditions:

$$d(p,q) = 0 \quad \text{if } p \text{ and } q \text{ coincide}; \tag{1.1.1}$$

$$d(p,q) > 0 \quad \text{if } p \text{ and } q \text{ are distinct}; \tag{1.1.2}$$

$$d(p,q) = d(q,p) \quad \text{for all } p,q \text{ in } S; \tag{1.1.3}$$

$$d(p,r) \leqslant d(p,q) + d(q,r) \quad \text{for all } p,q,r \text{ in } S. \tag{1.1.4}$$

Condition (1.1.4), whose antecedents go back at least to Euclid's Proposition I.20, is the *triangle inequality*.

Any function d satisfying (1.1.1)–(1.1.4) is a *metric* on S. (Occasionally it is convenient to drop (1.1.2), in which case d is a *pseudometric* on S.) A *metric space* is a pair (S,d) where S is a set and d is a metric on S.

In 1942 K. Menger, who had played a major role in the development of the theory of metric spaces (see [Menger 1928, 1930, 1932, 1954]), proposed a probabilistic generalization of this theory. Specifically, he proposed replacing the number $d(p,q)$ by a real function F_{pq} whose value $F_{pq}(x)$, for any real number x, is interpreted as the probability that the distance between p and q is less than x. Since probabilities can neither be negative nor be greater than 1, we have

$$0 \leqslant F_{pq}(x) \leqslant 1 \tag{1.1.5}$$

for every real x; and clearly we also have

$$F_{pq}(x) \leqslant F_{pq}(y) \tag{1.1.6}$$

whenever $x < y$. Hence F_{pq} is a *probability distribution function*.

Other conditions on the functions F_{pq} are also immediate. Thus, since distances cannot be negative, we have

$$F_{pq}(0) = 0. \tag{1.1.7}$$

Similarly, (1.1.1), (1.1.2), and (1.1.3) yield the following:

$$\text{If } p = q, \text{ then } F_{pq}(x) = 1 \quad \text{for all } x > 0. \tag{1.1.8}$$

$$\text{If } p \neq q, \text{ then } F_{pq}(x) < 1 \quad \text{for some } x > 0. \tag{1.1.9}$$

$$F_{pq} = F_{qp}. \tag{1.1.10}$$

In contradistinction, the probabilistic generalization of the triangle inequality (1.1.4) is quite another matter. Even at the outset Menger and A. Wald proposed different generalizations, and the study of alternative triangle inequalities has been a central theme in the development of the theory of probabilistic metric spaces. The major steps in this development are traced in the next few sections.

1.2. Menger, 1942

In his original paper, Menger [1942] defined a *statistical metric space*[1] as a set S together with an associated family of probability distribution functions F_{pq} satisfying (1.1.7)–(1.1.10) and the inequality

$$F_{pr}(x+y) \geqslant T(F_{pq}(x), F_{qr}(y)) \tag{1.2.1}$$

for all p, q, r in S and all real numbers x, y. Here T is a function from the closed unit square $[0,1] \times [0,1]$ into the closed unit interval $[0,1]$ satisfying

$$T(a,b) = T(b,a), \tag{1.2.2}$$

$$T(a,b) \leqslant T(c,d) \quad \text{whenever } a \leqslant c,\ b \leqslant d, \tag{1.2.3}$$

$$T(a,1) > 0 \quad \text{whenever } a > 0, \text{ and } T(1,1) = 1. \tag{1.2.4}$$

Given (1.2.2)–(1.2.4), the inequality (1.2.1) implies that our knowledge of the third side of a triangle depends in a symmetric manner on our knowledge of the other two sides; that it increases, or at least does not decrease, as our knowledge of these other sides increases; that if we have an upper bound for the length of one side and know something about the second, then we know something about the third; and that if we have upper bounds for the lengths of two sides, then we have an upper bound for the length of the third. In particular it follows that if there is a function d from $S \times S$ into the nonnegative reals such that

$$F_{pq}(x) = \begin{cases} 0, & x \leqslant d(p,q), \\ 1, & x > d(p,q), \end{cases} \tag{1.2.5}$$

[1] The change in adjective from "statistical" to "probabilistic" was made in 1964 in order to conform to the by then customary usage of these terms.

for all p, q in S and all real x, then (S, d) is a metric space. For let p, q, r be points of S and suppose that $d(p,q) < x$ and $d(q,r) < y$ for some $x, y > 0$. Then (1.2.5) yields $F_{pq}(x) = 1$ and $F_{qr}(y) = 1$, whence it follows from (1.2.4) and (1.2.1) that $F_{pr}(x+y) = 1$. By virtue of (1.2.5) this implies that $d(p,r) < x + y$, which in turn implies (1.1.4). The other conditions, (1.1.1)–(1.1.3), are trivial. Conversely, if a metric space is given and the functions F_{pq} are defined by (1.2.5), then it is immediate that (1.1.5)–(1.1.10) and (1.2.1) hold for any T satisfying (1.2.2)–(1.2.4). Thus ordinary metric spaces may be viewed as special cases of Menger's statistical metric spaces.

In addition to the basic definitions and the inequality (1.2.1), Menger also introduced a notion of betweenness that has some, but generally not all, of the properties of ordinary metric betweenness (see Sections 3.3 and 14.3).

1.3. Wald, 1943

Menger's paper was followed almost immediately by a paper of Wald [1943]. In his paper Wald stated that the inequality (1.2.1)

> has the drawback that it involves an unspecified function $T(a,b)$ and one can hardly find sufficient justification for a particular choice of this function. Furthermore the notion of "between" introduced by Menger on the basis of inequality (1.2.1) has the properties of the between relationship in metric spaces only under restrictive conditions on the distribution functions . . . .

Wald suggested replacing (1.2.1) by the inequality

$$F_{pr}(x) \geqslant (F_{pq} * F_{qr})(x), \tag{1.3.1}$$

where "*" denotes convolution, so that

$$(F_{pq} * F_{qr})(x) = \int_{-\infty}^{\infty} F_{pq}(x-y)\, dF_{qr}(y) = \int_{0}^{x} F_{pq}(x-y)\, dF_{qr}(y). \tag{1.3.2}$$

On the basis of (1.3.1) Wald introduced a notion of betweenness that has all the properties of ordinary betweenness (see Sections 3.3 and 14.1).

Wald's inequality has the following natural interpretation: The probability that the distance from p to r is less than x is at least as large as the probability that the sum of the distances from p to q and from q to r, regarded as independent, is less than x.

In his subsequent writings on the subject, Menger [1949, 1951abc, 1954] adopted Wald's inequality. This led to serious difficulties, for while it was relatively easy to derive properties of spaces satisfying (1.1.6)–(1.1.8) and (1.3.1), constructing bona fide examples was quite another matter. In retrospect, it is evident that the difficulties stem from two facts. First,

(1.3.1) is too strong in the sense that the structure it imposes is too similar to the usual metric space structure. Second, any Wald space can be viewed as a collection of pairwise independent random variables with the property that any subset of three or more is dependent in a rather complicated way. Except for some obvious cases, such collections are not easily constructed.

1.4. Developments, 1956–1960

Our collaboration on probabilistic metric spaces began in 1956. At first we too were impeded by the above-mentioned difficulties with (1.3.1). Our "breakthrough" came in 1957 when, in the course of studying certain particular spaces, we rediscovered Menger's inequality (1.2.1)—not at first in its general form, but rather via particular functions T satisfying (1.2.2)–(1.2.4). Among these were the following:

$$\begin{aligned} W(a,b) &= \mathrm{Max}(a+b-1,0), \\ \Pi(a,b) &= ab, \\ M(a,b) &= \mathrm{Min}(a,b), \\ M^*(a,b) &= \mathrm{Max}(a,b), \\ \Pi^*(a,b) &= a+b-ab, \\ W^*(a,b) &= \mathrm{Min}(a+b,1). \end{aligned} \tag{1.4.1}$$

If T is taken to be any one of these functions, then (1.2.1) not only tells us that the probability $F_{pr}(x+y)$ depends in some monotonic way on the probabilities $F_{pq}(x)$ and $F_{qr}(y)$, but also makes the nature of this dependence precise. For example, if $T = \Pi$, then (1.2.1) states that the probability that the distance from p to r is less than $x+y$ is at least as large as the joint probability that, independently, the distance from p to q is less than x and the distance from q to r is less than y. The other functions in (1.4.1) yield corresponding interpretations.

An announcement of our initial results appeared in our note [S² 1958], followed two years later by our paper [S² 1960]. In the first part of the latter we laid the foundation for much of the subsequent development of the theory. In order to attain an appropriate level of generality, we began by defining (what we now call) a *weak probabilistic metric space* as a set S together with an associated family of distribution functions F_{pq} satisfying (1.1.7)–(1.1.10) and the additional condition:

$$\text{If } F_{pq}(x) = 1 \text{ and } F_{qr}(y) = 1, \text{ then } F_{pr}(x+y) = 1. \tag{1.4.2}$$

The principal advantages of (1.4.2) are that it is implied by both (1.2.1) and (1.3.1) and that it is weak enough to be valid in a very large class of spaces, yet strong enough to imply that this class includes all ordinary metric spaces. Indeed, (1.4.2) can be regarded as a minimal generalization of the ordinary triangle inequality. Its principal disadvantage is that it is vacuous in all spaces in which the functions F_{pq}, for $p \neq q$, never attain the value 1.

Turning to the inequality (1.2.1), we proved three lemmas which showed, first, that if S has more than one point and $T \geqslant M^*$, then (1.2.1) cannot hold for all p, q, r in S; second, that if not all the functions F_{pq} are of the form (1.2.5) (i.e., if the space is not simply a metric space) and if (1.2.1) holds for a given T, then there is a number a in the open interval $(0, 1)$ such that $T(a, 1) \leqslant a$; and finally, that if in addition T is continuous, then a can be taken to be any number in the range of any F_{pq}, whence $T(F_{pq}(x), 1) \leqslant F_{pq}(x)$ for all p, q in S and any $x > 0$.

Motivated by these results, we replaced (1.2.4) by the stronger boundary condition

$$T(a, 1) = a \quad \text{for all } a \text{ in } [0, 1]. \tag{1.4.3}$$

Taken together, (1.2.2), (1.2.3), and (1.4.3) yield $T(a,b) \leqslant T(a,1) = a$ and $T(a,b) \leqslant T(1,b) = T(b,1) = b$ for all a, b in $[0,1]$. Hence every function T that satisfies these conditions also satisfies

$$T(a,b) \leqslant M(a,b) \quad \text{for all } a, b \text{ in } [0, 1]. \tag{1.4.4}$$

We also added the requirement that T be *associative*, i.e., that

$$T(T(a,b),c) = T(a,T(b,c)) \quad \text{for all } a, b, c \text{ in } [0, 1]. \tag{1.4.5}$$

This enabled us to extend (1.2.1) to a polygonal inequality. For if p, q, r, s are four points in S, and if $F_{pq}(x)$, $F_{qr}(y)$, and $F_{rs}(z)$ are given, then $F_{ps}(x+y+z)$ can be estimated in two ways: either by estimating $F_{pr}(x+y)$ and combining this estimate with $F_{rs}(z)$, or by combining $F_{pq}(x)$ with the estimate of $F_{qs}(y+z)$ (see Figure 1.4.1). Requiring that these estimates be consistent leads naturally to (1.4.5).

A function that satisfies (1.2.2), (1.2.3), (1.4.3), and (1.4.5) is a *triangular norm* (briefly, a *t-norm*). Only the first three functions listed in (1.4.1) are t-norms. They are in fact the most important t-norms, and the names

Figure 1.4.1

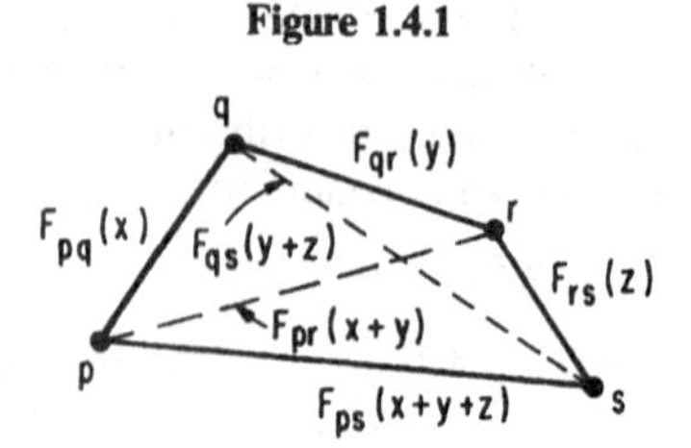

conferred upon them in (1.4.1) will be used consistently throughout this book. (The structure of t-norms and related associative functions is the subject of Chapter 5.)

After introducing the notion of a t-norm, we defined a *Menger space*, specifically, a *Menger space under* (a given t-norm) T, as a space in which the distribution functions F_{pq} satisfy (1.1.7)–(1.1.10) and (1.2.1) with the given t-norm T. We also defined a *Wald space* as one in which the functions F_{pq} satisfy (1.1.7)–(1.1.10) and (1.3.1). We then showed that every Wald space is a Menger space under Π (which implies that every Wald space satisfies (1.4.2) and is therefore a weak probabilistic metric space) and that the converse is not valid (see below and Section 8.4).

1.5. Some Examples

The rehabilitation of the Menger inequality enabled us and others attracted to the subject to introduce and study many classes of probabilistic metric spaces. To describe some of the more elementary types, it is convenient first to define the family of distribution functions ε_a for $-\infty < a < \infty$ by

$$\varepsilon_a(x) = \begin{cases} 0, & x \leqslant a, \\ 1, & x > a. \end{cases} \tag{1.5.1}$$

Now let (S, d) be a metric space and G a distribution function such that $G(0) = 0$ and $G(x) > 0$ for some $x > 0$. Then the *simple space* (S, d, G) consists of the set S and the family of distribution functions F_{pq} defined by

$$F_{pq} = \varepsilon_0 \quad \text{if } p = q, \tag{1.5.2}$$

$$F_{pq}(x) = G(x/d(p,q)) \quad \text{if } p \neq q. \tag{1.5.3}$$

If $G = \varepsilon_1$, then (1.5.3) reduces to (1.2.5), so that metric spaces are special cases of simple spaces. Otherwise, a simple space can be regarded as a metric space that is "smeared out" or randomized by the distribution function G.

In [S^2 1960] we showed that every simple space is a Menger space under any t-norm and that there are simple spaces that are not Wald spaces. Subsequently, in collaboration with T. Erber, we used simple spaces to construct a phenomenological theory that describes certain aspects of hysteresis in large-scale physical systems (see Section 8.5).

If we choose a nonnegative number α and replace (1.5.3) by

$$F_{pq}(x) = G\big(x/(d(p,q))^\alpha\big) \quad \text{if } p \neq q, \tag{1.5.4}$$

then we obtain the class of *α-simple spaces* [S^2 1963b]. For $0 < \alpha \leqslant 1$, an α-simple space is a simple space, but this is no longer the case for $\alpha > 1$. Indeed, for $\alpha > 1$, an α-simple space need not even be a weak probabilistic metric space, let alone a Menger space. For example, let S be the real line, $d(p,q) = |p - q|$, and U_1 the distribution function defined by

$$U_1(x) = \begin{cases} 0, & x \leqslant 0, \\ x, & 0 \leqslant x \leqslant 1, \\ 1, & 1 \leqslant x. \end{cases}$$

Then, in the corresponding 2-simple space, we have

$$F_{01}(1) = F_{12}(1) = U_1\left(\frac{1}{1^2}\right) = 1 \quad \text{but} \quad F_{02}(2) = U_1\left(\frac{2}{2^2}\right) = \frac{1}{2},$$

which contradicts (1.4.2). On the other hand, for each $\alpha > 1$ there exist α-simple spaces that are Menger spaces under appropriate t-norms; and it can be shown that every space that is a Menger space under a sufficiently well-behaved t-norm is topologically equivalent, with respect to a very natural topology, to an appropriate α-simple space (see Sections 8.6 and 12.3).

In [S² 1962] we introduced the class of *distribution-generated spaces* (see Chapter 10). The points of such a space are random vectors in Euclidean n-space E^n. The distance between two such points p, q is thus a random variable, whose distribution function is our F_{pq}. In particular, we investigated *C-spaces*. These are the distribution-generated spaces in which any two distinct points are independent and a spherically symmetric, unimodal, n-dimensional probability density g_p is associated with each (nonsingular) point p. Any such point p may generally be identified with g_p. Since each g_p can be visualized as a "cloud" in E^n, a C-space can be visualized as a set of clouds, each spherically symmetric and unimodal. One can also visualize a C-space as a set of "particles" p whose uncertain position in E^n is governed by the probability density g_p.

If every g_p is a normal density, then we have a *normal C-space*. Normal C-spaces may be viewed as the probabilistic analogs of Euclidean spaces. They arise frequently in multivariate analysis, notably in connection with problems of classification and discrimination (see Section 10.6). Moreover, most of the attempts at building uncertainty into an underlying geometry mentioned in Sections 1.0 and 10.6 have employed normal C-spaces. These spaces have also found application in psychology (see [Marley 1971]).

All C-spaces are weak probabilistic metric spaces and, under suitable conditions, are Menger spaces under W. In contrast, there are one-dimensional normal C-spaces that are not Menger spaces under Π, and hence not Wald spaces. When they exist, the means of the distribution functions F_{pq} are metrics on S. The resulting metric spaces are Euclidean in the large but discrete in the small, in the sense that for every point p there is a positive number t_p such that the sphere with center p and radius t_p contains no other points of the space (see Section 10.5).

In [S[2] 1973], we defined the class of *transformation-generated spaces*. These are obtained when one considers a metric space (S, d) endowed with a probability measure and a measure-preserving transformation ψ acting on S. It follows from the Birkhoff ergodic theorem that for any $x > 0$ and almost all p, q in S, there is a distribution function F_{pq} such that the fraction of times the distance between the points $\psi^n(p)$ and $\psi^n(q)$ is less than x converges to $F_{pq}(x)$ as $n \to \infty$. The resulting space is a Menger space under W on which ψ is probabilistic distance preserving (see Sections 11.1 and 11.2). The case when ψ is mixing is of particular interest, for then the distribution functions F_{pq} are independent of p and q, and, as we jointly with T. Erber [1973] showed, this fact can be used to study dispersive behavior and recurrence in statistical mechanics (see Section 11.3). It can also be used to develop efficient random number generators [Erber, Everett, and Johnson 1979].

1.6. Šerstnev, 1962

In 1962 A. N. Šerstnev introduced an inequality that includes all those previously proposed as special cases and is without doubt the appropriate probabilistic generalization of the ordinary triangle inequality (1.1.4).

Let us call a function F a *distance distribution function* (briefly, a *d.d.f.*) if F is a distribution function and $F(0) = 0$. Distance distribution functions are ordered by defining $F \leqslant G$ to mean $F(x) \leqslant G(x)$ for all $x > 0$. In particular, for $a \geqslant 0$, the functions ε_a defined in (1.5.1) are d.d.f.'s and $F \leqslant \varepsilon_0$ for all d.d.f.'s F. Following Šerstnev [1962, 1964a], we say that τ is a *triangle function* if τ assigns a d.d.f. to every pair of d.d.f.'s and satisfies the following conditions:

$$\tau(F, G) = \tau(G, F), \tag{1.6.1}$$

$$\tau(F, G) \leqslant \tau(H, K) \quad \text{whenever } F \leqslant H,\ G \leqslant K, \tag{1.6.2}$$

$$\tau(F, \varepsilon_0) = F, \tag{1.6.3}$$

$$\tau(\tau(F, G), H) = \tau(F, \tau(G, H)). \tag{1.6.4}$$

We then define a *probabilistic metric space* (briefly, a *PM space*) *under* a given triangle function τ to be a set S together with an associated family of distance distribution functions F_{pq} satisfying (1.1.7)–(1.1.10) and the *Šerstnev triangle inequality*

$$F_{pr} \geqslant \tau(F_{pq}, F_{qr}) \quad \text{for all } p, q, r \text{ in } S. \tag{1.6.5}$$

If all the foregoing conditions, with the possible exception of (1.1.9), hold, then the space is a *probabilistic pseudometric space* (briefly, a *PPM space*) *under* τ.

On comparing (1.6.5) with (1.1.4), we see that the essence of the passage from ordinary to probabilistic metric spaces lies in the replacement of real numbers, i.e., numerical distances, by distance distribution functions, and the replacement of the operation of addition of real numbers by a triangle function τ. Since the function τ in (1.6.5) is not further specified, many distinct and inequivalent triangle inequalities are possible. If τ is such that

$$\tau(\varepsilon_a, \varepsilon_b) \geqslant \inf\{\varepsilon_c \mid c < a + b\} \quad \text{for all } a, b > 0, \tag{1.6.6}$$

then (1.6.5) yields (1.4.2). When τ is convolution, then (1.6.5) is Wald's inequality (1.3.1). If the t-norm T is sufficiently well behaved and the function τ_T is defined by

$$(\tau_T(F,G))(x) = \sup\{T(F(u),G(v)) \mid u + v = x\}, \tag{1.6.7}$$

then τ_T is a triangle function and (1.6.5) with τ_T is equivalent to Menger's inequality (1.2.1) (see Section 8.2).

After 1963 the subject grew rapidly—so rapidly that each of the surveys [Onicescu 1964, Ch. VII; Schweizer 1967; Istrăţescu 1974] were dated soon after they appeared. In the following sections, we trace some of the facets of this development.

1.7. Random Metric Spaces

If one considers the notion of a PM space from the point of view of the standard measure-theoretic model of probability theory (see Section 2.3), then one is naturally inclined toward a different formulation of the subject, namely, one that begins with random variables on a given probability space. This approach leads to several closely related classes of PM spaces which are introduced below and described in detail in Chapter 9.

The first to look at the subject in this light was A. Špaček [1956], who considered a set S and the family $\mathfrak{R}$ of all real-valued functions defined on $S \times S$. The set $\mathfrak{M}$ of all metrics on S is a subset of $\mathfrak{R}$, and Špaček gave necessary and sufficient conditions that a probability measure P on $\mathfrak{R}$ satisfy $P(\mathfrak{M}) = 1$. Subsequently, Šerstnev [1967] showed that if $P(\mathfrak{M}) = 1$ and if, for any p, q in S and any real x, $D(p,q;x)$ is the set of all functions d in $\mathfrak{R}$ such that $d(p,q) < x$, then the functions F_{pq} defined by

$$F_{pq}(x) = P(D(p,q;x)) \tag{1.7.1}$$

are distribution functions satisfying (1.1.7)–(1.1.10), and the set S together with these functions F_{pq} is a Menger space under W. Šerstnev also gave a number of examples that clarify the relationship between PM spaces and Špaček's "random metrics" and concluded that "Although Špaček's approach appears classical from the point of view of the axiomatics of probability theory, it turns out to be restrictive in certain essential respects."

In his doctoral dissertation of 1965, R. R. Stevens modified Špaček's approach. Stevens [1968] started with a set S, a family of metrics $\mathfrak{D}$ on S, and a probability measure P on $\mathfrak{D}$ (rather than on the set $\mathfrak{R}$ described earlier). The idea behind this setup is that one has a set S and a collection $\mathfrak{D}$ of "measuring rods." One chooses a measuring rod d from $\mathfrak{D}$ "at random" and uses it to measure the distance between two given points p and q of S. Given this, Stevens used (1.7.1) to define the distribution functions F_{pq} and showed that the *metrically generated space* so obtained is a Menger space under W.

A few years later, H. Sherwood [1969] approached the subject from a different direction. Motivated by our work on distribution-generated spaces, he introduced the concept of an *E-space*. The points of an E-space are functions from a probability space $(\Omega, \mathscr{A}, P)$ into a metric space (M, d). For each pair (p, q) of functions in the space, the composite function $d(p, q)$ defined by

$$(d(p,q))(\omega) = d(p(\omega), q(\omega)) \quad \text{for all } \omega \text{ in } \Omega \tag{1.7.2}$$

is assumed to be a random variable on $(\Omega, \mathscr{A}, P)$. The function F_{pq} is taken to be the distribution function of this random variable, so that for any real x,

$$F_{pq}(x) = P\{\omega \text{ in } \Omega \mid (d(p,q))(\omega) < x\}. \tag{1.7.3}$$

Thus, by construction, $F_{pq}(x)$ is the probability that the distance between p and q is less than x. Sherwood showed that every E-space is a Menger space under W and that E-spaces are closely related to distribution-generated spaces, but that neither class includes the other.

Given an E-space, for each ω in Ω, the function d_ω defined on $S \times S$ by

$$d_\omega(p,q) = d(p(\omega), q(\omega)) \tag{1.7.4}$$

is a pseudometric on S—but generally not a metric, since $p(\omega) = q(\omega)$ does not imply $p = q$. On identification of d_ω with ω, it follows that

$$F_{pq}(x) = P\{d_\omega \mid d_\omega(p,q) < x\}; \tag{1.7.5}$$

and comparison of (1.7.5) with (1.7.1) shows that every E-space is a *pseudometrically generated space*. Thus Sherwood's E-spaces lead naturally to a generalization of Stevens's metrically generated spaces. But much more is true; for as Sherwood proved, every pseudometrically generated space can be realized as an E-space. Thus these two concepts are coextensive.

Sherwood also showed that if the points of a distribution-generated space are random variables on a common probability space (a strong assumption), then this space is an E-space.

There is yet another way to look at E-spaces. For each fixed pair of points p, q in S, let d_{pq} be the function defined on Ω by

$$d_{pq}(\omega) = d(p(\omega), q(\omega)). \tag{1.7.6}$$

Then d_{pq} is a random variable with distribution function F_{pq}, and the following conditions are satisfied:

$$P\{\omega \text{ in } \Omega \mid d_{pq}(\omega) = 0\} = 1 \quad \text{if } p = q, \tag{1.7.7}$$

$$P\{\omega \text{ in } \Omega \mid d_{pq}(\omega) = 0\} < 1 \quad \text{if } p \neq q, \tag{1.7.8}$$

$$P\{\omega \text{ in } \Omega \mid d_{pq}(\omega) = d_{qp}(\omega)\} = 1 \quad \text{for all } p, q, \tag{1.7.9}$$

$$P\{\omega \text{ in } \Omega \mid d_{pr}(\omega) \leqslant d_{pq}(\omega) + d_{qr}(\omega)\} = 1 \quad \text{for all } p, q, r. \tag{1.7.10}$$

Conditions (1.7.7)–(1.7.10) are closely related to those imposed by Špaček [1956, 1960], and a collection of random variables satisfying them is called a *random metric space*. Such spaces were studied by P. Calabrese in his doctoral dissertation in 1968 and, somewhat later, by J. B. Brown [1972]. Calabrese [1978] showed that only under stringent conditions on certain null sets is a random metric space derivable from an E-space in the manner described above. Brown's paper is devoted mainly to (1.7.10) and various weaker versions of this condition. Other alternatives to (1.7.10) have been considered by G. Sîmboan and R. Theodorescu [1962] (see also [Onicescu 1964, Ch. VII]).

Using an argument due independently to Šerstnev and to Stevens, it is easy to show that a random metric space determines a unique PPM space under τ_W. This immediately gives rise to the converse question: Given a PPM space, does there exist a random metric space that determines it? This question was raised and studied by the authors mentioned above. Collectively they have constructed a variety of counterexamples which show that unless the PPM space in question is a Menger space under M, the answer is "No." Thus the theory of random metric spaces and, a fortiori, the theory of E-spaces are a proper part of the theory of probabilistic metric spaces.

1.8. Topologies

A metric space (S, d) is endowed with a natural topology. This "metric topology" is essentially unique and may be defined by taking the sets

$$B(p, x) = \{q \text{ in } S \mid d(p, q) < x\} \tag{1.8.1}$$

for p in S and $x > 0$ as neighborhoods. In a PM space, neighborhoods and corresponding topological structures may be defined in many nonequivalent ways (see Section 13.1). We were aware of this from the outset. However, in [S^2 1960] we felt it best to begin by considering the "strongest" of these, i.e., the one that most resembles the standard metric space construction. Accordingly, for any ϵ, $\lambda > 0$, we defined the *ϵ, λ-neighborhood* of a point p in a PM space by

$$N_p(\epsilon, \lambda) = \{q \text{ in } S \mid F_{pq}(\epsilon) > 1 - \lambda\}. \tag{1.8.2}$$

The interpretation is that two points of a PM space are "near" when it is highly probable that the distance between them is small.

Note that in general the ϵ, λ-neighborhood of a point p in a simple space $(S, d; G)$ is an ordinary neighborhood of p in the metric space (S, d) (see Section 12.3).

Since 1960 many papers have been devoted to the study of the topological structures induced by the system of ϵ, λ-neighborhoods. In particular, it was shown by us jointly with E. Thorp [1960] that if T is a t-norm satisfying the condition

$$\sup\{T(x, x) \mid 0 < x < 1\} = 1, \tag{1.8.3}$$

then in any Menger space under T, the ϵ, λ-neighborhoods induce a topology that is metrizable. Subsequently, B. Morrel and J. Nagata [1978] showed that no condition weaker than (1.8.3) can guarantee that the ϵ, λ-neighborhoods induce a bona fide topology. Other topics that have been studied include convergence [S^2 1960; Anthony, Sherwood, and Taylor 1974], continuity of the probabilistic distance [S^2 1960; Schweizer 1966], measures of compactness [Bocşan and Constantin 1973, 1974], completion [Sherwood 1966; Muštari 1967; Nishiura 1970], contraction mappings [Sehgal 1966; Sehgal and Bharucha-Reid 1972; Sherwood 1971; Istrăţescu and Săcuiu 1973; Cain and Kasriel 1976] (see also [Bharucha-Reid 1976, p. 654]), entropy [Saleski 1974, 1975], and product and quotient spaces [Istrăţescu and Vaduva 1961; Schweizer 1964; Egbert 1968; Xavier 1968; Tardiff 1976; Radu 1977; Alsina 1978ab]. Under appropriate continuity conditions on the t-norms or triangle functions involved, the results obtained generally match the corresponding results for metric spaces. There is one striking exception: Sherwood [1971] showed that for virtually all t-norms except M, there is a PM space under that t-norm which is complete but admits a contraction map having no fixed point. These and related matters are the subject of Chapter 12.

In the definition of $N_p(\epsilon, \lambda)$ in (1.8.2) all positive values of ϵ and λ are allowed. This is equivalent to the assumption that statements about arbitrarily small distances can be made with probabilities arbitrarily close to 1. In many situations, such precision or such certainty is unrealistic. Thus one is naturally led to consider limitations on the possible values of ϵ and λ.

This was first done by Thorp [1962], who restricted ϵ and λ to a specified subset of the strip $(0, 1) \times (0, \infty)$ and showed that such a restriction leads naturally to the generalized topologies first defined by Fréchet [1917, 1928] and subsequently developed by A. Appert and Ky-Fan [1951].

R. T. Fritsche [1971] simplified Thorp's work by introducing the notion of a *profile function*. This is simply a fixed distance distribution function φ whose value $\varphi(x)$, for any $x > 0$, is interpreted as the maximum degree of confidence that can be assigned to statements about distances less than x. For example, if $\varphi = \varepsilon_a$, then nothing can be said about distances less than a; and if $\varphi(x) = b$ for all $x > 0$, where $0 \leqslant b \leqslant 1$, then statements about distances have at best a probability b of being valid. Fritsche's structures coincide with Thorp's when the subset of the strip $(0, 1) \times (0, \infty)$ is the set $\{(x, y) \mid 0 < y < \varphi(x)\}$; and Fritsche showed that there is virtually no loss of generality in restricting one's attention to profile functions.

If p and q are points of a PM space and if $F_{pq} \geqslant \varphi$, then F_{pq} cannot be effectively distinguished from ε_0, whence p and q are themselves indistinguishable relative to φ. This relation of indistinguishability leads naturally to Poincaré's paradox, for it is perfectly possible to have p indistinguishable from q and q indistinguishable from r, while p is distinguishable from r. Furthermore, in many PM spaces one can use this relation, in conjunction with certain properties of the triangle function, to define *degrees of distinguishability* and *degrees of indistinguishability*, both for points and for subsets of the space (see [Schweizer 1975b] and Section 13.2).

Thorp's and Fritsche's topological structures suffered from the defect that the intersection of two neighborhoods may not contain a neighborhood. This technical nuisance was overcome in 1974, at which time it was also noted that the closure spaces of E. Čech [1966] provide a better framework than the generalized topologies of Appert and Ky-Fan. The resultant theory was developed by R. M. Tardiff [1976] (see Section 13.1).

1.9. Tools

At virtually every stage in our development of the theory of PM spaces we have had either to construct new mathematical machinery or to modify existing machinery to suit our needs. This has often led us to mathematical problems that are of interest in their own right.

With our rediscovery of the Menger triangle inequality in 1957, we were immediately faced with two tasks: (1) to relate the geometric and probabilistic aspects of this inequality; (2) to develop a repertory of t-norms for use in (1.2.1). The first led us to the theory of copulas; the second to the structure and representation theory of semigroups on real intervals.

A *joint* (two-dimensional) *distribution function* is a function G from the extended plane into the interval [0, 1] that satisfies the following conditions:

i. $G(\infty, \infty) = 1$ and $G(-\infty, x) = G(x, -\infty) = G(-\infty, -\infty) = 0$;
ii. $G(x_1, y_1) - G(x_1 y_2) - G(x_2, y_1) + G(x_2, y_2) \geqslant 0$ whenever $x_1 \leqslant x_2$, $y_1 \leqslant y_2$.

The *margins* of G are the one-dimensional distribution functions F_1, F_2 defined by

$$F_1(x) = G(x, \infty), \qquad F_2(x) = G(\infty, x).$$

Joint distribution functions can be complicated objects. The fundamental fact, however, is that they are composites of simpler ones. Specifically, let G be a joint distribution function with margins F_1, F_2. Then there is a continuous function C from the unit square $[0, 1] \times [0, 1]$ onto the unit interval [0, 1] such that

$$G(x, y) = C(F_1(x), F_2(y)) \quad \text{for all } x, y. \tag{1.9.1}$$

We call such a function C a *copula* (because it links a joint distribution function to its margins). Conversely, if C is a copula and if F_1, F_2 are distribution functions, then the function G defined by (1.9.1) is a joint distribution function with margins F_1, F_2. Note the resemblance between the right-hand sides of (1.9.1) and (1.2.1).

Copulas were introduced (for the general n-dimensional case) in [Sklar 1959] and were discussed at greater length in [Sklar 1973] and [S^2 1974]. Their systematic use unifies and simplifies many topics in probability theory. For example, in [Schweizer and Wolff 1976, 1981] it is shown how they can be used to define very natural nonparametric measures of dependence for random variables.

Representation theorems for associative functions on real intervals date back to N. H. Abel [1826] (see [Aczél 1966, §6.2]). In [S^2 1961a] we used a representation theorem due to Aczél [1949] to construct many classes of t-norms virtually at will. We extended these results somewhat in [S^2 1963c]. Subsequently, C. H. Ling [1965] proved the following: Let T be a t-norm that is continuous and *Archimedean*, i.e., such that $T(x,x) < x$ for $0 < x < 1$. Then there are continuous nondecreasing functions f, g such that

$$T(x, y) = f(g(x) + g(y)) \tag{1.9.2}$$

for all x, y in [0, 1]. Using (1.9.2) and Theorem B of P. S. Mostert and A. L. Shields [1957], Ling characterized the structure of continuous t-norms. This in effect settles the question as far as PM spaces are concerned (see Sections 5.3 and 5.5).

The representation (1.9.2) has since been shown to hold under much weaker conditions (see [Krause 1981]). In particular, the requirement that T be continuous can be substantially weakened. The results are of interest both in the theory of functional equations and in the theory of topological semigroups.

Copulas and t-norms meet in the fact that the three most important t-norms, M, Π, and W, are also the three most important copulas. Moreover, (1.9.2) can be used to show that an Archimedean t-norm is a copula if and only if f is convex. This result can be extended to higher dimensions, and the extension has been used by C. H. Kimberling [1973b, 1974] to characterize sets of exchangeable random variables.

With the appearance of the work of Šerstnev, our attention shifted to problems concerning spaces of distribution functions and binary operations on them. Our first concern was to develop a repertory of triangle functions and to learn how to calculate with them.

A basic tool is a modified version of the Lévy metric introduced by D. A. Sibley [1971]. Convergence in this metric is equivalent to weak convergence (convergence in Lévy's original metric implies weak convergence, but not conversely). In this book we consistently use a slightly simplified version of this modified Lévy metric (see Section 4.2).

Many of the basic properties of the operations τ_T defined by (1.6.7) were established in [Šerstnev 1964a]. This work was carried further in [Schweizer 1975a], where, in addition, several other classes of operations on distribution functions were defined and studied. Since then, thanks primarily to the efforts of M. J. Frank and R. Moynihan, the subject has blossomed. Frank [1975, 1979] settled the difficult question of determining which members of various classes of generalized convolutions that are derived from copulas are associative. Moynihan [1975, 1978ab] studied the algebraic structure of the semigroups of distribution functions determined by operations of the type τ_T; and in his doctoral dissertation of 1975 and in [Moynihan 1980ab], he developed a theory of conjugate transforms for these semigroups. These transforms play the role of characteristic functions. Lastly, in [Frank and Schweizer 1979] several useful duality theorems for the operations τ_T are obtained. These matters are discussed in Chapter 7.

In their original papers Menger and Wald each introduced a notion of betweenness. With the sole exception of a paper by Sherwood [1970], in which he discussed betweenness in E-spaces, the subject lay dormant for many years, primarily because adequate mathematical tools were not available. The development of the above-mentioned machinery changed all this. Thus in [Moynihan and Schweizer 1979], both Menger's and Wald's notions of betweenness are developed within a common framework, and several other betweenness notions, related to conjugate transforms, are presented as well (see Chapter 14).

From a philosophical point of view, perhaps the most striking consequence of the development of the machinery is the following: An operation τ on distribution functions is derivable from a function V on random variables if, given any two distribution functions F, G, there exist random variables X, Y defined on a common probability space such that F is the distribution function of X, G the distribution function of Y, and $\tau(F, G)$ the distribution function of $V(X, Y)$. Thus convolution is derivable from addition (of independent random variables) and it is easily shown that τ_M is derivable from addition (of strictly dependent random variables). But if $T \neq M$, then, as was shown in [S^2 1974], the operation τ_T is not derivable from any function on random variables (see Section 7.6).

1.10. Postscript

Although we believe that the results presented in this book amply justify our original decision to define PM spaces in terms of distribution functions rather than random variables, we prefer to leave the last word on the general subject to M. Loève. In his classic work [Loève 1977, §10.2] he states

> the primary datum in random phenomena is not the probability space but the joint distributions of the families of random variables which describe the characteristics of the phenomena. . . .
>
> In particular, since in the numerical case a distribution is represented by the corresponding distribution functions, we can say that
>
> —*the probability-theoretic properties of a random variable X are those which can be expressed in terms of its distribution function F_X,*
> —*the probability-theoretic properties of a finite family $(X_1, X_2, \ldots, X_N)$ of random variables are those which can be expressed in terms of the joint distribution function $F_{X_1, X_2, \ldots, X_N}$,*
> —*the probability-theoretic properties of any family $(X_t, t \in T)$ of random variables are those which can be expressed in terms of the joint distribution functions of its finite subfamilies. . . .*
>
> It is important to realize fully that measurements of a stochastic variable are relative to the induced probability space; the original probability space is but a mathematical fiction. Yet it is basic, for it permits the use of a "common frame of reference" for the families of stochastic variables we investigate. . . . However, precisely because of the existence of a common frame of reference in the present setup, modern physics forces us to introduce a different setup.

This position is amplified in [Loève 1965, especially pp. 268–272].

2

Preliminaries

2.1. Sets and Functions

We take for granted all the usual apparatus of elementary set theory (see, e.g., [Foulis 1969] or [Halmos 1960]). We also assume familiarity with the elements of advanced calculus (see, e.g., [Apostol 1974]).

For a given set S we use the symbol $\mathcal{P}(S)$ to denote the *power set* of S, i.e., the set of all subsets of S. For any positive integer n, we use S^n to denote the nth *Cartesian power of* S, i.e., the set $S \times S \times \cdots \times S$ (n times).

We use R to denote the *extended reals*, i.e., the set of real numbers with the symbols $-\infty$, ∞ adjoined, and ordered via $-\infty < x < \infty$ for all real x. Thus R is a closed interval. Other symbols that we use to denote particular closed intervals are R^+ for $[0, \infty]$ and I for $[0, 1]$.

We also take for granted the notion of *function*. The *domain* of a function f, i.e., the set of arguments of f, is denoted by Dom f, and the *range* of f, i.e., the set of values of f, by Ran f. A function f is defined by specifying its domain, Dom f, and its value $f(x)$ for every x in Dom f. Two functions, f and g, are *equal* if and only if they are identical, i.e., if and only if Dom $f =$ Dom g and $f(x) = g(x)$ for all x in Dom $f =$ Dom g. For any set S, we denote the *identity function on* S by j_S. Thus

$$\text{Dom } j_S = S \quad \text{and} \quad j_S(x) = x \quad \text{for all } x \text{ in } S. \tag{2.1.1}$$

When there is no possibility of confusion, we omit the subscript.

The *composite* of the functions f and g is the function $f \circ g$ defined by

$$\begin{aligned} \operatorname{Dom}(f \circ g) &= \{x \mid x \text{ in } \operatorname{Dom} g \text{ and } g(x) \text{ in } \operatorname{Dom} f\}, \\ (f \circ g)(x) &= f(g(x)) \quad \text{for all } x \text{ in } \operatorname{Dom}(f \circ g). \end{aligned} \tag{2.1.2}$$

Since we admit a unique *empty function*, with empty domain and empty range, any two functions have a composite. Note that $f \circ j_{\operatorname{Dom} f} = j_{\operatorname{Ran} f} \circ f = f$ for any function f. Another straightforward consequence of (2.1.2) is that composition of functions is associative, i.e., that

$$(f \circ g) \circ h = f \circ (g \circ h) \tag{2.1.3}$$

for any three functions f, g, h. It follows that we can without ambiguity omit parentheses and write $f_1 \circ f_2 \circ \cdots \circ f_n$ for the composite of the functions $f_1, f_2, \ldots, f_n$.

The *iterates* of a function f are defined recursively by

$$f^1 = f, \quad f^{n+1} = f^n \circ f \quad \text{for } n \geqslant 1. \tag{2.1.4}$$

Since composition is associative, it follows that

$$f^{m+n} = f^m \circ f^n = f^n \circ f^m$$

and

$$(f^m)^n = f^{mn} = (f^n)^m \tag{2.1.5}$$

for all $m, n \geqslant 1$. When $\operatorname{Ran} f \subseteq \operatorname{Dom} f$, it is possible to extend (2.1.4) by defining f^0 to be $j_{\operatorname{Dom} f}$, in which case (2.1.5) is valid for all nonnegative integers m, n.

A function f is *idempotent* if $f^2 = f$. Idempotent functions are completely characterized by

Lemma 2.1.1. *A function f is idempotent if and only if $f \circ j_{\operatorname{Ran} f} = j_{\operatorname{Ran} f}$, i.e., if and only if the restriction of f to* Ran f *is the identity function on* Ran f.

PROOF. If $f \circ j_{\operatorname{Ran} f} = j_{\operatorname{Ran} f}$, then $f^2 = f \circ f = f \circ j_{\operatorname{Ran} f} \circ f = j_{\operatorname{Ran} f} \circ f = f$, whence f is idempotent.

Next, suppose f is idempotent, i.e., that $f(f(x)) = f(x)$ for every x in $\operatorname{Dom} f$. Then $\operatorname{Ran} f \subseteq \operatorname{Dom} f$. Now for any y in $\operatorname{Ran} f$ there is an x in $\operatorname{Dom} f$ such that $f(x) = y$. Thus $f(y) = f(f(x)) = f(x) = y$ for each y in $\operatorname{Ran} f$, which is equivalent to $f \circ j_{\operatorname{Ran} f} = j_{\operatorname{Ran} f}$. □

Definition 2.1.2. A function g is a *quasi-inverse* of a function f, and we write $g[Q]f$ if $g \circ j_{\operatorname{Ran} f}$, the restriction of g to $\operatorname{Ran} f$, is a right inverse of f whose range is equal to $\operatorname{Ran} g$; i.e., if

$$f \circ g \circ j_{\operatorname{Ran} f} = j_{\operatorname{Ran} f} \quad \text{and} \quad \operatorname{Ran}(g \circ j_{\operatorname{Ran} f}) = \operatorname{Ran} g. \tag{2.1.6}$$

For example, if $f(x) = x^2$ for all real x, and if $g_1(x) = \sqrt{x}$ and $g_2(x) = -\sqrt{x}$ for all $x \geqslant 0$, then both g_1 and g_2 are quasi-inverses of f, as are the functions g_3 and g_4 given by

$$g_3(x) = \begin{cases} \sqrt{x}, & x \geqslant 0, \\ x^4, & x < 0, \end{cases}$$

and

$$g_4(x) = \begin{cases} -\sqrt{x}, & x \geqslant 0, \\ x^3, & x < 0. \end{cases}$$

Note that x^4 (resp., x^3) can be replaced by any nonnegative (resp., nonpositive) function defined on any subset of $(-\infty, 0)$.

The next two lemmas give the basic properties of quasi-inverses. They may be proved by standard but tedious arguments. They are much easier to establish with the aid of techniques developed in our study of the algebra of functions [S² 1967], but since we do not use these techniques elsewhere in this book, we relegate the proofs to the end of this section.

Lemma 2.1.3. *If* $g[Q]f$, *then* $f[Q]g$. *Furthermore,* $\operatorname{Ran} f \subseteq \operatorname{Dom} g$ *and* $\operatorname{Ran} g \subseteq \operatorname{Dom} f$.

Lemma 2.1.4. *If* $g_1[Q]f_1$, $g_2[Q]f_2$ *and, in addition,* $\operatorname{Dom} f_1 \subseteq \operatorname{Ran} f_2$, *then* $(g_2 \circ g_1)[Q](f_1 \circ f_2)$.

If $g[Q]f$ and $\operatorname{Ran} f = \operatorname{Dom} g$, $\operatorname{Ran} g = \operatorname{Dom} f$, then f and g are *inverses* of each other in the usual sense. In this case we denote g by f^{-1}. In general, if $g[Q]f$, then the functions $g \circ j_{\operatorname{Ran} f}$ and $f \circ j_{\operatorname{Ran} g}$ are inverses. Since $g \circ j_{\operatorname{Ran} f}$ is the restriction of g to $\operatorname{Ran} f$ and since $f \circ j_{\operatorname{Ran} g}$ is a restriction of f whose range is equal to $\operatorname{Ran} f$, it follows that if $g[Q]f$, then g is an extension of the inverse h of some maximal invertible restriction of f, with the property that $\operatorname{Ran} g = \operatorname{Ran} h$ (see the example above). Inverses, when they exist, are unique. However, as is well known, only one-to-one functions have inverses. Quasi-inverses, on the other hand, exist for all functions —at least under the assumption of the Axiom of Choice. Indeed, it can be shown that the statement "every function has at least one quasi-inverse" is equivalent to the Axiom of Choice (cf. [S² 1965]). However, in many applications of the concept, including most of those we shall make, the existence of quasi-inverses can be demonstrated without an appeal to the Axiom of Choice. Thus the notion of a quasi-inverse is a useful generalization of that of an inverse.

Lastly, we note that every function f induces two associated set functions, one on the power set of its domain, the other on the power set of its range.

These functions are customarily also denoted by f and f^{-1} and are defined by

$$f(A) = \{y \text{ in } \operatorname{Ran} f \mid y = f(x) \text{ for } x \text{ in } A\} \tag{2.1.7}$$

for any subset A of Dom f, and

$$f^{-1}(B) = \{x \text{ in } \operatorname{Dom} f \mid f(x) = y \text{ for } y \text{ in } B\} \tag{2.1.8}$$

for any subset B of Ran f. The range of the set function f is $\mathcal{P}(\operatorname{Ran} f)$. The range of the set function f^{-1} is a subset of the $\mathcal{P}(\operatorname{Dom} f)$ and is equal to $\mathcal{P}(\operatorname{Dom} f)$ if and only if the point function f is one-to-one. The set functions f and f^{-1} are quasi-inverses.

PROOFS OF LEMMAS 2.1.3 AND 2.1.4. Following [S² 1967], we write Rf for $j_{\operatorname{Dom} f}$ and Lf for $j_{\operatorname{Ran} f}$. It is then easy to show that for all functions f, g we have (i) $RLf = LLf = Lf$; (ii) $Lf \circ f = f$; (iii) $L(g \circ f) = L(g \circ Lf)$; (iv) $R(f \circ g) \subseteq Rg$; (v) $R(g \circ Lf) \subseteq Rg$.

Now suppose $g[Q]f$. This translates into (vi) $f \circ g \circ Lf = Lf$ and (vii) $L(g \circ Lf) = Lg$. Using (ii) and (vi), we obtain

$$(g \circ f)^2 = g \circ f \circ g \circ f = g \circ f \circ g \circ Lf \circ f = g \circ Lf \circ f = g \circ f,$$

whence $g \circ f$ is idempotent. Then (vii), (iii), and Lemma 2.1.1 yield

$$\begin{aligned} g \circ f \circ Lg &= g \circ f \circ L(g \circ Lf) = g \circ f \circ L(g \circ f) \\ &= L(g \circ f) = L(g \circ Lf) = Lg, \end{aligned}$$

which is (vi) with f and g interchanged. Similarly, we have

$$L(f \circ Lg) = L(f \circ L(g \circ Lf)) = L(f \circ g \circ Lf) = LLf = Lf,$$

which is (vii) with f and g interchanged. Hence $f[Q]g$.

Next, using (i), (vi), (iv), and (v), we obtain

$$Lf = RLf = R(f \circ g \circ Lf) \subseteq R(g \circ Lf) \subseteq Rg,$$

which is equivalent to Ran $f \subseteq$ Dom g. Interchanging f and g yields $Lg \subseteq Rf$, i.e., Ran $g \subseteq$ Dom f. This proves Lemma 2.1.3.

For Lemma 2.1.4 we need the additional properties: (viii) if $Rf_1 \subseteq Lf_2$, then $L(f_1 \circ f_2) = Lf_1$; (ix) if $Lg_1 \subseteq Lf_2$, then $Lf_2 \circ g_1 = g_1$. Now suppose that f_1, g_1, f_2, and g_2 satisfy the hypotheses of Lemma 2.1.4. Then $Rf_1 \subseteq Lf_2$, and Lemma 2.1.3 yields $Lg_1 \subseteq Rf_1$, whence (viii) and (ix) yield $L(f_1 \circ f_2) = Lf_1$ and $Lf_2 \circ g_1 = g_1$. Hence we obtain

$$\begin{aligned} &f_1 \circ f_2 \circ g_2 \circ g_1 \circ L(f_1 \circ f_2) \\ &\quad = f_1 \circ f_2 \circ g_2 \circ g_1 \circ Lf_1 = f_1 \circ f_2 \circ g_2 \circ Lf_2 \circ g_1 \circ Lf_1 \\ &\quad = f_1 \circ Lf_2 \circ g_1 \circ Lf_1 = f_1 \circ g_1 \circ Lf_1 = Lf_1 = L(f_1 \circ f_2) \end{aligned}$$

and

$$L(g_2 \circ g_1 \circ L(f_1 \circ f_2)) = L(g_2 \circ g_1 \circ Lf_1) = L(g_2 \circ L(g_1 \circ Lf_1))$$
$$= L(g_2 \circ Lg_1) = L(g_2 \circ g_1).$$

Thus $g_2 \circ g_1[Q]f_1 \circ f_2$, which proves Lemma 2.1.4. □

2.2. Functions on Intervals

If f is a *real function*, i.e., if Dom f and Ran f are both subsets of R, and if c is an accumulation point of Dom f, then we denote the *left-hand limit* of f at c (if it exists) by $f(c-)$ and the *right-hand limit* of f at c (if it exists) by $f(c+)$. If c is in Dom f and $f(c-) = f(c)$, then f is *left continuous* at c; if $f(c+) = f(c)$, then f is *right continuous at* c; and f is *continuous at* c if $f(c+) = f(c-) = f(c)$.

Now let f be a real function defined on the closed interval $[a,b]$. Define $f(a-) = f(a)$, $f(b+) = f(b)$. Then f is *continuous on* $[a,b]$ if f is continuous at every point c in $[a,b]$. If c is in $[a,b]$ and f is not continuous at c, then f is *discontinuous at* c. In this case one of the following conditions holds:

i. $f(c+)$ and $f(c-)$ exist and $f(c+) = f(c-) \neq f(c)$. In this case c is a *removable discontinuity* of f.
ii. $f(c+)$ and $f(c-)$ exist, but $f(c+) \neq f(c-)$. In this case c is a *jump discontinuity* of f.
iii. Either $f(c+)$ or $f(c-)$ does not exist.

If f has only removable or jump discontinuities, then the values of f on any dense subset of $[a,b]$ including a and b uniquely determine $f(c-)$ and $f(c+)$ for all c in $[a,b]$.

Definition 2.2.1. Let f be a real function defined on $[a,b]$ (or on a dense subset of $[a,b]$, including a and b) having only removable or jump discontinuities. Then l^-f and l^+f are the functions defined on $[a,b]$ via

$$l^-f(x) = f(x-) \quad \text{and} \quad l^+f(x) = f(x+). \tag{2.2.1}$$

It is easy to show that l^-f is left continuous on $[a,b]$ and that l^+f is right continuous, whence

$$l^-l^-f = l^-f = l^-l^+f, \qquad l^+l^+f = l^+f = l^+l^-f. \tag{2.2.2}$$

Definition 2.2.2. A real function f is *nondecreasing* (resp., *strictly increasing*) on a subset S of its domain if $f(x) \leqslant f(y)$ (resp., $f(x) < f(y)$) for any x, y in S such that $x < y$. Similarly, f is *nonincreasing* (resp., *strictly decreasing*) on S if $f(x) \geqslant f(y)$ (resp., $f(x) > f(y)$) for any x, y in S such that

$x < y$. Finally, f is *monotonic* if it is either nondecreasing or nonincreasing.

If f is nondecreasing (resp., strictly increasing) on S, then $-f$ is nonincreasing (resp., strictly decreasing) on S. Thus we need only consider nondecreasing or strictly increasing functions.

If f is nondecreasing on $[a, b]$, then the functions $l^- f$ and $l^+ f$ defined in (2.2.1) are given by

$$l^- f(x) = \begin{cases} f(a), & x = a, \\ \sup\{f(t) | a \leqslant t < x\}, & a < x \leqslant b; \end{cases} \tag{2.2.3a}$$

$$l^+ f(x) = \begin{cases} \inf\{f(t) | x < t \leqslant b\}, & a \leqslant x < b, \\ f(b), & x = b. \end{cases} \tag{2.2.3b}$$

Furthermore, it is well known that if f is a monotonic function defined on $[a, b]$, then any discontinuity of f must be a jump discontinuity; that f can have at most denumerably many such discontinuities; and that f is completely determined by its values on any dense subset of $[a, b]$ that includes the points of discontinuity of f. Moreover, for $a \leqslant x < y \leqslant b$,

$$l^- f(x) \leqslant f(x) \leqslant l^+ f(x) \leqslant l^- f(y) \leqslant f(y) \leqslant l^+ f(y). \tag{2.2.4}$$

(For a discussion of the basic properties of monotonic functions, see [Boas 1972, §22]).

If f is defined and continuous on $[a, b]$, then Ran f is a closed interval. If f is defined and nondecreasing on $[a, b]$ and Ran f is an interval, then f is continuous on $[a, b]$.

Combining these facts with Lemma 2.1.1 yields

Lemma 2.2.3. *Let f be defined on $[a, d]$. Then f is continuous, nondecreasing, and idempotent if and only if there are points b, c with $a \leqslant b \leqslant c \leqslant d$ such that*

$$f(x) = \begin{cases} b, & x \text{ in } [a, b], \\ x, & x \text{ in } [b, c], \\ c, & x \text{ in } [c, d]. \end{cases} \tag{2.2.5}$$

Definition 2.2.4. Let f be defined on $[0, \infty)$. Then f is *subadditive* if

$$f(x + y) \leqslant f(x) + f(y) \quad \text{for all } x, y \text{ in } [0, \infty), \tag{2.2.6}$$

and f is *superadditive* if

$$f(x + y) \geqslant f(x) + f(y) \quad \text{for all } x, y \text{ in } [0, \infty). \tag{2.2.7}$$

(For a detailed treatment of subadditive functions, see [Hille and Philips 1957, Ch. 7].)

Definition 2.2.5. Let f be defined on an interval J that contains neither $-\infty$ nor ∞. Then f is *convex* on J if

$$f(\lambda x + (1-\lambda)y) \leqslant \lambda f(x) + (1-\lambda) f(y) \tag{2.2.8}$$

for all x, y in J and all λ in I; and f is *concave* on J if

$$f(\lambda x + (1-\lambda)y) \geqslant \lambda f(x) + (1-\lambda) f(y) \tag{2.2.9}$$

for all x, y in J and all λ in I.

A connection between the functions of Definitions 2.2.4 and 2.2.5 is provided by

Lemma 2.2.6. *Let f be defined on* $[0, \infty)$. *If f is concave and* $f(0) = 0$, *then f is subadditive. If f is convex and* $f(0) = 0$, *then f is superadditive.*

PROOF. Let x, y be in $[0, \infty)$. If $x + y = 0$, then $x = y = 0$ and, with $f(0) = 0$, both (2.2.6) and (2.2.7) are trivial. Hence we can assume that $x + y > 0$, whence

$$x = \left(\frac{x}{x+y}\right)(x+y) + \left(\frac{y}{x+y}\right)0$$

and

$$y = \left(\frac{x}{x+y}\right)0 + \left(\frac{y}{x+y}\right)(x+y).$$

If f is concave and $f(0) = 0$, we therefore obtain

$$f(x) \geqslant \frac{x}{x+y} f(x+y) + \frac{y}{x+y} f(0) = \frac{x}{x+y} f(x+y)$$

and similarly,

$$f(y) \geqslant \frac{y}{x+y} f(x+y).$$

Adding the inequalities yields (2.2.6). Similarly, if f is convex and $f(0) = 0$, we obtain (2.2.7). □

2.3. Probabilities, Integrals, Random Variables

We confine ourselves to the most basic definitions and results. Full details can be found in many places (e.g., [Loève 1977, Ch. I–III; Kingman and Taylor 1966, Ch. 3–11]).

Definition 2.3.1. Let Ω be a nonempty set. A *sigma field* on Ω is a family $\mathcal{Q}$ of subsets of Ω such that:

i. Ω is in $\mathcal{Q}$.
ii. If A is in $\mathcal{Q}$, then $\Omega \backslash A$, the complement of A, is in $\mathcal{Q}$.
iii. If A_n is in $\mathcal{Q}$ for $n = 1, 2, 3, \ldots$, then $\bigcup_{n=1}^{\infty} A_n$ is in $\mathcal{Q}$.

Note that the *empty set* $\varnothing = \Omega \backslash \Omega$ is in $\mathcal{Q}$. If A and B are in $\mathcal{Q}$, then $A \cup B$ is in $\mathcal{Q}$, whence $A \cap B$ and $B \backslash A$ also are in $\mathcal{Q}$. If A_n is in $\mathcal{Q}$ for $n = 1, 2, 3, \ldots$, then $\bigcap_{n=1}^{\infty} A_n$ is in $\mathcal{Q}$.

If Ω is a closed interval, then there is a unique smallest sigma field on Ω that contains all the subintervals of Ω; this is the *Borel field* on Ω, and its members are the *Borel sets* in Ω.

Definition 2.3.2. Let $\mathcal{Q}$ be a sigma field on a set Ω. A function f from Ω into R is *measurable with respect to* $\mathcal{Q}$ if, for every x in R, the inverse image $f^{-1}[-\infty, x)$ of the interval $[-\infty, x)$ is in $\mathcal{Q}$. If f is measurable with respect to the Borel field on a closed interval, then f is *Borel measurable*.

Clearly, any function that is nondecreasing or nonincreasing on a closed interval is Borel measurable.

Definition 2.3.3. A *probability space* is a triple $(\Omega, \mathcal{Q}, P)$, where Ω is a nonempty set, $\mathcal{Q}$ is a sigma field on Ω, and P is a function from $\mathcal{Q}$ into I such that the following conditions hold:

i. $P(\Omega) = 1$ and $P(\varnothing) = 0$. (2.3.1)
ii. If $\{A_n\}$ is a sequence of pairwise disjoint sets in $\mathcal{Q}$, then

$$P\left(\bigcup_{n=1}^{\infty} A_n\right) = \sum_{n=1}^{\infty} P(A_n). \tag{2.3.2}$$

The function P is a *probability measure* on Ω; P is *complete* if, whenever A in $\mathcal{Q}$ is such that $P(A) = 0$, then any subset of A is also in $\mathcal{Q}$.

An elementary consequence of this definition is

Lemma 2.3.4. *Let $(\Omega, \mathcal{Q}, P)$ be a probability space. Then for any A, B in $\mathcal{Q}$, we have*

i. *If $A \subseteq B$, then*

$$P(B) = P(A) + P(B \backslash A), \tag{2.3.3}$$

whence

$$P(A) \leqslant P(B), \tag{2.3.4}$$

so that P is nondecreasing on $\mathscr{Q}$.

ii. $P(\Omega \backslash A) = 1 - P(A)$. (2.3.5)

iii. $P(A \cup B) + P(A \cap B) = P(A) + P(B)$. (2.3.6)

iv. *If W and M are as in* (1.4.1), *then*

$$W(P(A), P(B)) \leqslant P(A \cap B) \leqslant M(P(A), P(B)). \tag{2.3.7}$$

v. *If P is complete, $A \subseteq C \subseteq B$, and $P(A) = P(B)$, then C is in $\mathscr{Q}$, and*

$$P(C) = P(A) = P(B). \tag{2.3.8}$$

PROOF. The proofs of (2.3.3)–(2.3.6) and (2.3.8) are standard exercises. The right-hand inequality in (2.3.7) follows from (2.3.4) and the left-hand inequality from the fact that $P(A \cap B) \geqslant 0$ and

$$\begin{aligned} P(A \cap B) &= P(A) + P(B) - P(A \cup B) \geqslant P(A) + P(B) - P(\Omega) \\ &= P(A) + P(B) - 1. \end{aligned}$$

□

If Ω is a closed interval $[a, d]$ and F is a nondecreasing function on Ω with $F(a) = 0$ and $F(d) = 1$, then F defines a unique probability measure P_F, called the *Lebesgue–Stieltjes F-measure* on Ω, with the following properties.

i. Every subinterval of Ω, and therefore every Borel set in Ω, is in $\operatorname{Dom} P_F$.

ii. If $a \leqslant b \leqslant c \leqslant d$, then

$$\begin{aligned} P_F[b, c] &= F(c+) - F(b-), \\ P_F[b, c) &= F(c-) - F(b-), \\ P_F(b, c] &= F(c+) - F(b+), \\ P_F(b, c) &= \operatorname{Max}(F(c-) - F(b+), 0). \end{aligned} \tag{2.3.9}$$

iii. P_F is complete.

If $\Omega = I$ and F is the identity function on I, then P_F is *Lebesgue measure* on I; in this case, we write λ for P_F and denote the corresponding probability space by (I, λ).

Definition 2.3.5. Let $(\Omega, \mathscr{Q}, P)$ be a probability space. Then $L_1^+(\Omega)$ is the set of all functions f from Ω into R that are measurable with respect to $\mathscr{Q}$ and satisfy

$$P\{\omega \text{ in } \Omega \mid f(\omega) < 0\} = 0.$$

Since a probability measure, like any function, determines its domain, it is always possible—and often convenient—to speak of a function as being *P-measurable*, rather than *measurable with respect to the sigma field* $\mathcal{Q} = \operatorname{Dom} P$. Similarly, any property that holds everywhere on Ω except possibly on a set of probability 0 is said to hold *almost everywhere* (*a.e.*) or *almost surely* (*a.s.*). Thus the elements of $L_1^+(\Omega)$ are the P-measurable functions that are a.s. nonnegative.

Any probability measure P determines a corresponding integral. When it exists, the integral of a P-measurable function f with respect to P over a set A in $\operatorname{Dom} P$ is denoted by $\int_A f dP$. If f is in $L_1^+(\Omega)$, then $\int_A f dP$ always exists (though it may be infinite) and has the following properties:

i. If $P(A) = 0$, then $\int_A f dP = 0$. (2.3.10)

ii. If c_1 and c_2 are constants such that $c_1 \leqslant f \leqslant c_2$ on A, then

$$c_1 P(A) \leqslant \int_A f dP \leqslant c_2 P(A). \tag{2.3.11}$$

iii. If $A = \bigcup_{n=1}^{\infty} A_n$, where the sets A_n in $\mathcal{Q}$ are pairwise disjoint, then

$$\int_A f dP = \sum_{n=1}^{\infty} \int_{A_n} f dP. \tag{2.3.12}$$

Upon taking f to be the constant function of value 1, it follows from (2.3.11) that

$$\int_A dP = P(A) \quad \text{for any } A \text{ in } \mathcal{Q}. \tag{2.3.13}$$

When dealing with a Lebesgue-Stieltjes measure P_F, we generally write $\int_A f dF$ rather than $\int_A f dP_F$.

Definition 2.3.6. A *random variable* (briefly, an *r.v.*) on a probability space $(\Omega, \mathcal{Q}, P)$ is a P-measurable function on Ω.

Random variables are often denoted by X, Y, etc. For any random variable X on Ω and any Borel-measurable function g, the *expectation* of $g(X)$ (when it exists) is the number $E(g(X))$ given by the integral

$$E(g(X)) = \int_{\Omega} g(X)\, dP. \tag{2.3.14}$$

2.4. Binary Operations

Definition 2.4.1. A *binary operation* on a nonempty set S is a function T from $S \times S$ into S, i.e., a function T with $\operatorname{Dom} T = S \times S$ and $\operatorname{Ran} T \subseteq S$. A *binary system* is a pair (S, T) where S is a nonempty set and T is a binary operation on S. If the restriction to $S \times S$ of a function T is a binary operation on S, then we say that S is *closed under* T.

Thus ordinary addition, subtraction, and multiplication are binary operations on $(-\infty, \infty)$, while $[0, \infty)$ is closed under both addition and multiplication, but not under subtraction. Division is not a binary operation on $(-\infty, \infty)$, but $(0, \infty)$ is closed under division.

Definition 2.4.2. Let T be a binary operation on S. For any a in S, the *vertical section of T at a* is the function v_a from S into S defined by

$$v_a(x) = T(a, x); \tag{2.4.1}$$

and the *horizontal section of T at a* is the function h_a from S into S defined by

$$h_a(x) = T(x, a). \tag{2.4.2}$$

The *diagonal section of T* is the function δ from S into S defined by

$$\delta(x) = T(x, x). \tag{2.4.3}$$

Definition 2.4.3. Let T be a binary operation on S. An element a of S is a *left null element* of T if $T(a, x) = v_a(x) = a$ for all x in S; a *right null element* of T if $T(x, a) = h_a(x) = a$ for all x in S; and a *null element* of T if it is both a left and a right null element of T. Correspondingly, an element a of S is a *left identity* of T if $v_a(x) = x$ for all x in S; a *right identity* of T if $h_a(x) = x$ for all x in S; and an *identity* of T if it is both a left and a right identity of T. An element a of S is *idempotent* under T if $T(a, a) = \delta(a) = a$, i.e., if a is a fixed point of δ.

Thus (left or right) null elements and (left or right) identities are idempotent elements.

Lemma 2.4.4. *If a is a left null element and b a right null element of T, then $a = b$. If a is a left identity and b a right identity of T, then $a = b$.*

It follows that a binary operation can have at most one null element and at most one identity. Idempotent elements that are neither null elements nor identities can of course be much more numerous.

Definition 2.4.5. Let T_1 be a binary operation on S_1, and T_2 a binary operation on S_2. If h is a function such that $\operatorname{Dom} h = S_1$, $\operatorname{Ran} h = S_2$, and

$$h(T_1(x, y)) = T_2(h(x), h(y)) \tag{2.4.4}$$

for all x, y in S_1, then h is a *homomorphism from* (S_1, T_1) *to* (S_2, T_2). A homomorphism h with an inverse h^{-1} is an *isomorphism.*

Evidently, if h is an isomorphism from (S_1, T_1) to (S_2, T_2), then h^{-1} is an isomorphism from (S_2, T_2) to (S_1, T_1), and (S_1, T_1) and (S_2, T_2) are *isomorphic*.

An immediate consequence of Definition 2.4.5 is

Lemma 2.4.6. *Let h be a homomorphism from (S_1, T_1) to (S_2, T_2). If a is, respectively, a left null element, right null element, null element, left identity, right identity, identity, or idempotent element of T_1, then $h(a)$ is, respectively, a left null element, right null element, null element, left identity, right identity, identity, or idempotent element of T_2. If T_1 is commutative, i.e., if $T_1(x, y) = T_1(y, x)$ for all x, y in S_1, then T_2 is also commutative. Similarly, if T_1 is associative, i.e., satisfies* (1.4.5) *for all x, y, z in S_1, then so is T_2.*

3

Metric and Topological Structures

3.1. Metric and Related Spaces

We begin with a summary of basic definitions and results. Details and further information may be found, e.g., in [Apostol 1974, §§3.13–3.16, 4.2–4.4; Blumenthal 1953, Ch. I; Sierpiński 1956].

Definition 3.1.1. A *metric space* is a pair (S, d) where S is a nonempty set and d, the *distance function* or *metric* of the space, is a function from $S \times S$ into $[0, \infty)$ satisfying the following conditions:

i. $d(p,q) = 0 \quad \text{if } p = q,$ (3.1.1)

ii. $d(p,q) > 0 \quad \text{if } p \neq q,$ (3.1.2)

iii. $d(p,q) = d(q,p) \quad \text{for all } p, q \text{ in } S,$ (3.1.3)

iv. $d(p,r) \leqslant d(p,q) + d(q,r) \quad \text{for all } p, q, r \text{ in } S.$ (3.1.4)

The function d is a *pseudometric* if it satisfies (3.1.1), (3.1.3), and (3.1.4); a *semimetric* if it satisfies (3.1.1), (3.1.2), and (3.1.3); a *nonsymmetric metric* if it satisfies (3.1.1), (3.1.2), and (3.1.4); an *extended real-valued metric* if it is allowed to assume the value ∞; and an *ultrametric* if, instead of (3.1.4), it satisfies the stronger condition

$$d(p,r) \leqslant \operatorname{Max}(d(p,q), d(q,r)) \quad \text{for all } p, q, r \text{ in } S. \tag{3.1.5}$$

The triangle inequality (3.1.4) is symmetric in $d(p,q)$, $d(q,r)$, and $d(p,r)$.

For using (3.1.3), we have

$$d(p,q) \leqslant d(p,r) + d(r,q) = d(p,r) + d(q,r)$$

and

$$d(q,r) \leqslant d(q,p) + d(p,r) = d(p,q) + d(p,r).$$

These inequalities and (3.1.4) may be combined into the single inequality

$$D(p,q,r) \leqslant 0 \tag{3.1.6}$$

where

$$D(p,q,r) = \begin{vmatrix} 0 & 1 & 1 & 1 \\ 1 & 0 & (d(p,q))^2 & (d(p,r))^2 \\ 1 & (d(p,q))^2 & 0 & (d(q,r))^2 \\ 1 & (d(p,r))^2 & (d(q,r))^2 & 0 \end{vmatrix}$$

is a *Cayley-Menger determinant* (see [Blumenthal 1961, §VII.1]). Such determinants play an essential role in the metric characterization of subsets of Euclidean and Hilbert spaces (see [Menger 1928; Blumenthal 1953, §§40–42; Menger 1954, Ch. I, §3]).

A pseudometric space (S,d) can always be condensed to a metric space by means of the following well-known construction: Let $\sim$ be the equivalence relation on S defined by

$$p \sim q \quad \text{if and only if} \quad d(p,q) = 0. \tag{3.1.7}$$

Let p^* be the equivalence class containing p, i.e., $p^* = \{p' \text{ in } S \mid p' \sim p\}$, and let S^* be the set of all such equivalence classes. Then the expression

$$d^*(p^*,q^*) = d(p,q)$$

defines a function d^* on $S^* \times S^*$ and (S^*,d^*) is a metric space, the *quotient space of the pseudometric space* (S,d).

Given a metric on S, or even a nonnegative function d merely satisfying (3.1.1) and (3.1.3), we can introduce a notion of *convergence* for sequences in S via

$$p_n \to p \quad \text{if and only if} \quad d(p_n, p) \to 0. \tag{3.1.8}$$

In general, only the complete set of conditions (3.1.1)–(3.1.4) will ensure that this convergence behaves properly; e.g., if d is only a pseudometric, then a sequence of points in S may converge to two or more distinct points. Similarly, semimetrics abound with pathologies (see [Blumenthal 1953, Ch. I]). On the other hand, we have the following:

Theorem 3.1.2. *Let d be a pseudometric on the set S and let convergence of sequences in S be defined by* (3.1.8). *Then d is a continuous function on $S \times S$, i.e.,*

$$\text{If } p_n \to p \text{ and } q_n \to q, \text{ then } d(p_n, q_n) \to d(p,q). \tag{3.1.9}$$

For future comparison with Theorem 12.2.2, we give the very simple proof of (3.1.9). For any $n \geqslant 1$, two applications of (3.1.4) yield

$$\begin{aligned} d(p_n, q_n) &\leqslant d(p_n, p) + d(p, q_n) \\ &\leqslant d(p_n, p) + d(p,q) + d(q, q_n). \end{aligned} \tag{3.1.10}$$

Similarly, using (3.1.3), we obtain

$$d(p,q) \leqslant d(p_n, p) + d(p_n, q_n) + d(q, q_n). \tag{3.1.11}$$

Combining (3.1.10) and (3.1.11), we have

$$|d(p_n, q_n) - d(p,q)| \leqslant d(p_n, p) + d(q_n, q) \tag{3.1.12}$$

for every $n \geqslant 1$, from which (3.1.9) follows immediately.

It should be noted that under the hypotheses of Theorem 3.1.2, the function d is *uniformly continuous* in the sense that the estimate in (3.1.12) is independent of p and q. It should further be noted that neither (3.1.1) nor (3.1.2) was used in the proof of Theorem 3.1.2. Without these conditions, however, the theorem may be vacuous in the sense that there may be no convergent sequences in S. For example, without (3.1.1), a constant or eventually constant sequence need not converge.

A *Cauchy sequence* is a sequence $\{p_n\}$ such that $d(p_n, p_m) \to 0$ as $m, n \to \infty$. Every convergent sequence is a Cauchy sequence; but the converse may not be valid. If it is (i.e., if every Cauchy sequence in a metric space (S,d) is convergent), then (S,d) is *complete*. Also, a subset B of a metric space (S,d) is *compact* if every infinite sequence in B has a convergent subsequence whose limit is in B.

If (S,d) is a metric space, then for any p in S and any $x > 0$, the *open ball* with *center* p and *radius* x is the set $B(p,x) = \{q \mid d(p,q) < x\}$. A point p is an *accumulation point* of a subset A of S if every open ball $B(p,x)$ contains at least one point of A distinct from p. The union of A and the set of its accumulation points is the *closure* of A and is denoted by $\bar{A}$. The set A is *closed* if $A = \bar{A}$. The set A is *open* if its complement is closed—or, equivalently, if for every p in A there is an $x > 0$ such that $B(p,x)$ is in A. The set A is *dense* in S if $\bar{A} = S$; and finally, the space (S,d) is *separable* if S contains a denumerable dense subset.

A function f from S into S is *continuous at a point* p in S if for every $\eta > 0$ there is a $\lambda(\eta, p) > 0$ such that

$$d(f(p), f(q)) < \eta \quad \text{whenever} \quad d(p,q) < \lambda(\eta, p); \tag{3.1.13}$$

f is *continuous on a subset* A of S if (3.1.13) holds for all points p in A; and f is *uniformly continuous on* A if $\lambda(\eta, p)$ is independent of p. Similarly, a binary operation T on S is continuous at a point (p_1, p_2) in $S \times S$ if for every $\eta > 0$ there is a $\lambda(\eta, (p_1, p_2))$ such that

$$d(T(p_1, p_2), T(q_1, q_2)) < \eta \tag{3.1.14}$$

whenever $d(p_1, q_1) < \lambda$ and $d(p_2, q_2) < \lambda$. Continuity and uniform continuity of T on a subset A of $S \times S$ are defined similarly. Furthermore, T is *continuous in each place* on $S \times S$ if all vertical sections v_a and all horizontal sections h_a are continuous functions from S into S. Clearly, if T is continuous, then T is continuous in each place; the converse is false. If J is an interval and T is a binary operation on J, then T is *nondecreasing in each place* if all vertical sections v_a and all horizontal sections h_a are nondecreasing on J.

The following theorem is easily established.

Theorem 3.1.3. *Let T be a binary operation on the interval J. If T is continuous in each place and nondecreasing in each place, then T is continuous on $J \times J$.*

3.2. Isometries, Homotheties, Metric Transforms

Let (S_1, d_1) and (S_2, d_2) be metric spaces and suppose that φ is a one-to-one function from S_1 onto S_2. If

$$d_2(\varphi(p), \varphi(q)) = d_2(\varphi(r), \varphi(s)) \quad \text{whenever} \quad d_1(p, q) = d_1(r, s), \tag{3.2.1}$$

then there is a real function f such that

$$d_2(\varphi(p), \varphi(q)) = f(d_1(p, q)) \quad \text{for all } p, q \text{ in } S. \tag{3.2.2}$$

If there is an $a > 0$ such that $f(x) = ax$ for all x in $\operatorname{Ran} d_1$, then φ is a *homothety* from (S_1, d_1) to (S_2, d_2). If $a = 1$, so that f is the identity function on $\operatorname{Ran} d_1$, then φ is an *isometry* from $(S_1 d_1)$ to (S_2, d_2). If φ is a homothety (isometry) from (S_1, d_1) to (S_2, d_2), then φ^{-1} is a homothety (isometry) from (S_2, d_2) to (S_1, d_1) and we say that (S_1, d_1) and (S_2, d_2) are *homothetic* (*isometric*). Since homothety and isometry are both equivalence relations in the class of metric spaces, homothetic and isometric spaces can be regarded as spaces with identical metric structures.

In the other direction, if we start with a metric space (S_1, d_1), a one-to-one function φ from S_1 onto a set S_2, and a real function f, then we can use (3.2.2) to define d_2. In this case (3.2.1) follows automatically, but d_2 may not be a metric on S_2. If we want d_2 to be a metric, then we must impose appropriate conditions on f. To do this we begin with

Definition 3.2.1. A function f from $[0, \infty)$ into $[0, \infty)$ is a *weak metric transform* if, whenever (S_1, d_1) is a metric space, φ is a one-to-one function from S_1 onto a set S_2, and d_2 is defined via (3.2.2), then (S_2, d_2) is a metric space; f is a *metric transform* if it is a weak metric transform and

$$d_1(p,q) < d_1(r,s) \quad \text{implies} \quad d_2(\varphi(p),\varphi(q)) \leqslant d_2(\varphi(r),\varphi(s)) \tag{3.2.3}$$

for all p, q, r, s in S.

Note that in view of (3.1.6), weak metric transforms can be succinctly described as functions that preserve the sign of every Cayley-Menger determinant $D(p,q,r)$.

Familiar examples of metric transforms are the function $f(x) = x/1 + x$ and the power functions x^α for $0 < \alpha \leqslant 1$.

Theorem 3.2.2. *A function f from $[0, \infty)$ into $[0, \infty)$ is a weak metric transform if and only if it satisfies the following conditions:*

i. $f(0) = 0,$ (3.2.4)

ii. $f(x) > 0$ *for* $x > 0,$ (3.2.5)

iii. $f(z) \leqslant f(x) + f(y)$ *whenever* $|x - y| \leqslant z \leqslant x + y.$ (3.2.6)

Furthermore, f is a metric transform if and only if f is a weak metric transform and is nondecreasing.

PROOF. Suppose that f is a weak metric transform. Let (S_1, d_1) be the Euclidean plane and φ the identity function on S_1. Then (3.2.2) becomes

$$d_2(p,q) = f(d_1(p,q)), \tag{3.2.7}$$

whence for any p in S_1 the requirement that d_2 be a metric yields

$$0 = d_2(p,p) = f(d_1(p,p)) = f(0),$$

which is (3.2.4). Next, for any $x > 0$ there are points p, q in S_1 for which $d_1(p,q) = x$. Hence we have

$$0 < d_2(p,q) = f(d_1(p,q)) = f(x),$$

which is (3.2.5). Finally, if x, y, $z \geqslant 0$ are any numbers such that $|x - y| \leqslant z \leqslant x + y$, then there is a triangle (p,q,r) in S_1 with $d_1(p,q) = x$, $d_1(q,r) = y$, $d_1(p,r) = z$. But

$$d_2(p,r) \leqslant d_2(p,q) + d_2(q,r),$$

and this, by virtue of (3.2.7), yields (3.2.6).

In the other direction, suppose f is a function satisfying (3.2.4), (3.2.5), and (3.2.6). Let (S_1, d_1) be an arbitrary metric space and φ an arbitrary one-to-one map of S_1 onto a set S_2. Define d_2 via (3.2.2). Then for any p in S_2 we have

$$d_2(p,p) = f\big(d_1(\varphi^{-1}(p), \varphi^{-1}(p))\big) = f(0) = 0.$$

Next, for any distinct points p, q in S_2, $\varphi^{-1}(p)$ and $\varphi^{-1}(q)$ are distinct points in S_1, whence $d_1(\varphi^{-1}(p), \varphi^{-1}(q)) > 0$ and

$$d_2(p,q) = f\big(d_1(\varphi^{-1}(p), \varphi^{-1}(q))\big) > 0.$$

The symmetry condition for d_2 follows automatically from (3.2.2) and the symmetry of d_1. Lastly, if p, q, r are any three points in S_2, then (3.1.4) and (3.1.3) imply that the distances $x = d_1(\varphi^{-1}(p), \varphi^{-1}(q))$, $y = d_1(\varphi^{-1}(q), \varphi^{-1}(r))$, and $z = d_1(\varphi^{-1}(p), \varphi^{-1}(r))$ satisfy the condition $|x - y| \leqslant z \leqslant x + y$, whence (3.2.6) yields the triangle inequality for d_2. Thus d_2 is a metric and the proof of the first part of the theorem is complete.

Turning to metric transforms proper, we observe first that if f is a nondecreasing weak metric transform, then (3.2.3) follows immediately from (3.2.2). In the other direction, there are metric spaces (e.g., the Euclidean line) such that for any numbers x, y with $0 \leqslant x < y$ there are points p, q, r, s in the space such that $d_1(p,q) = x$, $d_1(r,s) = y$. Hence if f is a metric transform, then (3.2.2) and (3.2.3) yield

$$\begin{aligned} f(x) = f(d_1(p,q)) &= d_2(\varphi(p), \varphi(q)) \leqslant d_2(\varphi(r), \varphi(s)) \\ &= f(d_1(r,s)) = f(y), \end{aligned}$$

whence f is nondecreasing. This completes the proof. □

There are weak metric transforms that are not monotonic and hence not metric transforms,—for example, the function f defined by

$$f(x) = \begin{cases} x, & 0 \leqslant x \leqslant 2, \\ 4 - x, & 2 \leqslant x \leqslant 3, \\ 1, & 3 \leqslant x. \end{cases}$$

Upon comparing (3.2.6) with (2.2.6), we see that (3.2.6) is a stronger condition than subadditivity. On the other hand, if a function f defined on $[0, \infty)$ is both subadditive and nondecreasing, then for any x, y, z such that $x, y \geqslant 0$ and $0 \leqslant z \leqslant x + y$, we have

$$f(z) \leqslant f(x + y) \leqslant f(x) + f(y),$$

which is even stronger than (3.2.6). Hence we have

Theorem 3.2.3. *A function f from $[0, \infty)$ into $[0, \infty)$ is a metric transform if and only if it satisfies (3.2.4) and (3.2.5) and is nondecreasing and subadditive.*

An appeal to Lemma 2.2.6 then yields the following:

Corollary 3.2.4. *If f is a function from* $[0, \infty)$ *into* $[0, \infty)$ *that is nondecreasing, concave, and satisfies* (3.2.4) *and* (3.2.5), *then f is a metric transform.*

The converse of Corollary 3.2.4 is not valid: There are metric transforms that are not concave—for example, the function f defined by

$$f(x) = \begin{cases} \sqrt{x}, & 0 \leqslant x \leqslant 1, \\ x, & 1 < x. \end{cases}$$

A continuous metric transform is a *scale function* in the sense of W. A. Wilson [1935]. On the other hand, there are metric transforms that are not continuous, hence not scale functions. The simplest example is the restriction to R^+ of the function ε_0 defined in (1.5.1), which is discontinuous at 0. Note that ε_0 transforms any metric (or even any semimetric) space (S, d) into the *discrete* metric space (S, d_0), where $d_0(p,q) = 0$ if $p = q$, and $d_0(p,q) = 1$ if $p \neq q$.

3.3. Betweenness

In the first of his four "Untersuchungen über allgemeine Metrik," Menger [1928] showed that the notion of betweenness could be defined in arbitrary metric spaces as follows:

Definition 3.3.1. Let (S, d) be a metric space. If p, q, r are points of S, then q is *between* p and r, and we write $\langle pqr \rangle$, if and only if p, q, r are distinct and

$$d(p,r) = d(p,q) + d(q,r). \tag{3.3.1}$$

The following statements are immediate consequences of the definition.

i. If $\langle pqr \rangle$, then $\langle rqp \rangle$. (3.3.2)
ii. If $\langle pqr \rangle$, then neither $\langle qpr \rangle$ nor $\langle prq \rangle$. (3.3.3)
iii. If $\langle pqr \rangle$ and $\langle prs \rangle$, then both $\langle pqs \rangle$ and $\langle qrs \rangle$. (3.3.4)
iv. If $p \neq r$, then the set $A(p,r) = \{q \mid \langle pqr \rangle\} \cup \{p, r\}$ is closed. (3.3.5)

In studying the geometric structure of a metric space, it is important to know whether the space admits segments joining all pairs of points. Here a *segment* joining the distinct points p, q in the metric space (S, d) is a subset of S that contains p and q and is isometric to a Euclidean line segment of length $d(p,q)$. Menger showed that a metric space admits a segment joining each pair of distinct points if the space is complete and has the following property: For any two distinct points p, r, there is a point q such that $\langle pqr \rangle$ (for details, see [Blumenthal 1953, Section 14]).

A natural question is, what metric transforms preserve betweenness? The answer is given by

Theorem 3.3.2. *A metric transform f preserves betweenness if and only if $f(x) = ax$ for some $a > 0$. Hence any two metric spaces connected via (3.2.2) by a betweenness-preserving metric transform are homothetic. Conversely, homotheties preserve betweenness.*

3.4. Minkowski Metrics

The following procedure, due to H. Minkowski [1910, Kap. 1, §2], leads to a class of spaces that play a prominent role in our phenomenological model of hysteresis. Let n be a fixed positive integer, $R_0 = (-\infty, \infty)$, and d_E the Euclidean metric on R_0^n. Let z designate a fixed point in R_0^n, and let $\mathfrak{M}$ be a set of points in R_0^n with the property that each ray issuing from z intersects $\mathfrak{M}$ in precisely one point distinct from z, i.e., $\mathfrak{M}$ is the boundary of a *star-shaped* region. Then for every ordered pair (p, q) of distinct points of R_0^n there is a unique point r in $\mathfrak{M}$ such that the directed segment $\overrightarrow{zr}$ is parallel to and has the same sense as the directed segment $\overrightarrow{pq}$. Now define the function $d_{\mathfrak{M}}$ induced by the *Minkowski indicatrix* $\mathfrak{M}$ by setting $d_{\mathfrak{M}}(p, p) = 0$ and

$$d_{\mathfrak{M}}(p, q) = d_E(p, q)/d_E(z, r) \tag{3.4.1}$$

when p and q are distinct. It is clear that $d_{\mathfrak{M}}$ always satisfies (3.1.1) and (3.1.2), and that it satisfies (3.1.3) if and only if $\mathfrak{M}$ is symmetric about z. As for the triangle inequality, it is well known (see e.g., [Valentine 1964, p. 32]) that $d_{\mathfrak{M}}$ satisfies this condition if and only if $\mathfrak{M}$ is the boundary of a convex body. Thus $d_{\mathfrak{M}}$ is a metric if and only if $\mathfrak{M}$ is the boundary of a convex body symmetric about z.

In particular, if our space is the Cartesian plane, then we have the indicatrices $\mathfrak{M}_\alpha$, $\alpha > 0$, given by the equations

$$|x|^\alpha + |y|^\alpha = 1. \tag{3.4.2}$$

These are all symmetric, but convex only for $\alpha \geqslant 1$. For $\alpha = 1$ the indicatrix $\mathfrak{M}_\alpha$ is a square with vertices at the points $(\pm 1, 0)$, $(0, \pm 1)$; for $\alpha = 2$ it is the unit circle; while as $\alpha \to \infty$ the curves $\mathfrak{M}_\alpha$ approach the curve $\mathfrak{M}_\infty$, whose equation is $\operatorname{Max}(|x|, |y|) = 1$, i.e., the square with vertices at the points $(\pm 1, \pm 1)$. The corresponding metrics d_α are the well-known two-dimensional l_α-metrics, which are given by

$$d_\alpha((x_1, y_1), (x_2, y_2)) = (|x_1 - x_2|^\alpha + |y_1 - y_2|^\alpha)^{1/\alpha} \quad (3.4.3)$$

for $1 \leqslant \alpha < \infty$, and

$$d_\infty((x_1, y_1), (x_2, y_2)) = \mathrm{Max}(|x_1 - x_2|, |y_1 - y_2|). \quad (3.4.4)$$

3.5. Topological Structures

Definition 3.5.1. Let S be a nonempty set. For each p in S let $\mathcal{V}_p$ be a nonempty family of subsets of S such that:

N1. p is in V for every V in $\mathcal{V}_p$.

Then each V in $\mathcal{V}_p$ is a *neighborhood of* p, $\mathcal{V}_p$ is a *neighborhood system at* p, and the family $\mathcal{V} = \bigcup_{p \text{ in } S} \mathcal{V}_p$ is a *neighborhood system for* S.

Definition 3.5.2. Two neighborhood systems $\mathcal{V}$ and $\mathcal{W}$ for the same set S are *equivalent at a point* p in S if for each V in $\mathcal{V}_p$ there is a W in $\mathcal{W}_p$ such that $W \subseteq V$, and conversely. The neighborhood systems $\mathcal{V}$ and $\mathcal{W}$ are *equivalent* if they are equivalent at each point of S. A neighborhood system $\mathcal{V}$ for S satisfies the *first axiom of countability* (briefly, is *first countable*) if there is a neighborhood system $\mathcal{W}$ such that $\mathcal{W}$ is equivalent to $\mathcal{V}$, and $\mathcal{W}_p$ is denumerable for each p in S.

In addition to N1, a neighborhood system $\mathcal{V}$ for S may satisfy one or more of the following conditions:

N2. For each U and V in $\mathcal{V}_p$, there is a W in $\mathcal{V}_p$ such that $W \subseteq U \cap V$.
N3. If V is a neighborhood of p, and q is in V, then there is a neighborhood W of q such that $W \subseteq V$.
N4. If $p \neq q$, then there is a V in $\mathcal{V}_p$ and a W in $\mathcal{V}_q$ such that $V \cap W = \varnothing$.

Definition 3.5.3. If $\mathcal{V}$ is a neighborhood system for S, then the pair $(S, \mathcal{V})$ is a *V-space*. If $\mathcal{V}$ also satisfies N2, then $(S, \mathcal{V})$ is a *V_D-space*.

The concepts of V-space, V_D-space, and related neighborhood spaces are due to Fréchet [1917, 1928]. A detailed analysis of such spaces is given in [Appert and Ky-Fan 1951], and a clear exposition of the theory of V-spaces may be found in Chapter 1 of [Sierpiński 1956].

Definition 3.5.4. Let $(S, \mathcal{V})$ be a V-space, p a point in S, and A a subset of S. Then p is *$\mathcal{V}$-contiguous to A* if and only if $A \cap V \neq \varnothing$ for every V in $\mathcal{V}_p$.

It is immediate that if $\mathcal{V}$ and $\mathcal{W}$ are equivalent neighborhood systems, then p is $\mathcal{V}$-contiguous to A if and only if p is $\mathcal{W}$-contiguous to A.

Definition 3.5.5. If $(S, \mathcal{V})$ is a V-space, then c is the function from $\mathcal{P}(S)$ into $\mathcal{P}(S)$ defined by

$$c(A) = \{ p \mid p \text{ is } \mathcal{V}\text{-contiguous to } A \}. \tag{3.5.1}$$

Lemma 3.5.6. *If* $(S, \mathcal{V})$ *is a* $\mathcal{V}$*-space and* c *is defined by* (3.5.1), *then*

i. $c(\emptyset) = \emptyset,$ (3.5.2)

ii. $A \subseteq c(A)$ *for all* A *in* $\mathcal{P}(S)$, (3.5.3)

iii. $c(A) \cup c(B) \subseteq c(A \cup B)$ *for all* A, B *in* $\mathcal{P}(S)$. (3.5.4)

If $(S, \mathcal{V})$ *is a* V_D*-space, then*

iv. $c(A) \cup c(B) = c(A \cup B)$ *for all* A, B *in* $\mathcal{P}(S)$. (3.5.5)

It follows from (3.5.3) that $c(S) = S$, and from (3.5.4) that c is *nondecreasing*, i.e., that

$$c(A) \subseteq c(B) \quad \text{whenever} \quad A \subseteq B. \tag{3.5.6}$$

Definition 3.5.7. Let S be a nonempty set. A *Fréchet closure operation for S* is a function c from $\mathcal{P}(S)$ into $\mathcal{P}(S)$ satisfying (3.5.2), (3.5.3), and (3.5.4); if, in addition, c satisfies (3.5.5), then c is a *Čech closure operation*. A *closure space* is a pair (S, c) where c is a Čech closure operation for S. If (S, c) is a closure space and $A \subseteq S$, then $c(A)$ is the *c-closure of A*. The set A is *c-closed* if $c(A) = A$; and A is *c-open* if $S \setminus A$, the complement of A, is c-closed. The set $\operatorname{int} A = S \setminus c(S \setminus A)$ is the *c-interior* of A.

Note that in any closure space, the intersection of any collection, and the union of any finite number, of c-closed sets is c-closed.

Closure spaces appear in the work of Fréchet and Appert and Ky-Fan; but the name, as well as the systematic approach to topological structures from this point of view, is due to E. Čech [1966]. They are directly related to V-spaces in the following manner:

Definition 3.5.8. Let (S, c) be a closure space. For any p in S, a *c-neighborhood of p* is a subset V of S such that p is in $\operatorname{int} V$; the *c-neighborhood system at p* is the family $\mathcal{V}_p^c$ of all c-neighborhoods of p; and the *c-neighborhood system for S* is the family $\mathcal{V}^c = \bigcup_{p \text{ in } S} \mathcal{V}_p^c$.

Theorem 3.5.9. *If (S, c) is a closure space and $\mathcal{V}^c$ is the c-neighborhood system for S, then $(S, \mathcal{V}^c)$ is a V_D-space. In the other direction, if $(S, \mathcal{V})$ is a V_D-space and c is defined by* (3.5.1), *then (S, c) is a closure space and the neighborhood systems $\mathcal{V}$ and $\mathcal{V}^c$ are equivalent.*

PROOF. Let (S, c) be a closure space and p any point in S. Since $\operatorname{int} S = S$, it follows that p is in $\operatorname{int} S$, whence $\mathcal{V}_p^c$ is nonempty. Now suppose that p is in $\operatorname{int} V$ for some subset V of S. Then p is not in $c(S \setminus V)$, whence by (3.5.3) p is not in $S \setminus V$. Thus p is in V and $\mathcal{V}^c$ satisfies N1. As for N2, let U and V be in $\mathcal{V}_p^c$. Then p is in $(\operatorname{int} U) \cap (\operatorname{int} V)$ and using (3.5.5) we have

$$\begin{aligned}(\operatorname{int} U) \cap (\operatorname{int} V) &= S \setminus (c(S \setminus U) \cup c(S \setminus V))\\ &= S \setminus c((S \setminus U) \cup (S \setminus V))\\ &= \operatorname{int}(U \cap V).\end{aligned}$$

Hence p is in $\operatorname{int}(U \cap V)$, whence $U \cap V$ is in $\mathcal{V}_p^c$ and N2 is satisfied. Thus $(S, \mathcal{V}^c)$ is a V_D-space.

Conversely, if $(S, \mathcal{V})$ is a V_D-space and c is defined by (3.5.1), then it follows immediately from Lemma 3.5.6 that (S, c) is a closure space.

Next, let p be in S and V be in $\mathcal{V}_p$. Since $V \cap (S \setminus V) = \emptyset$, it follows from Definitions 3.5.4 and 3.5.5 that p is not in $c(S \setminus V)$. Hence p is in $\operatorname{int} V$, whence V is in $\mathcal{V}_p^c$. Thus any set V in $\mathcal{V}_p$ contains a set (V itself) in $\mathcal{V}_p^c$. In the other direction, let V be in $\mathcal{V}_p^c$. Then p is in $\operatorname{int} V = S \setminus c(S \setminus V)$. Therefore p is not in $c(S \setminus V)$ and, again by Definitions 3.5.4 and 3.5.5, there is a set U in $\mathcal{V}_p$ such that $U \cap (S \setminus V) = \emptyset$. It follows that $U \subseteq V$. Thus any set in $\mathcal{V}_p^c$ contains a set in $\mathcal{V}_p$. Hence $\mathcal{V}$ and $\mathcal{V}^c$ are equivalent at p, and since p is arbitrary, $\mathcal{V}$ and $\mathcal{V}^c$ are equivalent. □

Definition 3.5.10. If c is a Čech closure operation for a set S and if

$$c(c(A)) = c(A) \quad \text{for all } A \text{ in } \mathcal{P}(S), \tag{3.5.7}$$

i.e., if c is *idempotent*, then c is a *Kuratowski closure operation* (cf. [Kelley 1955, p. 43]).

Lemma 3.5.11. *If $\mathcal{V}$ is a neighborhood system for S that satisfies* N2 *and* N3, *and if c is defined by* (3.5.1), *then c is a Kuratowski closure operation for S. Conversely, let c be a Kuratowski closure operation for S and let $\mathcal{V}^c$ be the c-neighborhood system for S. If V is in $\mathcal{V}^c$, then* $\operatorname{int} V$ *is in $\mathcal{V}^c$ and the family $\{\operatorname{int} V \mid V \text{ in } \mathcal{V}^c\}$ is a neighborhood system for S that is equivalent to $\mathcal{V}^c$ and satisfies* N2 *and* N3.

Definition 3.5.12. Let S be a nonempty set. A *topology* for S is a family $\mathfrak{T}$ of subsets of S that includes S and $\varnothing$ and is closed under arbitrary unions and finite intersections. A subset A of S is *open* (*relative to* $\mathfrak{T}$) if and only if A is in $\mathfrak{T}$; A is *closed* (*relative to* $\mathfrak{T}$) if and only if $S \setminus A$ is open. The pair $(S, \mathfrak{T})$ is a *topological space.*

Lemma 3.5.13. *Let $(S, \mathfrak{T})$ be a topological space. Then $\mathfrak{T} \setminus \{\varnothing\}$ is a neighborhood system for S that satisfies* N2 *and* N3. *Conversely, if $\mathfrak{V}$ is a neighborhood system for S that satisfies* N2 *and* N3, *then the family $\mathfrak{T}$ consisting of $\varnothing$ and all unions of elements of $\mathfrak{V}$ is a topology for S, and $(S, \mathfrak{T})$ is a topological space.*

Lemma 3.5.14. *Let $(S, \mathfrak{T})$ be a topological space. Then the function c defined by* (3.5.1) *is a Kuratowski closure operation for S and the family of all c-open sets in S coincides with $\mathfrak{T}$. Conversely, let c be a Kuratowski closure operation for S and define $\mathfrak{T}$ to be the family of c-open sets. Then $(S, \mathfrak{T})$ is a topological space.*

The condition N4 is the *Hausdorff separation property*; and a *Hausdorff space* is a topological space $(S, \mathfrak{T})$ in which the family $\mathfrak{T}$ satisfies N4. If (S, d) is a metric space, then the family $\mathfrak{B}_d$ of all open balls of S satisfies N1–N4. Thus a metric d on a set S determines a unique topology $\mathfrak{T}_d$ for S. This is the *metric topology*. Lastly, a topological space $(S, \mathfrak{T})$ is *metrizable* if and only if there exists a metric d on S such that the neighborhood systems $\mathfrak{T}$ and $\mathfrak{B}_d$ are equivalent.

Returning to closure spaces, we have

Theorem 3.5.15. *Let (S, c) be a closure space. Define a function c^K from $\mathfrak{P}(S)$ into $\mathfrak{P}(S)$ by taking $c^K(A)$ to be the intersection of all the c-closed subsets of S containing A. Then c^K is a Kuratowski closure operation. In particular, if A is c-closed, then $c^K(A) = A$, whence A is c^K-closed.*

It follows that every closure space (S, c) has a natural Kuratowski closure operation and hence a topology associated with it. In this topology the closed sets are precisely those subsets of S that are c-closed.

Since $c^K(A)$ is c-closed and $A \subseteq c^K(A)$, we have

$$c(A) \subseteq c\big(c^K(A)\big) = c^K(A) \quad \text{for every } A \text{ in } \mathfrak{P}(S).$$

The inclusion can be proper as the following example shows: Let S be the set $\{p,q,r\}$ and let

$$c(\varnothing)=\varnothing,\quad c\{p\}=\{p,r\},\quad c\{q\}=\{q,p\},\quad c\{r\}=\{r,q\},$$
$$c\{p,q\}=c\{q,r\}=c\{p,r\}=c(S)=S.$$

Then c is a closure operation, and the only c-closed subsets of S are $\varnothing$ and S itself. In particular, $c\{p\}\neq c^K\{p\}$.

We conclude with one further example. Let (S,d) be a metric space and θ a nonnegative number, which may be interpreted as a threshold. For any p in S and any $t>0$, define the (θ,t)-neighborhood of p to be the set

$$N_p(\theta,t)=\{q\mid d(p,q)<\theta+t\};$$

define the θ-neighborhood system at p to be the family

$$\mathfrak{N}_p^\theta=\{N_p(\theta,t)\mid t>0\},$$

and the θ-neighborhood system for S to be the union $\mathfrak{N}^\theta=\bigcup_{p\text{ in }S}\mathfrak{N}_p^\theta$. It is readily seen that $(S,\mathfrak{N}^\theta)$ is a V_D-space. Hence for every value of θ, (3.5.1) yields a closure operation c^θ for S. The c^0-closure operation is a Kuratowski closure operation and yields the metric topology; but for $\theta\neq0$ this is generally not the case. For example, if $S=(-\infty,\infty)^n$ and d is the usual Euclidean metric on S, then for any positive threshold θ, the only c^θ-closed subsets of S are $\varnothing$ and S itself. Hence the topology determined by the Kuratowski closure $c^{\theta K}$ is the *indiscrete* topology which lumps together all the points of S. In contrast, the c^θ closure itself fails to distinguish only those points whose separation falls within the threshold θ. Otherwise, if $d(p,q)>\theta$, then there is a (θ,t)-neighborhood of p that does not contain q and a (θ,t)-neighborhood of q that does not contain p; while if $d(p,q)>2\theta$, then there are (θ,t)-neighborhoods of p and of q that are disjoint. We note that such structures have been considered in connection with the idea of an "elementary length" in physics (see, e.g., [Gudder 1968]).

4
Distribution Functions

4.1. Spaces of Distribution Functions

Definition 4.1.1. A *distribution function* (briefly, a *d.f.*) is a nondecreasing function F defined on R, with $F(-\infty) = 0$ and $F(\infty) = 1$.

There is nothing probabilistic about this definition, yet distribution functions usually occur in contexts where statements about probabilities are made. Such statements involve a "quantity" X and an element x in R and take the following forms: "The probability that X is less than x is $l^- F(x)$"; "The probability that X is not greater than x is $l^+ F(x)$"; "The probability that X is equal to x is $l^+ F(x) - l^- F(x)$."

Since such statements involve only $l^- F$ and $l^+ F$ and not F itself directly, and since each of $l^+ F$, $l^- F$ determines the other via (2.2.2), it is customary to normalize d.f.'s by requiring them to be either left or right continuous. We normalize d.f.'s by taking them to be left continuous on $(-\infty, \infty)$. Accordingly, we make the following:

Definition 4.1.2. The set of all distribution functions that are left continuous on $(-\infty, \infty)$ will be denoted by Δ. The subset of Δ consisting of those elements F such that $l^+ F(-\infty) = F(-\infty) = 0$ and $l^- F(\infty) = F(\infty) = 1$ will be denoted by $\mathscr{D}$.

Since the appearance of the epoch-making book by A. N. Kolmogorov [1933], the "quantity" X referred to above is customarily taken to be a random variable defined on some probability space (see Section 2.3). Accordingly, we have

Definition 4.1.3. Let X be a random variable on the probability space $(\Omega, \mathcal{A}, P)$. Then F_X is the function on R defined by $F_X(-\infty) = 0$, $F_X(\infty) = 1$, and

$$F_X(x) = P\{\omega \text{ in } \Omega \mid X(\omega) < x\} \quad \text{for } -\infty < x < \infty. \tag{4.1.1}$$

It is readily verified that F_X is indeed a distribution function. On occasion we also denote F_X by $df(X)$.

In much of the literature on probability theory, it is assumed that $P\{\omega \text{ in } \Omega \mid -\infty < X(\omega) < \infty\} = 1$, i.e., that F_X is in $\mathcal{D}$. However, there are situations (see, e.g., [Feller 1966, p. 360; Hammersley 1974, p. 652] and Sections 12.4, 12.6, and 13.1) where $P\{X(\omega) = \infty\} > 0$. In this case, the random variable X is *defective*.

For a further discussion of distribution functions and random variables, see Sections 10 and 11 of [Loève 1977]. For an analysis of the meaning of the term "quantity" used loosely above, see [Menger 1956].

The elements of Δ are partially ordered via

$$F \leqslant G \quad \text{if and only if} \quad F(x) \leqslant G(x) \quad \text{for all } x \text{ in } R. \tag{4.1.2}$$

Definition 4.1.4. For any a in R, ε_a, the *unit step at* a, is the function in Δ given by

$$\varepsilon_a(x) = \begin{cases} 0, & -\infty \leqslant x \leqslant a, \\ 1, & a < x \leqslant \infty, \end{cases} \quad \text{for } -\infty \leqslant a < \infty, \tag{4.1.3a}$$

$$\varepsilon_\infty(x) = \begin{cases} 0, & -\infty \leqslant x < \infty, \\ 1, & x = \infty. \end{cases} \tag{4.1.3b}$$

For any a, b such that $-\infty < a < b < \infty$, U_{ab}, the *uniform distribution on* $[a,b]$, is the function in $\mathcal{D}$ given by:

$$U_{ab}(x) = \begin{cases} 0, & x \text{ in } [-\infty, a], \\ \dfrac{x-a}{b-a}, & x \text{ in } [a,b], \\ 1, & x \text{ in } [b, \infty]. \end{cases} \tag{4.1.4}$$

We shall usually abbreviate U_{0b} to U_b.

Note that $\varepsilon_a \leqslant \varepsilon_b$ if and only if $b \leqslant a$; that ε_a is in $\mathcal{D}$ if $-\infty < a < \infty$; and that $\varepsilon_{-\infty}$ is the maximal element, and ε_∞ the minimal element, of Δ. Moreover, if X is an r.v. such that $df(X) = \varepsilon_a$, then $X = a$ a.s.

Using expectations, the following is easily established (for an alternate proof, see Lemma 4.1 in [Calabrese 1978]).

Lemma 4.1.5. *Let X, Y be random variables on the same probability space. If $X \leqslant Y$ a.s., $df(X)$ and $df(Y)$ are both in $\mathfrak{D}$, and $df(X) = df(Y)$, then $X = Y$ a.s.*

Finally, note that if F is in Δ and if A is any bounded subset of R, then

$$\sup\{F(x) \mid x \text{ in } A\} = F(\sup\{x \mid x \text{ in } A\}). \tag{4.1.5}$$

4.2. The Modified Lévy Metric

In 1925 P. Lévy introduced (by use, rather than explicit definition) a metric for the set Δ. The explicit definition came in 1935 in a note by Lévy appended to [Fréchet 1936] (see also [Gnedenko and Kolmogorov 1954, §9]). A useful modification of the Lévy metric was introduced by D. A. Sibley [1971], and it is a slight modification of this modified Lévy metric that is best suited to our purposes.

Definition 4.2.1. Let F and G be in Δ, let h be in $(0, 1]$, and let $(F, G; h)$ denote the condition

$$F(x-h) - h \leqslant G(x) \leqslant F(x+h) + h \quad \text{for all } x \text{ in } \left(-\frac{1}{h}, \frac{1}{h}\right). \tag{4.2.1}$$

The *modified Lévy distance* is the function d_L defined on $\Delta \times \Delta$ by

$$d_L(F, G) = \inf\{h \mid \text{both } (F, G; h) \text{ and } (G, F; h) \text{ hold}\}. \tag{4.2.2}$$

Note that for any F and G in Δ, both $(F, G; 1)$ and $(G, F; 1)$ hold, whence the set in (4.2.2) is nonempty and $d_L(F, G) \leqslant 1$. Furthermore we have

Lemma 4.2.2. *If $d_L(F, G) = h > 0$, then both $(F, G; h)$ and $(G, F; h)$ hold.*

PROOF. For any $t > 0$, let J_t denote the interval $(-1/t, 1/t)$ and note that if $0 < s < t \leqslant 1$, then $J_t \subseteq J_s$.

Let x be in J_h. Since J_h is open, there is a y in J_h such that $y < x$ and a $t > 0$ such that y is in J_{h+t}. Since $d_L(F, G) = h$, it follows that $(F, G; h + t)$ holds. Therefore,

$$F(y-h-t) - (h+t) \leqslant G(y) \leqslant F(y+h+t) + (h+t).$$

Letting t decrease to 0 and using the left-continuity of F yields

$$F(y-h) - h \leqslant G(y) \leqslant F((y+h)+) \leqslant F(x+h) + h.$$

Next, letting y increase to x and using the left-continuity of F and G yields

$$F(x-h)-h \leqslant G(x) \leqslant F(x+h)+h$$

for all x in J_h. Thus $(F,G;h)$ holds. Interchanging the roles of F and G yields $(G,F;h)$ and completes the proof. □

Theorem 4.2.3. *The function d_L is a metric on Δ.*

PROOF. Conditions (3.1.1) and (3.1.3) for a metric are trivial. Next, if $d_L(F,G)=0$, then (4.2.1) and (4.1.5) yield $l^-F(x) \leqslant G(x)$ and $l^-G(x) \leqslant F(x)$ for all x in $(-\infty,\infty)$, whence by left-continuity, $F=G$ and (3.1.2) holds. To establish the triangle inequality

$$d_L(F,H) \leqslant d_L(F,G)+d_L(G,H) \tag{4.2.3}$$

for any distinct F, G, H in Δ, let $\alpha=d_L(F,G)>0$ and $\beta=d_L(G,H)>0$. If $\alpha+\beta \geqslant 1$, then (4.2.3) holds trivially. Assume therefore that $\alpha+\beta<1$ and let x be in $J_{\alpha+\beta}$. Then it easily follows that both $x-\beta$ and $x+\beta$ are in J_α. Using this fact together with Lemma 4.2.2 and (4.2.1), we then have

$$\begin{aligned} F(x-\alpha-\beta)-\alpha-\beta &\leqslant G(x-\beta)-\beta \leqslant H(x) \\ &\leqslant G(x+\beta)+\beta \leqslant F(x+\alpha+\beta)+\alpha+\beta, \end{aligned}$$

whence $(F,H;\alpha+\beta)$ holds. Similarly, since $x-\alpha$ and $x+\alpha$ are both in J_β, it follows that $(H,F;\ \alpha+\beta)$ also holds. Consequently, $d_L(F,H) \leqslant \alpha+\beta=d_L(F,G)+d_L(G,H)$ and the proof is complete. □

The significance of the modified Lévy metric lies in its connection with weak convergence. There are several equivalent definitions of this notion. The most convenient for our purposes is

Definition 4.2.4. A sequence $\{F_n\}$ of d.f.'s *converges weakly* to a d.f. F (and we write $F_n \xrightarrow{w} F$) if and only if the sequence $\{F_n(x)\}$ converges to $F(x)$ at each continuity point x of F.

Note that convergence at each point of continuity of F is not equivalent to convergence at every point of R. For example, $U_{a-(1/n),a} \xrightarrow{w} \varepsilon_a$, but for all n, $U_{a-(1/n),a}(a)=1 \neq 0=\varepsilon_a(a)$.

Theorem 4.2.5. *Let $\{F_n\}$ be a sequence of functions in Δ, and let F be in Δ. Then $F_n \xrightarrow{w} F$ if and only if $d_L(F_n,F) \to 0$.*

PROOF. Suppose $d_L(F_n,F) \to 0$ and let x be a point of continuity of F. We can assume x to be in $(-\infty,\infty)$, since convergence at $-\infty$ or ∞ is automatic. Then for all sufficiently small positive h, the interval $(x-h,x+h)$

is contained in the interval $(-1/h, 1/h)$; for any such h, $(F_n, F; h)$ holds for sufficiently large n, whence we have

$$F(x-2h) - h \leqslant F_n(x-h) \leqslant F_n(x) \leqslant F_n(x+h) \leqslant F(x+2h) + h.$$

Since h can be arbitrarily small, and F is continuous at x, it follows that $F_n(x) \to F(x)$.

In the other direction, suppose that $F_n \xrightarrow{w} F$. Let h satisfy $0 < h \leqslant 1$. Since the set of continuity points of F is dense in R, there is a finite set $A = \{c_0, c_1, \ldots, c_p\}$ of these continuity points such that $c_0 \leqslant -1/h$, $c_p \geqslant 1/h$, and $c_{m-1} < c_m \leqslant c_{m-1} + h$ for $m = 1, 2, \ldots, p$. Since A is finite, it follows that for sufficiently large n we have

$$|F_n(c_m) - F(c_m)| \leqslant h \quad \text{at every } c_m.$$

Now let x be any point in $(-1/h, 1/h)$. Then x is in one of the intervals $[c_{m-1}, c_m]$, and we have

$$\begin{aligned} F(x-h) - h &\leqslant F(c_{m-1}) - h \leqslant F_n(c_{m-1}) \leqslant F_n(x) \leqslant F_n(c_m) \\ &\leqslant F(c_m) + h \leqslant F(x+h) + h, \end{aligned}$$

which is $(F_n, F; h)$. Interchanging F_n and F in the last string of inequalities yields $(F, F_n; h)$. It follows that $d_L(F_n, F) \to 0$, and this completes the proof. □

Helly's weak compactness theorem (see [Loève 1977, §11.2]) states that every infinite sequence of d.f.'s contains a weakly convergent subsequence. Combined with Theorem 4.2.5 this yields

Corollary 4.2.6. *The metric space* (Δ, d_L) *is compact, and hence complete.*

Lévy's original definition differed from Definition 4.2.1 in requiring the inequalities in (4.2.1) to hold for all x in $(-\infty, \infty)$. This does not affect the first half of the proof of Theorem 4.2.5. However, it invalidates the argument in the second half. Indeed, the "only if" part of the statement of the theorem does not hold for the original Lévy metric. For example, $\varepsilon_n \xrightarrow{w} \varepsilon_\infty$ but the sequence $\{\varepsilon_n\}$ does not converge to ε_∞ in the original Lévy metric. On the other hand, as is easily seen, for any $a > 0$, $d_L(\varepsilon_a, \varepsilon_\infty) = \operatorname{Min}(1, 1/a)$, whence $d_L(\varepsilon_n, \varepsilon_\infty) \to 0$.

It should also be noted that our definition of d_L differs slightly from Sibley's. Our alteration amounts to a slight change in the lengths of the intervals on which the inequalities in (4.2.1) are required to hold. This change does not affect the validity of any previously established results; it does, however, simplify many definitions and proofs. A comparison of Sibley's metric with the original Lévy metric may be found in [Sibley 1971] (see also [Schweizer 1975a]).

4.3. The Space of Distance Distribution Functions

In the sequel we shall be interested almost exclusively in the subset of Δ consisting of those elements F that satisfy $F(0) = 0$. Since any such F is automatically equal to 0 on the half-line $[-\infty, 0]$, it is convenient to consider F as a function whose domain is R^+ rather than R, while continuing to refer to F as a distribution function. There are numerous technical advantages in this; and since the set of all such functions F is a closed subspace of (Δ, d_L), nothing is lost. Accordingly, we make the following:

Definition 4.3.1. A *distance distribution function* (briefly, a *d.d.f.*) is a nondecreasing function F defined on R^+ that satisfies $F(0) = 0$ and $F(\infty) = 1$, and is left continuous on $(0, \infty)$. The set of all d.d.f's will be denoted by Δ^+; and the set of all F in Δ^+ for which $l^- F(\infty) = F(\infty) = 1$ by $\mathcal{D}^+$. A d.d.f. is *strict* if it is continuous and strictly increasing on R^+.

Note that a strict or merely strictly increasing d.d.f. is invertible in Δ^+, but not in Δ.

For any $a \geqslant 0$, we shall view the unit step function ε_a as belonging to either Δ or Δ^+, as the case may be. The set Δ^+ inherits the ordering defined by (4.1.2), but now ε_0 is the maximal element of Δ^+, while ε_∞ remains the minimal element.

The function d_L is clearly a metric on Δ^+; and again $d_L(F, F_n) \to 0$ if and only if $F_n \xrightarrow{w} F$, so that the metric space (Δ^+, d_L) is compact and complete. In addition, it follows from (4.2.1), (4.2.2), and the fact that $F(0) = G(0) = 0$ that if we let $[F, G; h]$ denote the condition

$$G(x) \leqslant F(x+h) + h \quad \text{for } x \text{ in } (0, 1/h), \tag{4.3.1}$$

then for any F, G in Δ^+ and h in $(0, 1]$,

$$d_L(F, G) = \inf\{h \mid \text{both } [F, G; h] \text{ and } [G, F; h] \text{ hold}\}. \tag{4.3.2}$$

We also have the following results, which will be needed later:

Lemma 4.3.2. *If F, G are in Δ^+ and $d_L(F, G) = h > 0$, then both $[F, G; h]$ and $[G, F; h]$ hold.*

Lemma 4.3.3. *For any F in Δ^+,*

$$\begin{aligned} d_L(F, \varepsilon_0) &= \inf\{h \mid [F, \varepsilon_0; h] \text{ holds}\} \\ &= \inf\{h \mid F(h+) > 1 - h\}; \end{aligned} \tag{4.3.3}$$

and for any $t > 0$,

$$F(t) > 1 - t \quad \text{iff} \quad d_L(F, \varepsilon_0) < t. \tag{4.3.4}$$

Geometrically, $d_L(F, \varepsilon_0)$ is the abscissa of the point of intersection of the line $y = 1 - x$ and the graph of F (completed, if necessary, by the addition of vertical segments at discontinuities). From this we immediately obtain

Lemma 4.3.4. *If F and G are in Δ^+ and $F \leqslant G$, then $d_L(G, \varepsilon_0) \leqslant d_L(F, \varepsilon_0)$.*

Finally, we also have

Lemma 4.3.5. *The supremum of any set of d.f.'s in Δ^+ is in Δ^+.*

On the other hand, the infimum of a set of d.f.'s in Δ^+, although nondecreasing, need not be left continuous, hence not in Δ^+: Consider, for example, the sequence $\{\varepsilon_{a+1/n}\}$.

4.4. Quasi-Inverses of Nondecreasing Functions

Let f be nondecreasing on a closed interval $[a, b]$ and let y be a point in $[f(a), f(b)]$. If the set $f^{-1}(y)$ is not empty, then it is an interval (which may be a single point). If it is empty, then $\sup\{x \mid f(x) < y\} = \inf\{x \mid f(x) > y\}$ and f has a jump discontinuity at $\sup\{x \mid f(x) < y\}$. These observations motivate the following:

Definition 4.4.1. Let f be nondecreasing on $[a, b]$. Then $Q(f)$ is the set of functions f^* defined on $[f(a), f(b)]$ via

i. $f^*(f(a)) = a$ and $f^*(f(b)) = b$. (4.4.1)

ii. If y is in Ran f, then

$$f^*(y) \text{ is in } f^{-1}(y). \tag{4.4.2}$$

iii. If y is not in Ran f, then

$$f^*(y) = \sup\{x \mid f(x) < y\} = \inf\{x \mid f(x) > y\}. \tag{4.4.3}$$

Note that as an immediate consequence of the conditions (4.4.1) and (4.4.2), we have

$$f(f^*(y)) = y \quad \text{for all } y \text{ in Ran } f. \tag{4.4.4}$$

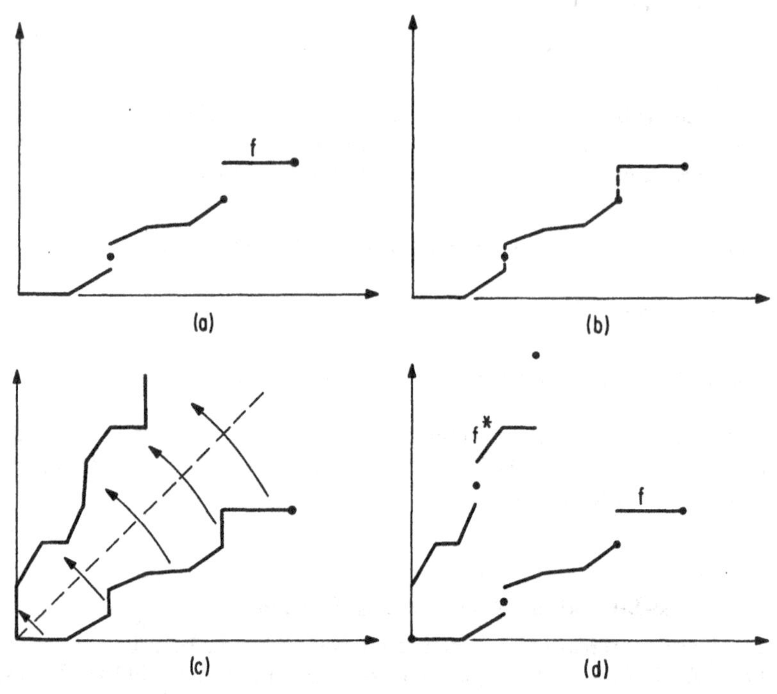

Figure 4.4.1. **(a)** The graph of f. **(b)** Vertical segments added (step i). **(c)** Reflection in the line $y = x$ (step ii). **(d)** Vertical segments deleted from the reflected curve (step iii). Comparison of the graphs of f and f^*.

The structure of the functions in $Q(f)$ can be characterized geometrically as follows (see Figure 4.4.1).

i. Consider the graph of f and complete it to a continuous curve by adding the vertical segments joining the points $(x, f(x-))$ and $(x, f(x+))$ at each discontinuity x of f.
ii. Reflect the resulting curve in the line $y = x$.
iii. Delete all but one point from any vertical segment contained in the reflected curve.

The resulting set is the graph of a function in $Q(f)$. All functions in $Q(f)$ are obtainable in this way.

It follows, either from the geometric construction or by a direct argument, that all functions f^* in $Q(f)$ are nondecreasing and coincide except on an at most denumerable set of discontinuities. There is a unique function $f^\wedge$ in $Q(f)$ that is left continuous on $(f(a), f(b))$ and a unique function $f^\vee$ that is right continuous on $(f(a), f(b))$. These are given, respectively, by

$$f^\wedge(y) = \sup\{x \mid f(x) < y\} \quad \text{and} \quad f^\vee(y) = \inf\{x \mid f(x) > y\}. \tag{4.4.5}$$

Furthermore, for all f^* in $Q(f)$,

$$\begin{gathered} f^\wedge \leqslant f^* \leqslant f^\vee, \\ l^- f^* = f^\wedge, \quad \text{and} \quad l^+ f^* = f^\vee \quad \text{on } (f(a), f(b)). \end{gathered} \tag{4.4.6}$$

Theorem 4.4.2. *Let f be nondecreasing on (a,b). Then we have*

i. *If f is left continuous on (a,b), then $f(f^*(y)) \leqslant y$ for all y in $[f(a), f(b)]$ and all f^* in $Q(f)$; and $f^\vee\,[Q]f$ (see Definition 2.1.2).* (4.4.7)
ii. *If f is right continuous on (a,b), then $f(f^*(y)) \geqslant y$ for all y in $[f(a), f(b)]$ and all f^* in $Q(f)$; and $f^\wedge\,[Q]f$.* (4.4.8)
iii. *If f is continuous on $[a,b]$, then each f^* in $Q(f)$ is strictly increasing on $[f(a), f(b)]$, and $f^*[Q]f$.* (4.4.9)
iv. *If f is strictly increasing on $[a,b]$, then $Q(f)$ consists of a single continuous function f^*, and $f^*[Q]f$.* (4.4.10)

PROOF. Let y be in $(f(a), f(b))$ and let $\bar{x} = \sup\{x \mid f(x) \leqslant y\}$. Let f be left continuous. Then $f(\bar{x}) \leqslant y$, and $f(\bar{x}) = y$ iff y is in $\operatorname{Ran} f$. Thus, using (4.4.5) and (4.4.6), for any f^* in $Q(f)$ we have

$$\begin{aligned} f(f^*(y)) &\leqslant f(f^\vee(y)) = f(\inf\{x \mid f(x) > y\}) \\ &= f(\sup\{x \mid f(x) \leqslant y\}) = f(\bar{x}) \leqslant y. \end{aligned}$$

Combined with (4.4.4), this yields $f(f^*(y)) \leqslant y$ for all y in $[f(a), f(b)]$. If y is not in $\operatorname{Ran} f$, then using (4.4.3) we have

$$f^\vee(y) = \sup\{x \mid f(x) \leqslant y\} = \bar{x} = f^\vee(f(\bar{x})),$$

whence the range of the restriction of $f^\vee$ to $\operatorname{Ran} f$ is $\operatorname{Ran} f^\vee$. Combined with (4.4.4), this yields $f^\vee\,[Q]f$ and establishes (4.4.7). Similar arguments establish (4.4.8).

Next, if f is continuous, then $\operatorname{Ran} f$ is the entire interval $[f(a), f(b)]$. Hence, if f^* is in $Q(f)$, then $f^*[Q]f$; and if y_1, y_2 are such that $f^*(y_1) = f^*(y_2)$, then $y_1 = f(f^*(y_1)) = f(f^*(y_2)) = y_2$. Thus f^* is one to one and, being nondecreasing, is strictly increasing. This proves (4.4.9).

Finally, if f is strictly increasing, then it follows that $\sup\{x \mid f(x) < y\} = \inf\{x \mid f(x) > y\}$ for all y in $(f(a), f(b))$. Hence $f^\wedge = f^\vee$, whence (4.4.6) shows that $Q(f)$ consists of a single function f^*, which being both left and right continuous, is continuous. By (4.4.9), $f[Q]f^*$, whence $f^*[Q]f$ and the theorem is proved. □

Corollary 4.4.3. *If f, g are nondecreasing on (a,b) and if $f \geqslant g$, then $f^{\wedge} \leqslant g^{\wedge}$ and $f^{\vee} \leqslant g^{\vee}$.*

Corollary 4.4.4. *If f is nondecreasing and left continuous on (a,b), then for any x in (a,b) and any y in $(f(a), f(b))$ we have:*

i. *If $f(x) \geqslant y$, then $f^{\wedge}(y) \leqslant x$.* (4.4.11)

ii. *If $f(x) > y$, then $f^{\wedge}(y) < x$.* (4.4.12)

iii. *If $f^{\wedge}(y) < x$, then $f(x) \geqslant y$.* (4.4.13)

iv. *If $f^{\wedge}(y) \leqslant x$, then $f(x+h) \geqslant y$ for any $h > 0$.* (4.4.14)

Combining Lemma 4.3.3 and Corollary 4.4.4 yields

Lemma 4.4.5. *For any F in Δ^+ and any t in $(0,1)$, we have*

i. *If $d_L(F,\varepsilon_0) < t$, then $F^{\wedge}(1-t) < t$.* (4.4.15)
ii. *If $F^{\wedge}(1-t) < t$, then $d_L(F,\varepsilon_0) \leqslant t$.*

If F is in Δ, then $Q(F)$ consists of nondecreasing functions defined on I. Thus any F^* in $Q(F)$ is a Lebesgue-measurable function, hence a random variable on (I,λ). If F is in $\mathfrak{D}$, then F^* is finite a.e., hence a nondefective r.v. on (I,λ). The distribution function of F^* is F, i.e., for all x in $(-\infty,\infty)$ we have

$$\lambda\{\omega \mid \omega \text{ in } I,\ F^*(\omega) < x\} = F(x). \tag{4.4.16}$$

A standard result (see, e.g., [Kingman and Taylor 1966, Th. 6.11]) is

Lemma 4.4.6. *If F is in Δ, and g is a Borel-measurable function from R into R, then*

$$\int_R g\,dF = \int_I (g \circ F^*)\,d\lambda \quad \text{for any } F^* \text{ in } Q(F), \tag{4.4.17}$$

in the sense that if either side exists, then so does the other and the two are equal.

In particular, if k is a nonnegative integer and if $m^{(k)}(F)$, the kth *moment* of F, which is defined by

$$m^{(k)}(F) = \int_R x^k\,dF(x), \tag{4.4.18}$$

exists, then

$$m^{(k)}(F) = \int_0^1 (F^*(t))^k\,dt \tag{4.4.19}$$

for any F^* in $Q(F)$. Note that in view of (2.3.14), the right-hand side of (4.4.19) is the expectation of the kth power of F^*, considered as a random variable on (I, λ).

Various standard moment-convergence theorems (see [Loève 1977, §11.4]) are easy consequences of (4.4.19). For example we have

Lemma 4.4.7. *Let F_n, $n = 1, 2, \ldots$, and F be in $\mathfrak{D}$. If $F_n \xrightarrow{w} F$ and the sequence $\{(F_n^*)^k\}$ is dominated on I by an integrable function, then the moments $m^{(k)}(F_n)$ and $m^{(k)}(F)$ exist, and*

$$\lim_{n\to\infty} m^{(k)}(F_n) = m^{(k)}(F).$$

If F is in Δ^+, then for any β in $(0, \infty)$ we can define the *moment* $m^{(\beta)}(F)$ *of order* β of F. This always exists, but may be infinite, and is given by

$$m^{(\beta)}F = \int_{R^+} x^\beta \, dF(x) = \int_0^1 \left(F^\wedge(t)\right)^\beta dt. \tag{4.4.20}$$

In addition, if we define

$$m^{(0)}F = 1 - F(0+) \tag{4.4.21}$$

for all F in Δ^+, then it follows from (4.4.20) that

$$m^{(0)}F = \lim_{\beta\to 0} m^{(\beta)}F$$

whenever there is a $\beta > 0$ such that $m^{(\beta)}F < \infty$. Similarly, for any F in Δ^+ we have

$$\lim_{\beta\to\infty} \left(m^{(\beta)}F\right)^{1/\beta} = F^\wedge(1-). \tag{4.4.22}$$

The following simple fact, which follows immediately from (4.4.12), will also be needed in the sequel.

Lemma 4.4.8. *If $\{F_n\}$ is a sequence in Δ^+ that converges weakly to ε_0, then $F_n^\wedge(t) \to 0$ for every t in $[0, 1)$.*

Finally, (see the lemma in [Brown 1972]) we have

Lemma 4.4.9. *Let F_n, $n = 1, 2 \ldots$, and F be in $\mathfrak{D}$. If $F_n \xrightarrow{w} F$, and F_n^* is in $Q(F_n)$ for all n, then the sequence $\{F_n^*\}$ converges almost everywhere to a function in $Q(F)$.*

By Egorov's theorem (see [Kingman and Taylor 1966, Th. 7.1]), the phrase "almost everywhere" in Lemma 4.4.9 can be replaced by "almost uniformly."

5
Associativity

5.1. Associative Binary Operations

A binary operation T on a set S (see Definition 2.4.1) is *associative* if

$$T(T(x, y), z) = T(x, T(y, z)) \quad \text{for all } x, y, z \text{ in } S. \tag{5.1.1}$$

If T is an associative binary operation on S, then the pair (S, T) is a *semigroup*.

The following simple but elegant characterization of associative binary operations is due to D. Zupnik.

Theorem 5.1.1. *A binary operation T on S is associative if and only if every vertical section of T commutes with every horizontal section of T, i.e., if and only if*

$$v_a \circ h_b = h_b \circ v_a \quad \textit{for all } a, b \textit{ in } S. \tag{5.1.2}$$

As noted in Section 2.4, associativity is preserved under homomorphisms. Thus if (S_1, T_1) is a semigroup and h is a homomorphism from (S_1, T_1) to (S_2, T_2), then (S_2, T_2) is also a semigroup. Another useful result is

Lemma 5.1.2. *Let T be an associative binary operation on S. If the element a in S is idempotent under T, then both v_a and h_a are idempotent functions.*

PROOF. For any x in S, we have

$$\begin{aligned} v_a^2(x) &= v_a(v_a(x)) = v_a(T(a,x)) = T(a,T(a,x)) \\ &= T(T(a,a),x) = T(a,x) = v_a(x). \end{aligned}$$

Hence v_a is idempotent. Similarly, h_a is idempotent. □

Definition 5.1.3. Let T be a binary operation on S and let x be an element of S. The *T-powers* of x are the elements of S given recursively by

$$x^1 = x \quad \text{and} \quad x^{n+1} = T(x^n, x) \tag{5.1.3}$$

for all positive integers n.

Since both addition and multiplication are commutative operations on the integers, a familiar induction yields

$$x^{m+n} = T(x^m, x^n) = T(x^n, x^m) \tag{5.1.4a}$$

and

$$x^{mn} = (x^m)^n = (x^n)^m \tag{5.14b}$$

for all positive integers m, n and all x in S.

If T is associative and has an identity element e, then (5.1.3) can be extended to the nonnegative integers by defining $x^0 = e$ for all x except left and right null elements of T. When this is done, then (except for null elements) (5.1.4) extends to all nonnegative integers m, n.

Finally (see (2.4.3)), another simple induction yields

$$\delta^n(x) = x^{2^n} \tag{5.1.5}$$

for all integers $n \geqslant 0$ and all x in S.

5.2. Generators and Ordinal Sums

There are many ways of constructing new semigroups from old. Two such ways are presented in this section. We begin with

Theorem 5.2.1. *Let (S, T_1) be a semigroup and let f, g be functions such that:*

i. Ran $g \subseteq S$;
ii. *for any u, v in* Ran g, *$T_1(u,v)$ is in* Dom f;
iii. $f[Q]g$.

Suppose further that either

iv. *$T_1(g(x), g(y))$ is in* Ran g *for all x, y in* Dom g; *or*

v. *there is an element a in* Ran f *such that* $f(T_1(u,v)) = a$ *whenever* $f(u) = a$, *or* $f(v) = a$, *or* u, v *are in* Ran g *but* $T_1(u,v)$ *is not in* Ran g.

Define a binary operation T on Dom g *by*

$$T(x,y) = f(T_1(g(x), g(y))). \tag{5.2.1}$$

Then (Dom g, T) *is a semigroup.*

PROOF. For arbitrary x, y, z in Dom g, let $u = g(x)$, $v = g(y)$, $w = g(z)$. Then we have

$$T(T(x,y),z) = f(T_1(g(f(T_1(u,v))),w)) \tag{5.2.2}$$

and

$$T(x,T(y,z)) = f(T_1(u, g(f(T_1(v,w))))). \tag{5.2.3}$$

If $T_1(u,v)$ is in Ran g, then since $f[Q]g$, we have

$$g(f(T_1(u,v))) = (g \circ f)(T_1(u,v)) = T_1(u,v),$$

whence (5.2.2) reduces to

$$T(T(x,y),z) = f(T_1(T_1(u,v),w)). \tag{5.2.4}$$

If $T_1(u,v)$ is not in Ran g, then assumption v holds. Since a is in Ran f and $f[Q]g$, we have $f(g(a)) = a$, whence $f(T_1(g(a),w)) = a$. But since u, v are in Ran g and $T_1(u,v)$ is not, we also have $f(T_1(u,v)) = a$, whence $f(T_1(T_1(u,v),w)) = a$. Combining the last three equalities, we obtain

$$f(T_1(g(f(T_1(u,v))),w)) = f(T_1(g(a),w)) = a = f(T_1(T_1(u,v),w)),$$

which, when combined with (5.2.2), yields (5.2.4). Thus (5.2.4) holds in all cases.

In a similar manner, we obtain

$$T(x,T(y,z)) = f(T_1(u,T_1(v,w))),$$

which, combined with (5.2.4) and the associativity of T_1, yields the fact that T is associative. □

Theorem 5.2.1 is a generalization of Theorems 4 and 5 in [S^2 1963c]. It may be remarked that in general neither f nor g in Theorem 5.2.1 is a homomorphism in either direction. However, it is readily verified that when assumption iv holds, then T', the restriction of T_1 to Ran $g \times$ Ran g, is a binary operation on Ran g and g is a homomorphism from (Dom g, T) to (Ran g, T'). Also, it is easily seen that if f and g are inverses, then assumption iv holds and (Dom g, T) and (Ran g, T') are isomorphic.

Definition 5.2.2. Let S, T_1, f, g be as in Theorem 5.2.1 and let T be defined by (5.2.1). Then T is *generated from T_1 by the pair* (f, g).

By virtue of Lemma 2.1.4, we therefore obtain

Lemma 5.2.3. *Let T_1 be generated from T_0 by the pair (f_1, g_1) and T_2 from T_1 by the pair (f_2, g_2). Then T_2 is generated from T_0 by the pair $(f_2 \circ f_1, g_1 \circ g_2)$.*

The second topic of this section is the ordinal sum construction.

Definition 5.2.4. Let $\{(S_\alpha, T_\alpha)\}$ be a family of binary systems indexed by a set $\mathcal{I}$ of real numbers. For all α, β, γ in $\mathcal{I}$, let these binary systems satisfy the following compatibility conditions:

i. If $\alpha < \beta < \gamma$ and $S_\alpha \cap S_\gamma$ is nonempty, then $S_\beta = S_\alpha \cap S_\gamma$.
ii. If $\alpha < \beta$ and x is in $S_\alpha \cap S_\beta$, then x is the unique element in $S_\alpha \cap S_\beta$ and is the identity of T_α and the null element of T_β.

Finally, let S be the union of all the sets S_α and let T be defined on S as follows: For x in S_α and y in S_β,

$$T(x, y) = \begin{cases} x, & \alpha < \beta, \\ T_\alpha(x, y), & \alpha = \beta, \\ y, & \alpha > \beta. \end{cases} \tag{5.2.5}$$

(Note that the compatibility conditions guarantee that (5.2.5) is consistent, whence T is a well-defined binary operation on S.) Then the binary system (S, T) is the *ordinal sum* of the family of binary systems $\{(S_\alpha, T_\alpha)\}$.

Now suppose that each T_α is associative. Let x be in S_α, y in S_β, and z in S_γ. Then consideration of the various possibilities determined by the ordering of α, β, and γ shows that $T(T(x, y), z) = T(x, T(y, z))$ in all cases. This proves

Theorem 5.2.5. *An ordinal sum of associative binary systems is itself associative.*

Special cases of Theorem 5.2.5 go back at least to the work of A. C. Climescu [1946]. The term "ordinal sum" is due to A. H. Clifford [1954].

5.3. Associative Functions on Intervals

Let T be an associative binary operation on the closed interval $[a, e]$. We begin this section by studying some simple consequences of several conditions that may be imposed on T.

Lemma 5.3.1. *Let $[a,e]$ be a closed interval and let T be an associative binary operation on $[a,e]$. Suppose that T satisfies the following two conditions:*

i. *T is nondecreasing in each place, i.e.,*

$$T(x_1, y_1) \leqslant T(x_2, y_2) \tag{5.3.1}$$

for all x_1, x_2, y_1, y_2 in $[a,e]$ such that $x_1 \leqslant x_2$ and $y_1 \leqslant y_2$.

ii. *The endpoint e is an identity of T, i.e.,*

$$T(x,e) = T(e,x) = x \quad \text{for all } x \text{ in } [a,e]. \tag{5.3.2}$$

Then T also satisfies the following four conditions:

iii. *All vertical sections v_x, all horizontal sections h_x, and the diagonal δ of T are nondecreasing functions on $[a,e]$.*

iv. *The endpoint a is a null element of T; i.e.,*

$$T(a,x) = T(x,a) = a \quad \text{for all } x \text{ in } [a,e]. \tag{5.3.3}$$

v. $T(x, y) \leqslant \mathrm{Min}(x, y)$ *for all x, y in $[a,e]$.* (5.3.4)

vi. $\delta(x) \leqslant x$ *for all x in $[a,e]$.* (5.3.5)

Since δ is nondecreasing, both $l^-\delta$ and $l^+\delta$ are well-defined functions on $[a,e]$. By virtue of (2.2.4), an immediate consequence of (5.3.5) is

vii. If $l^-\delta(x) = x$ for some x in $[a,e]$, then $\delta(x) = x$. (5.3.6)

The symbol l^- in (5.3.6) cannot be replaced by l^+. However, we can impose the condition:

viii. If $l^+\delta(x) = x$ for some x in $[a,e]$, then $\delta(x) = x$. (5.3.7)

This condition, together with (5.3.6), is sufficient to ensure that the set of idempotents E of T is closed in $[a,e]$. Since a and e are in E, the complement of E in $[a,e]$ is an open subset of $[a,e]$ and may be uniquely expressed as a countable union of pairwise disjoint open intervals. This proves

Lemma 5.3.2. *Let T be such that* (5.3.1), (5.3.2), *and* (5.3.7) *hold. If $\delta(x) < x$ for a particular x in $[a,e]$, then there is a unique open subinterval (a_x, e_x) of $[a,e]$ containing x such that $\delta(a_x) = a_x$, $\delta(e_x) = e_x$, and $\delta(y) < y$ for all y in (a_x, e_x).*

Lemma 5.3.3. *Let T satisfy* (5.3.1), (5.3.2), *and the condition:*

ix. *If $T(x,x) = x$, then v_x and h_x are continuous.* (5.3.8)

Then for any such idempotent x, v_x and h_x are given by

$$v_x(y) = h_x(y) = \begin{cases} y, & a \leqslant y \leqslant x, \\ x, & x \leqslant y \leqslant e. \end{cases} \tag{5.3.9}$$

PROOF. By Lemma 5.1.2, v_x and h_x are idempotent, while by (5.3.1) and (5.3.8) they are, respectively, nondecreasing and continuous. Thus $\operatorname{Ran} v_x$ is a closed interval (since v_x is continuous) that contains a (since $v_x(a) = a$) and x (since $v_x(e) = x$), but no points outside $[a, x]$ (since v_x is nondecreasing). Hence $\operatorname{Ran} v_x = [a, x]$. Finally, an appeal to Lemma 2.2.3 (with $b = v_x(a) = a$ and $c = v_x(x) = x$) yields (5.3.9) for v_x. Similar arguments yield (5.3.9) for h_x. □

Since the conclusions established in Lemmas 5.3.2 and 5.3.3 are needed in the subsequent development, we make the following:

Definition 5.3.4. Let $[a, e]$ be a closed interval. Then $\mathcal{O}[a, e]$ is the class of all associative binary operations T on $[a, e]$ that satisfy (5.3.1), (5.3.2), (5.3.7), and (5.3.8).

It follows that any T in $\mathcal{O}[a, e]$ also satisfies (5.3.3), (5.3.4), (5.3.5), (5.3.6), and (5.3.9).

Lemma 5.3.5. *Let T in $\mathcal{O}[a, e]$ be such that $\delta(x) < x$ for all x in (a, e). Then for any x in (a, e), we have:*

i. *If $\delta^m(x) > a$ for some integer m, then*

$$\delta^n(x) < \delta^m(x) \quad \textit{for all } n > m. \tag{5.3.10}$$

ii. *If $\delta^m(x) = a$ for some integer m, then*

$$\delta^n(x) = a \quad \textit{for all } n \geqslant m. \tag{5.3.11}$$

iii. $$\lim_{n\to\infty} \delta^n(x) = \inf\{\delta^n(x)\} = a. \tag{5.3.12}$$

PROOF. From (5.3.5), for all x in $[a, e]$ we have $\delta^n(x) \leqslant \delta^{m+1}(x)$ whenever $n > m$. Suppose that x is in (a, e) and $\delta^m(x) > a$. Then $a < \delta^m(x) < x < e$, whence for $n > m$,

$$\delta^n(x) \leqslant \delta^{m+1}(x) = \delta(\delta^m(x)) < \delta^m(x),$$

which is (5.3.10). Next, (5.3.11) follows at once from the fact that $\delta(a) = a$. Finally, set $b = \inf\{\delta^n(x) \mid n \geqslant 1\}$. Then, by (5.3.5) and (5.3.11), $l^+\delta(b) = b$, whence (5.3.7) yields $\delta(b) = b$. But $b < e$, and therefore $b = a$, which is (5.3.12). □

Definition 5.3.6. Let T be an associative binary operation on $[a,e]$ satisfying (5.3.1) and (5.3.2). Then T is *Archimedean* if for any x, y in (a,e), there is a positive integer m such that $x^m < y$.

Lemma 5.3.7. *Let T be in $\mathfrak{S}[a,e]$. Then T is Archimedean if and only if $\delta(x) < x$ for all x in (a,e).*

PROOF. If $\delta(x) < x$ for all x in (a,e) and y, z are in (a,e), then by (5.3.12) and (5.1.5) there is an integer n such that $y^{2^n} = \delta^n(y) < z$. Hence T is Archimedean. If, on the contrary, there is an x in (a,e) such that $\delta(x) = x$, then $\delta^n(x) = x$ for all n, whence $x^m = x$ for all m, so that T is not Archimedean. □

The proof of Lemma 5.3.7 shows that a necessary condition for T to be Archimedean is $\delta(x) < x$ for all x in (a,e). This condition is not sufficient. For example, let T be the function defined on $I \times I$ by

$$T(x,y) = \begin{cases} 0, & x \text{ in } \left[0,\tfrac{1}{2}\right], y \text{ in } \left[0,\tfrac{1}{2}\right], \\ x, & x \text{ in } \left[0,\tfrac{1}{2}\right], y \text{ in } \left(\tfrac{1}{2},1\right], \\ y, & x \text{ in } \left(\tfrac{1}{2},1\right], y \text{ in } \left[0,\tfrac{1}{2}\right], \\ 2xy - x - y + 1, & x \text{ in } \left(\tfrac{1}{2},1\right], y \text{ in } \left(\tfrac{1}{2},1\right]. \end{cases} \tag{5.3.13}$$

It is easily verified that T is associative and satisfies (5.3.1) and (5.3.2), and that $\delta(x) < x$ for all x in $(0,1)$. However, if x is any number in $(\frac{1}{2},1)$, then $x^m = 2^{m-1}(x-\frac{1}{2})^m + \frac{1}{2}$, whence $x^m > \frac{1}{2}$ for all m. Thus T is not Archimedean.

Note that since $l^+\delta(\frac{1}{2}) = \frac{1}{2} \neq 0 = \delta(\frac{1}{2})$, T fails to satisfy (5.3.7) and so is not in $\mathfrak{S}[0,1]$.

The principal result of this section is

Theorem 5.3.8. *Let T be in $\mathfrak{S}[a,e]$. Then we have*

i. *If $\delta(x) < x$ for all x in (a,e), then T is Archimedean on $[a,e]$, and conversely.*
ii. *If $\delta(x) = x$ for all x in $[a,e]$, then T is the restriction of* Min *to $[a,e]$, and conversely.*
iii. *Otherwise, the semigroup $([a,e],T)$ is an ordinal sum of semigroups $\{(S_\alpha, T_\alpha)\}$ where each S_α is a closed subinterval of $[a,e]$, each T_α is in $\mathfrak{S}[S_\alpha]$, and T_α is either Archimedean on S_α or the restriction of* Min *to S_α.*

PROOF. Parts i and ii are restatements of Lemmas 5.3.7 and 5.3.3, respectively. As regards Part iii, first suppose that x in (a,e) is such that $\delta(x) < x$ and let (a_x, e_x) be the associated open interval containing x. Then if follows from Lemma 5.3.2 that the interval $[a_x, e_x]$ is closed under T and from Lemmas 5.3.2 and 5.3.3, taken together, that the restriction of T to the closed interval $[a_x, e_x]$ is an Archimedean function in $\mathfrak{G}[a_x, e_x]$. Similarly, if x in $[a,e]$ is such that $\delta(x) = x$, then there is a maximal closed subinterval $[b,c]$ of $[a,e]$ (which may consist of only a single point) containing x and such that the restriction of T to $[b,c]$ is Min.

Thus all points in $[a,e]$ belong to an interval of one type or the other. We can index these intervals, say, by their midpoints. Lemma 5.3.3 shows that the compatibility condition of Definition 5.2.4 is satisfied. It only remains to show that (5.2.5) holds when x and y are in different subintervals. Let ι_x and ι_y denote the indices of the subintervals containing x and y, respectively. If $\iota_x < \iota_y$ then there is a point c such that $x \leqslant c \leqslant y$ and $\delta(c) = c$. Thus we have

$$x = h_c(x) = T(x,c) \leqslant T(x,y) \leqslant T(x,e) = x, \tag{5.3.14}$$

whence $T(x,y) = x$. Similarly, if $\iota_y < \iota_x$, then $T(x,y) = y$. So (5.2.5) holds and $([a,e], T)$ is an ordinal sum. The converse statement is trivial. □

Theorem 5.3.8 extends Theorem B of Mostert and Shields [1957] from continuous to not necessarily continuous associative operations. The situation described in the theorem is illustrated in Figure 5.3.1. Here T is

Figure 5.3.1. An ordinal sum.

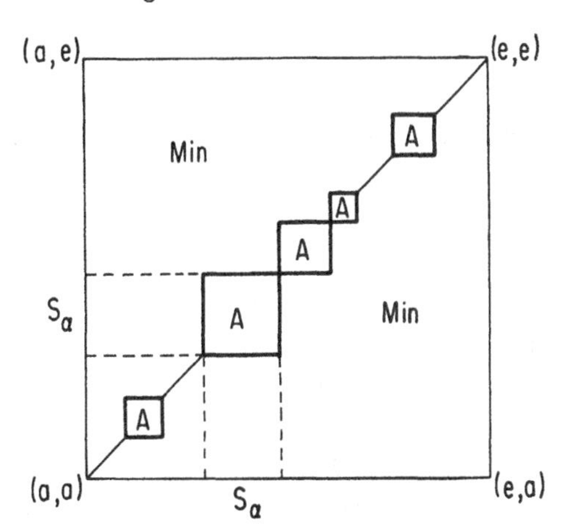

Archimedean on the squares labeled "A" and, in view of Theorem 5.3.8 and (5.2.5), equal to Min everywhere else in $[a, e]^2$.

5.4. Representation of Archimedean Functions

Theorem 5.3.8 characterizes the functions in $\Theta[a, e]$ in terms of restrictions of Min and Archimedean functions. As regards the restrictions of Min, there is nothing more to be said at this point. There is, however, much more to be said about the structure of Archimedean functions. We begin by stating a representation theorem that is basic to much of the subsequent development of the theory.

Theorem 5.4.1. *Let T be an Archimedean function in $\Theta[a, e]$. Suppose that T satisfies the following conditions:*

i. *There is a number e' in $[a, e)$ such that for all x in $[e', e]$,*

$$\lim_{y \to e} T(x, y) = x; \tag{5.4.1}$$

ii. *δ is continuous on $[a, e]$.* (5.4.2)

Then T admits the representation

$$T(x, y) = f(g(x) + g(y)) \tag{5.4.3}$$

for all x, y in $[a, e]$, where f and g have the following properties:

iii. *g is a function from $[a, e]$ into R^+ such that*

$$g \text{ is continuous, strictly decreasing, and } g(e) = 0; \tag{5.4.4}$$

iv. *f is a function from R^+ onto $[a, e]$ such that*

$$\begin{cases} f(0) = e, \\ f \text{ is continuous and strictly decreasing on } [0, g(a)], \\ f(x) = a \quad \text{for all } x \geqslant g(a); \end{cases} \tag{5.4.5}$$

v. $g[Q]f$. (5.4.6)

Corollary 5.4.2. *If T satisfies the hypotheses of Theorem 5.4.1, then T is continuous on $[a, e]^2$ and commutative, i.e.,*

$$T(x, y) = T(y, x) \tag{5.4.7}$$

for all x, y in $[a, e]$. Moreover, if $a \leqslant x_1 < x \leqslant e$, $a \leqslant y_1 < y \leqslant e$, and $T(x, y) > a$, then $T(x_1, y) < T(x, y)$ and $T(x, y_1) < T(x, y)$; i.e., T is strictly increasing in each place on the set $\{(x, y) \mid T(x, y) > a\}$. Lastly, *there is a c in* $[a, e]$ *such that* $\delta(x) = a$ *for* x *in* $[a, c]$, *while* δ *is strictly increasing on* $[c, e]$.

Corollary 5.4.3. *If* T *satisfies the hypotheses of Theorem* 5.4.1 *and is strictly increasing in each place on* $(a, e]^2$, *then* $g(a) = \infty$, *whence* $f = g^{-1}$ *and*

$$T(x, y) = g^{-1}(g(x) + g(y)) \tag{5.4.8}$$

for all x, y *in* $[a, e]$.

Combining Corollary 5.4.2 with Theorem 5.3.8 and (5.2.5) yields

Corollary 5.4.4. *Any continuous function in* $\mathfrak{G}[a, e]$ *is commutative.*

The representation (5.4.8) for associative functions has a long and distinguished history. It was first obtained, under considerably stronger assumptions, by N. H. Abel [1826]. It is, in fact, the subject of the first paper he published in Crelle's journal. Subsequently, it was established under successively weaker hypotheses by L. E. J. Brouwer [1909], É. Cartan [1930], and J. Aczél [1949]. The representation (5.4.3) is due to C. H. Ling [1965]. The hypotheses of Theorem 5.4.1 are considerably weaker than Ling's and are due to G. Krause. They are still not the weakest known [Krause 1981].

The proof of Theorem 5.4.1 is too long and involved to be presented in detail here. We shall instead give an indication of the main ideas. To begin with, note that any binary system $([a, e], T)$ is isomorphic, via a continuous and strictly increasing isomorphism, to a binary system $([a_1, e_1], T_1)$ where $-\infty < a_1 < e_1 < \infty$. Hence in proving Theorem 5.4.1 we can confine our attention to intervals $[a, e]$ in which both endpoints are finite (indeed, we could work throughout with the interval $[0, 1]$). Given this, the first part of the argument is a direct proof of Corollary 5.4.2. The second part begins with

Lemma 5.4.5. *A necessary condition for* (5.4.3) *to hold with* f *and* g *satisfying* (5.4.4)–(5.4.6) *is*

$$f(\alpha + \beta) = T(f(\alpha), f(\beta)) \quad \textit{for all } \alpha, \beta \leqslant g(a) \textit{ in } R^+. \tag{5.4.9}$$

Now let γ be the inverse of the restriction of δ to the interval $[c, e]$ of Corollary 5.4.2. Clearly, γ is continuous and strictly increasing, and $\gamma(x) > x$ for all x in (a, e). Next, fix a number $d < e$ such that $\delta(d) > a$. Then for all nonnegative integers m, n, the expressions $(\gamma^n(d))^m$ (see (5.1.3)) depend only on the binary rationals $m/2^n$ and not on their particular representations. Consequently, we can define a function f on the binary rationals in R^+ by

$$f(m/2^n) = (\gamma^n(d))^m \tag{5.4.10}$$

for all nonnegative integers m, n. We then have

Lemma 5.4.6. *The function f defined by* (5.4.10) *satisfies* (5.4.9) *for all binary rationals α, β in R^+.*

The proof of Theorem 5.4.1 is then completed by showing that f can be extended to a function defined on all of R^+ and that this extended function satisfies (5.4.5) and (5.4.9). We define g as the inverse of the restriction of f to the largest closed interval on which f is strictly decreasing. Then (5.4.4) and (5.4.6) are satisfied, and we can finally convert (5.4.9) to (5.4.3).

The converse of Theorem 5.4.1 is

Theorem 5.4.7. *If f and g are functions satisfying* (5.4.4), (5.4.5), *and* (5.4.6), *and T is defined by* (5.4.3), *then T is an Archimedean function in $\mathcal{O}[a,e]$ that also satisfies the conclusions of Corollary* 5.4.2.

Theorem 5.4.7 is easy to establish, since, apart from associativity, all the desired properties of T are straightforward consequences of (5.4.3) and the properties of f and g. As for the associativity of T, this follows from Theorem 5.2.1, since addition is an associative operation on R^+ and the functions f and g behave properly.

Theorem 5.4.1 establishes the existence of the representation (5.4.3). Uniqueness up to a multiplicative constant follows from

Theorem 5.4.8. *Let T satisfy the conditions of Theorem* 5.4.1. *Suppose T is generated from addition on R^+ by (f_1, g_1) and by (f_2, g_2), where both pairs satisfy* (5.4.4)–(5.4.6), *so that*

$$T(x, y) = f_1(g_1(x) + g_1(y)) = f_2(g_2(x) + g_2(y)) \tag{5.4.11}$$

for all x, y in $[a,e]$. Then there is an $\alpha > 0$ such that

$$g_2(x) = \alpha g_1(x) \quad \textit{for all } x \textit{ in } [a,e], \tag{5.4.12}$$

and

$$f_2(x) = f_1(\alpha x) \quad \textit{for all } x \textit{ in } R^+. \tag{5.4.13}$$

PROOF. From (5.4.11) and (5.4.4) it follows that $g_1(x) + g_1(y) \geqslant g_1(a)$ if and only if $g_2(x) + g_2(y) \geqslant g_2(a)$. Hence in view of (5.4.6), for any x, y in

$[a,e]$ such that $g_1(x)+g_1(y) \leqslant g_1(a)$, we have

$$g_2(f_2(g_2(x)+g_2(y))) = g_2(x)+g_2(y) = h(g_1(x)+g_1(y))$$

where $h = g_2 \circ f_1$. Setting $u = g_1(x)$ and $v = g_1(y)$ yields

$$h(u+v) = h(u)+h(v) \tag{5.4.14}$$

for all u, v in R^+ such that $u+v \leqslant g_1(a)$. The function h is nonnegative, and it is well known that the nonnegative solutions of (5.4.14) are all of the form $h(t) = \alpha t$, with $\alpha \geqslant 0$ (see [Aczél 1966, §2.1]). Since h is strictly increasing, $\alpha > 0$. Thus, for x, y in $[a,e]$, $g_2(x) = g_2(f_1(g_1(x))) = h(g_1(x)) = \alpha g_1(x)$, which is (5.4.12); and (5.4.13) follows at once. The converse is a routine calculation. □

5.5. Triangular Norms, Additive and Multiplicative Generators

Definition 5.5.1. A *triangular norm* (briefly, a *t-norm*) is an associative binary operation on I that satisfies (5.3.1), (5.3.2), and (5.4.7); i.e., that is commutative, nondecreasing in each place, and such that $T(a,1) = a$ for all a in I.

A t-norm may be visualized as a surface over the unit square that contains the skew quadrilateral whose vertices are $(0,0,0)$, $(1,0,0)$, $(1,1,1)$, and $(0,1,0)$ (see Figure 5.5.1). The geometric interpretations of the conditions other than associativity are evident; associativity, on the other hand, seems to have no simple geometric interpretation (see Problem 5.8.2).

As (5.3.13) shows, there are t-norms that are not in $\mathcal{O}[0,1]$. However, nearly all t-norms of interest to us, in particular all continuous t-norms, are

Figure 5.5.1. Graph of a t-norm.

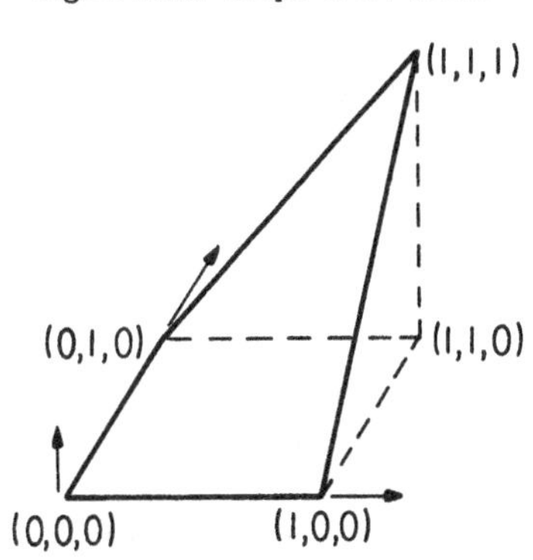

in $\mathcal{O}[0, 1]$. In the other direction, any function in $\mathcal{O}[0, 1]$ that is commutative is a t-norm. Thus, by virtue of Theorem 5.3.8 and Corollary 5.4.2, all continuous functions in $\mathcal{O}[0, 1]$ are t-norms.

An immediate consequence of Theorems 5.4.1 and 5.4.7 is

Theorem 5.5.2. *A t-norm T is continuous and Archimedean if and only if it admits the representation* (5.4.3), *i.e., if and only if*

$$T(x, y) = f(g(x) + g(y)) \tag{5.5.1}$$

where

i. *g is a continuous and strictly decreasing function from I into R^+ with* $g(1) = 0$;
ii. *f is a continuous function from R^+ onto I that is strictly decreasing on* $[0, g(0)]$ *and such that* $f(x) = 0$ *for all* $x \geqslant g(0)$ *(see Figure* 5.5.2);
iii. $f[Q]g$.

Definition 5.5.3. A t-norm T is *strict* if it is continuous on I^2 and strictly increasing in each place on $(0, 1]^2$.

If T is strict, then $x^n \to 0$ for any $x < 1$. Thus a strict t-norm is Archimedean. Furthermore, as a consequence of Corollary 5.4.3 and Theorem 5.4.7, we have

Corollary 5.5.4. *A t-norm T is strict if and only if it admits the representation* (5.4.8), *i.e., if and only if*

$$T(x, y) = g^{-1}(g(x) + g(y)) \tag{5.5.2}$$

Figure 5.5.2. Additive generators of an Archimedean t-norm.

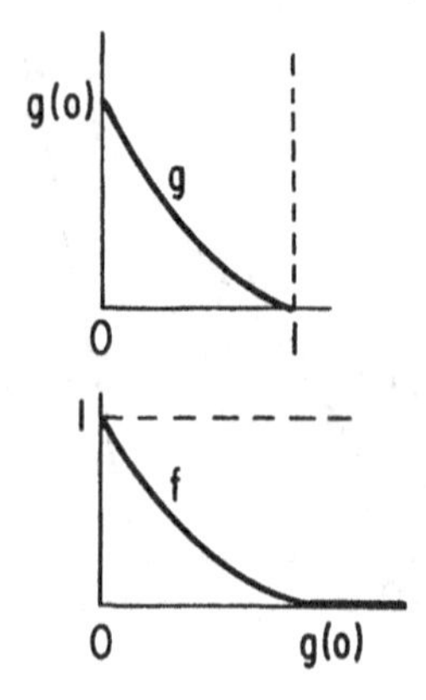

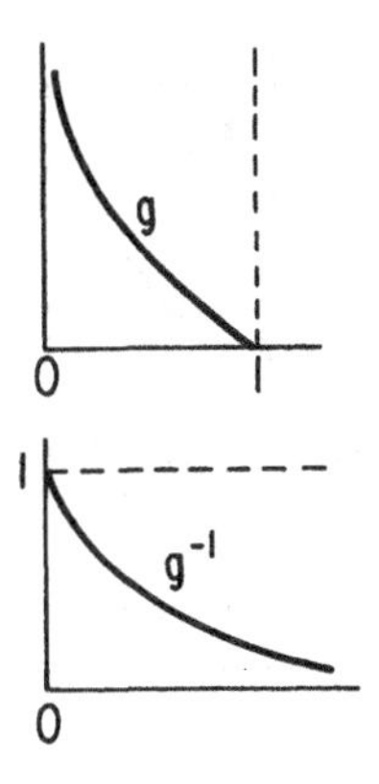

Figure 5.5.3. Additive generators of a strict t-norm.

where g is a continuous and strictly decreasing function from I onto R^+, with $g(0) = \infty$ and $g(1) = 0$ (see Figure 5.5.3).

Hence if T is strict, then the semigroups (I, T) and $(R^+, +)$ are isomorphic. In particular, if in (5.5.2) we take $g(x) = -\log x$ (with $g(0) = \infty$), then $g^{-1}(x) = e^{-x}$ (with $e^{-\infty} = 0$) and we obtain the strict t-norm Π defined by

$$\Pi(x, y) = xy \quad \text{for all } x, y \text{ in } I. \tag{5.5.3}$$

Now let T be any continuous Archimedean t-norm. Then, upon noting that

$$u + v = -\log(e^{-u}e^{-v}) \tag{5.5.4}$$

for all u, v in R^+, and substituting (5.5.4) into (5.5.1), with $u = g(x)$ and $v = g(y)$, we obtain

$$T(x, y) = h(k(x)k(y)) \tag{5.5.5}$$

for all x, y in I, where

$$h(x) = f(-\log x) \quad \text{and} \quad k(x) = \exp(-g(x)). \tag{5.5.6}$$

Hence we have

Theorem 5.5.5. *A t-norm T is continuous and Archimedean if and only if it admits the representation* (5.5.5) *where:*

i. *k is a continuous and strictly increasing function from I into I, with $k(1) = 1$;*

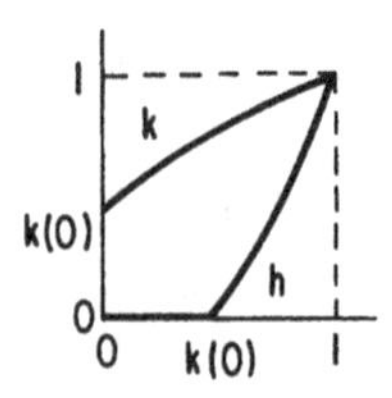

Figure 5.5.4. Multiplicative generators of an Archimedean t-norm.

ii. *h is a continuous function from I onto I that is strictly increasing on $[k(0), 1]$ and such that $h(x) = 0$ for all x in $[0, k(0)]$ (see Figure 5.5.4);*
iii. *$h[Q]k$.*

A t-norm T is strict if and only if it admits the representation

$$T(x, y) = k^{-1}(k(x)k(y)) \tag{5.5.7}$$

for all x, y in I, where k is a continuous and strictly increasing function from I onto I, so that $k(0) = 0$ and $k(1) = 1$ (see Figure 5.5.5).

Definition 5.5.6. Let T be a continuous Archimedean t-norm. Then we say that T is *additively generated* by the pair (f, g) of functions in (5.5.1) and *multiplicatively generated* by the pair (h, k) of functions in (5.5.5). We refer to f as an *outer additive generator* and to g as an *inner additive generator*. If T is strict, then the function g in (5.5.2) is an *additive generator* of T. *Multiplicative generators* are defined analogously.

Since additive and multiplicative generators are connected via (5.5.6), the multiplicative analog of Theorem 5.4.8 is

Theorem 5.5.7. *Let (h_1, k_1) and (h_2, k_2) be two pairs of multiplicative generators of the same continuous Archimedean t-norm T. Then there is an $\alpha > 0$ such that*

$$h_2(x) = h_1(x^{\alpha}) \quad \text{and} \quad k_2(x) = (k_1(x))^{\alpha} \tag{5.5.8}$$

Figure 5.5.5. Multiplicative generators of a strict t-norm.

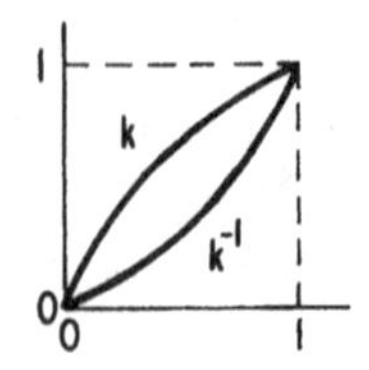

for all x in I. Conversely, if (h_1, k_1) are multiplicative generators of T and (h_2, k_2) are defined by (5.5.8), then (h_2, k_2) are also multiplicative generators of T.

Returning to additive generators, if T is additively generated by (f, g), then (5.4.4)–(5.4.6) yield

$$g(f(u)) = \begin{cases} u, & u \text{ in } [0, g(0)], \\ g(0), & u \text{ in } [g(0), \infty], \end{cases} \tag{5.5.9}$$

whence $g(f(u)) \leqslant u$ for all u in R^+, and $f(g(x)) = x$ for all x in I. These observations lead to

Lemma 5.5.8. *Let T_1 and T_2 be continuous Archimedean t-norms, additively generated by (f_1, g_1) and (f_2, g_2), respectively. Then*

$$T_1(x, y) \leqslant T_2(x, y) \quad \text{for all } x, y \text{ in } I \tag{5.5.10}$$

if and only if $g_1 \circ f_2$ is subadditive; and equality holds if and only if $g_1 \circ f_2$ is linear.

PROOF. For x, y in I, set $u = g_2(x)$, $v = g_2(y)$. Then we have $x = f_2(g_2(x)) = f_2(u)$, $y = f_2(g_2(y)) = f_2(v)$. Thus (5.5.10) is equivalent to

$$f_1[g_1(f_2(u)) + g_1(f_2(v))] \leqslant f_2(u + v) \tag{5.5.11}$$

for all u, v in $[0, g_2(0)]$. Applying g_1 to both sides yields

$$g_1(f_2(u + v)) \leqslant g_1(f_2(u)) + g_1(f_2(v)) \tag{5.5.12}$$

for all u, v in $[0, g_2(0)]$. Now it is easy to see that (5.5.12) extends to all u, v in R^+, whence $g_1 \circ f_2$ is subadditive. In the other direction, if $g_1 \circ f_2$ is subadditive, then (5.5.12) holds, and applying f_1 to both sides yields (5.5.11). The proof is completed by observing that in view of Theorem 5.4.8, $T_1 = T_2$ if and only if $g_1 \circ f_2$ is linear. □

It is convenient to extend the ordering of t-norms implicit in Lemma 5.5.8 to all binary operations on I, as follows:

Definition 5.5.9. Let T_1 and T_2 be binary operations on I. Then we say that T_1 is *weaker* than T_2, or T_2 *stronger* than T_1, and we write $T_1 < T_2$ or $T_2 > T_1$, if (5.5.10) is satisfied and if $T_1 \neq T_2$, i.e., if there is at least one pair (x, y) in $I \times I$ with $T_1(x, y) < T_2(x, y)$. We write $T_1 \leqslant T_2$, or $T_2 \geqslant T_1$, if T_2 is stronger than or equal to T_1.

Thus Lemma 5.5.8 can be paraphrased in the form: "T_2 is stronger than T_1 if and only if $g_1 \circ f_2$ is a nonlinear subadditive function."

Definition 5.5.10. A t-norm T is *positive* if $T(x, y) > 0$ for all x, y in $(0, 1]$.

Thus, for example, all strict t-norms are positive.

5.6. Examples

The most important t-norms are the functions M, Π, and W introduced in (1.4.1). The t-norm M is defined by

$$M(x, y) = \mathrm{Min}(x, y) \tag{5.6.1}$$

for all x, y in I. It is continuous and positive but not Archimedean. Thus M does not admit any representation of the form (5.4.3) with f and g satisfying (5.4.4)–(5.4.6). In fact, more can be said: V. I. Arnold and A. A. Kyrilov have shown that $M(x, y)$ cannot be represented in the form $f(g(x) + h(y))$ for any continuous (not necessarily monotonic) functions f, g, h (see [Arnold 1957]); and Ling [1965] has shown that $M(x, y)$ cannot be represented in the form $f(g(x) + g(y))$ with $f[Q]g$ for any strictly monotonic (not necessarily continuous) function g. The graph of M is the surface consisting of the two triangles whose vertices are $(0,0,0)$, $(1,1,1)$, $(1,0,0)$ and $(0,0,0)$, $(1,1,1)$, $(0,1,0)$, respectively; the graph of δ is the line segment joining $(0,0,0)$ to $(1,1,1)$ (see Figure 5.6.1).

The t-norm Π is defined by (5.5.3); as pointed out, it is strict, hence positive; it is additively generated by $g(x) = -\log x$, and multiplicatively generated by $k(x) = x$. The graph of Π is a portion of the hyperbolic paraboloid $z = xy$. The diagonal section δ is a parabolic arc (see Figure 5.6.2).

Figure 5.6.1. The t-norm M.

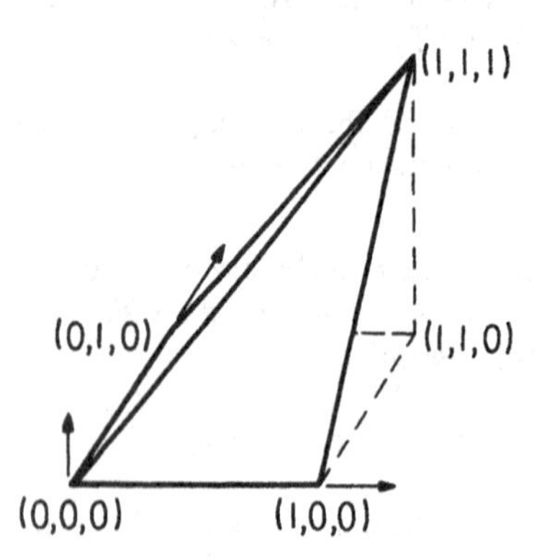

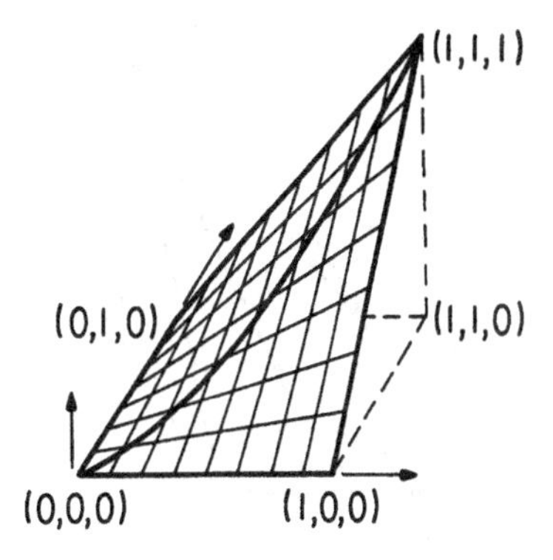

Figure 5.6.2. The t-norm Π.

The t-norm W is given by

$$W(x, y) = \text{Max}(x + y - 1, 0) \tag{5.6.2}$$

for all x, y in I; it is continuous and Archimedean, but not positive and a fortiori not strict. It has additive generators f_w and g_w given by

$$\begin{aligned} f_w(x) &= \text{Max}(1 - x, 0) \quad \text{for all } x \text{ in } R^+, \\ g_w(x) &= 1 - x \quad \text{for all } x \text{ in } I; \end{aligned} \tag{5.6.3}$$

and multiplicative generators h_w and k_w given by

$$h_w(x) = \text{Max}(1 + \log x, 0), \qquad k_w(x) = e^{x-1} \tag{5.6.4}$$

for all x in I. The graph of W is the surface consisting of the two triangles whose vertices are $(0, 0, 0)$, $(1, 0, 0)$, $(0, 1, 0)$ and $(1, 1, 1)$, $(1, 0, 0)$, $(0, 1, 0)$, respectively. Here $\delta(x) = \text{Max}(2x - 1, 0)$, and the graph of δ is a polygonal arc (see Figure 5.6.3).

It is easy to verify that

$$M(x, y) - \Pi(x, y) = \Pi(x, 1 - y) - W(x, 1 - y) \tag{5.6.5}$$

for all x, y in I, whence the graph of W is a twisted reflection of the graph of M in the graph of Π.

A fourth t-norm that merits a name is the t-norm Z defined by

$$Z(x, y) = \begin{cases} x, & x \text{ in } I, y = 1, \\ y, & x = 1, y \text{ in } I, \\ 0, & x \text{ in } [0, 1), y \text{ in } [0, 1); \end{cases} \tag{5.6.6}$$

Z is Archimedean but not continuous, and Figure 5.5.1 will serve to illustrate its graph. It is the minimal t-norm, just as M is the maximal

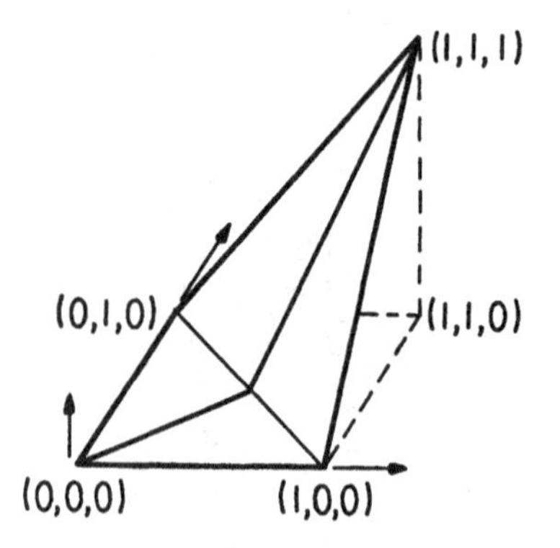

Figure 5.6.3. The t-norm W.

t-norm (by (5.3.4)). Thus

$$Z \leqslant T \leqslant M \quad \text{for any } t\text{-norm } T. \tag{5.6.7}$$

In particular, the four t-norms introduced above are ordered by

$$Z < W < \Pi < M. \tag{5.6.8}$$

By using (5.5.1) or (5.5.5), we can construct entire families of t-norms at will. As an example, consider the one-parameter family of pairs of functions (h_p, k_p) defined on I by

$$h_p(x) = (\mathrm{Max}(1 + p\log x, 0))^{1/p}, \quad k_p(x) = \exp\left(\frac{1}{p}(x^p - 1)\right), \quad p \neq 0,$$

$$h_0(x) = x = \lim_{p\to 0} h_p(x), \quad k_0(x) = x = \lim_{p\to 0} k_p(x).$$

It is readily verified that for all p in $(-\infty, \infty)$, h_p and k_p are multiplicative generators. The corresponding t-norms T_p are given by

$$\begin{aligned} T_p(x, y) &= (\mathrm{Max}(x^p + y^p - 1, 0))^{1/p}, \quad p \neq 0, \\ T_0(x, y) &= \Pi(x, y) = xy. \end{aligned} \tag{5.6.9}$$

The t-norm T_p is strict if and only if $p \leqslant 0$. Note that T_{-1} is given by

$$T_{-1}(x, y) = \frac{xy}{x + y - xy},$$

that $T_1 = W$, and that

$$\lim_{p\to -\infty} T_p(x, y) = M(x, y), \qquad \lim_{p\to\infty} T_p(x, y) = Z(x, y).$$

Thus each of the t-norms Z, W, Π, M is either a member of the family $\{T_p\}$ or a limit of members of this family.

Another family of t-norms, whose significance is discussed in Sections 6.4 and 7.4, was introduced by M. J. Frank [1979]. Here the additive generators are defined by

$$f_s(x) = \frac{\log(1 + (s-1)e^{-x})}{\log s}, \qquad g_s(x) = -\log\left(\frac{s^x - 1}{s - 1}\right) \tag{5.6.10}$$

for s in $(0, \infty)$, $s \neq 1$. The corresponding t-norms T_s are given by

$$T_s(x, y) = \log\left(1 + \frac{(s^x - 1)(s^y - 1)}{s - 1}\right) \Big/ \log s \tag{5.6.11}$$

for s in $(0, \infty)$, $s \neq 1$. All the t-norms T_s are strict, and we have

$$\begin{aligned} \lim_{s \to 0} T_s(x, y) &= M(x, y), \\ \lim_{s \to 1} T_s(x, y) &= \Pi(x, y), \\ \lim_{s \to \infty} T_s(x, y) &= W(x, y). \end{aligned} \tag{5.6.12}$$

5.7. Conorms and Composition Laws

Definition 5.7.1. Let T be a binary operation on I. Then T^* is the function given by

$$T^*(x, y) = 1 - T(1 - x, 1 - y) \tag{5.7.1}$$

for all x, y in I. If T is a t-norm, then T^* is the *t-conorm* of T. A function Y is a t-conorm if there is a t-norm T such that $Y = T^*$.

Clearly, T^* is itself a binary operation on I, and $T^{**} = T$. The graphs of T and T^* are reflections of each other in the center of the unit cube. Moreover, Theorem 5.2.1 with $f(x) = g(x) = 1 - x$ shows that T^* is associative if and only if T is. Hence it is a straightforward matter to verify

Lemma 5.7.2. *A binary operation Y on I is a t-conorm if and only if it is associative, commutative, and nondecreasing in each place, with $Y(0, a) = a$ and $Y(1, a) = 1$ for each a in I. If Y is a t-conorm, then Y^* is a t-norm.*

Lemma 5.7.3. *Let T_1, T_2 be binary operations on I. Then $T_1 < T_2$ if and only if $T_2^* < T_1^*$.*

The respective t-conorms of the t-norms M, Π, W, and Z are given by

$$M^*(x, y) = \mathrm{Max}(x, y), \tag{5.7.2}$$

$$\Pi^*(x, y) = x + y - xy, \tag{5.7.3}$$

$$W^*(x, y) = \mathrm{Min}(x + y, 1), \tag{5.7.4}$$

$$Z^*(x, y) = \begin{cases} x, & x \text{ in } I, y = 0, \\ y, & x = 0, y \text{ in } I, \\ 1, & x \text{ in } (0, 1], y \text{ in } (0, 1]. \end{cases} \tag{5.7.5}$$

Furthermore, it follows from (5.6.7) and Lemma 5.7.2 that every t-conorm Y satisfies $M^* \leqslant Y \leqslant Z^*$; and since $M < M^*$, for any t-norm T we have

$$Z \leqslant T \leqslant M < M^* \leqslant T^* \leqslant Z^*. \tag{5.7.6}$$

In particular

$$Z < W < \Pi < M < M^* < \Pi^* < W^* < Z^*. \tag{5.7.7}$$

If a t-norm is Archimedean, strict, etc., then we say that its t-conorm is Archimedean, strict, etc. Combining (5.7.1) with Theorem 5.5.2 and Corollary 5.5.4, we therefore obtain

Theorem 5.7.4. *A t-conorm Y is continuous and Archimedean if and only if Y admits the representation*

$$Y(x, y) = f^*(g^*(x) + g^*(y)) \tag{5.7.8}$$

for all x, y in I, where:

i. *g^* is a continuous and strictly increasing function from I into R^+, with $g^*(0) = 0$;*
ii. *f^* is a continuous function from R^+ onto I, strictly increasing on $[0, g^*(1)]$, with $f^*(0) = 0$ and $f^*(x) = 1$ for all $x \geqslant g^*(1)$;*
iii. *$f[Q]g$.*

If, in addition, $g^*(1) = \infty$ (in which case f^* and g^* are inverses of each other), then Y is strict, and conversely.

If (f, g) are additive generators of a t-norm and (f^*, g^*) are the corresponding additive generators of its t-conorm, then these functions are related by

$$\begin{aligned} f^*(x) &= 1 - f(x) \quad \text{for all } x \text{ in } R^+, \\ g^*(x) &= g(1 - x) \quad \text{for all } x \text{ in } I. \end{aligned} \tag{5.7.9}$$

Another class of functions closely related to t-norms was defined by

J. Kampé de Fériet and B. Forte [1967] in their axiomatic development of information theory.

Definition 5.7.5. Let L be an associative binary operation on R^+. Then L is a *composition law* if it satisfies (5.3.1), (5.3.2), and (5.4.7).

We immediately have

Lemma 5.7.6. *Let L be an associative binary operation on R^+. Then L is a composition law if and only if, for any continuous and strictly increasing function ψ from I onto R^+ (whence $\psi(0) = 0$ and $\psi(1) = \infty$), the function T defined by*

$$T(x, y) = \psi^{-1}(L(\psi(x), \psi(y)) \tag{5.7.10}$$

for all x, y in I is a t-norm.

Reversing (5.7.10) we have

$$L(x, y) = \psi\big(T(\psi^{-1}(x), \psi^{-1}(y)\big), \tag{5.7.11}$$

where ψ^{-1} is a continuous and strictly increasing function from R^+ onto I; and we can then use (5.7.11) to generate composition laws from t-norms. For example:

i. Take $T = W$ and define ψ and ψ^{-1} by

$$\psi(x) = -\frac{1}{c}\log(1 - x), \qquad \psi^{-1}(x) = 1 - e^{-cx},$$

where c is a fixed positive number. Then (5.7.11) yields

$$L_w(x, y) = \operatorname{Max}\left(-\frac{1}{c}\log(e^{-cx} + e^{-cy}), 0\right) \tag{5.7.12}$$

for all x, y in R^+; L_w is the *Wiener–Shannon law* with parameter c.

ii. Take $T = \Pi$, and ψ, ψ^{-1} as in i above. Then L is given by

$$L(x, y) = -\frac{1}{c}\log(e^{-cx} + e^{-cy} - e^{-c(x+y)}). \tag{5.7.13}$$

iii. Take $T = \Pi$, and define ψ and ψ^{-1} by

$$\psi(x) = \frac{-1}{\log x}, \qquad \psi^{-1}(x) = \exp\left(-\frac{1}{x}\right).$$

Then (5.7.11) yields

$$H(x, y) = \frac{xy}{x + y} \tag{5.7.14}$$

for all x, y in R^+; H is the *hyperbolic law.*

iv. Take $T = M$, and any ψ satisfying the conditions of Lemma 5.7.6. Then L is Min on R^+, which is the *Inf-law*.

If we replace the t-norm T in (5.7.11) by its conorm T^* we obtain an associative binary operation K on R^+ that satisfies (5.3.1) and (5.4.7) but for which 0 (rather than ∞) is the identity and ∞ (rather than 0) is the null element. If $T = M$, then $K(x, y) = \text{Max}(x, y)$ for all x, y in R^+. If T is Archimedean and additively generated by (f, g), then

$$K(x, y) = \varphi^{-1}(\varphi(x) + \varphi(y)) \tag{5.7.15}$$

for all x, y in R^+, where $\varphi(x) = g(1 - \psi^{-1}(x))$ is a continuous, strictly increasing function from R^+ to $[0, g(0)]$, with $\varphi(0) = 0$ and $\varphi(\infty) = g(0)$. If $\psi^{-1}(x) = 1 - e^{-x}$ and $g(x) = -\alpha \log x$, where $\alpha > 0$, then $\varphi(x) = x^\alpha$ and

$$K_\alpha(x, y) = (x^\alpha + y^\alpha)^{1/\alpha} \tag{5.7.16}$$

for all x, y in R^+. Note that $K_1 = \text{Sum}$, where

$$\text{Sum}(x, y) = x + y. \tag{5.7.17}$$

For $\alpha \geqslant 1$, the binary operations K_α obtained in this way are used, among other things, to define the Cartesian product of a finite number of metric spaces (see Section 12.7).

5.8. Open Problems

All proofs of the representation (5.4.3), from Abel's through Ling's, have assumed that the function T is continuous on $[a, e] \times [a, e]$. As pointed out in Section 5.4, G. Krause has succeeded in showing that these conditions can be weakened to the assumptions that δ is continuous on $[a, e]$ and that T has some continuity properties in a neighborhood of the point (e, e) (e.g., (5.4.1)). It would be desirable to remove this last assumption, and this leads to

Problem 5.8.1. Suppose T is an Archimedean function in $\mathcal{O}[a, e]$ and that the diagonal δ, given by $\delta(x) = T(x, x)$, is continuous on $[a, e]$. Is T necessarily continuous on $[a, e] \times [a, e]$?

The following results may have a bearing on this problem: (1) C. H. Kimberling [1973a] has shown that a continuous Archimedean t-norm T is not uniquely determined by its diagonal δ; (2) Frank [1980] has shown that T is uniquely determined by δ and an appropriately chosen segment of any vertical or horizontal section.

Suppose that T is a continuous t-norm (or, more generally, a continuous two-place function that admits the representation (5.4.8) on some appropriate region in R^2). The conditions (5.3.1), (5.3.2), and (5.4.7) have easily visualized geometric consequences for the surface $z = T(x, y)$. There seems to be no such simple geometric interpretation of the associativity condition (5.1.1). However, as we noted in [S² 1961a], there is some evidence to support the conjecture that if the surface is sufficiently smooth, then it is isothermal.

Recall that a surface is isothermal if its lines of curvature form an isothermally orthogonal net, i.e., if there exists a parametrization, say in terms of u, v, such that the lines of curvature are the curves u = constant, v = constant, and such that the first fundamental form is given by $ds^2 = \lambda(u,v)(du^2 + dv^2)$ (see [Darboux 1915, Ch. XI]). Our conjecture is based on the following facts.

i. The three systems of curves x = constant, y = constant, and $z = T(x, y)$ = constant form a hexagonal web on the surface (see [Blaschke and Bol 1938; Aczél 1965]).
ii. A surface is isothermal if and only if the lines of curvature and any one-parameter family of curves that cuts them isogonally form a hexagonal web [Thomsen 1927].
iii. The surface $z = xy$ is isothermal.
iv. The surface $z = T(x, y)$ is obtained from the surface $z = xy$ by means of the simple transformation (5.5.7).

Thus we have

Problem 5.8.2. Is every sufficiently smooth associative surface isothermal?

An affirmative answer to this question would yield an interesting differential-geometric characterization of associativity.

More generally, we have

Problem 5.8.3. What differential-geometric properties are preserved by transformations $S \to T$ of the form $T(x, y) = f^{-1}(S(f(x), f(y)))$ or $T(x, y) = f^{-1}(S(g(x), h(y)))$? In short, what differential-geometric properties are preserved by isotopies?

6
Copulas

6.1. Fundamental Properties of n-Increasing Functions

Along with one-dimensional distribution functions, we also need to consider their higher-dimensional analogs, joint distribution functions, and a class of related functions called copulas. To do this, it is convenient to extend certain notions from R to R^n. Thus, given points $\mathbf{u} = (u_1, \ldots, u_n)$, and $\mathbf{v} = (v_1, \ldots, v_n)$ in R^n, we write $\mathbf{u} \leqslant \mathbf{v}$ if $u_1 \leqslant v_1$, $u_2 \leqslant v_2, \ldots, u_n \leqslant v_n$; and $\mathbf{u} < \mathbf{v}$ if $u_1 < v_1$, $u_2 < v_2, \ldots, u_n < v_n$. We also write $\mathbf{u} + \mathbf{v}$ for the point $(u_1 + v_1, \ldots, u_n + v_n)$, $c\mathbf{u}$, where c is in $(-\infty, \infty)$, for the point $(cu_1, \ldots, cu_n)$, and $|\mathbf{u}|$ for the number $(u_1^2 + \cdots + u_n^2)^{1/2}$.

Definition 6.1.1. Let n be a positive integer. An *n-interval* is the Cartesian product of n real intervals, and an *n-box* the Cartesian product of n closed intervals. If J is the n-interval $J_1 \times J_2 \times \cdots \times J_n$, where for $m = 1, 2, \ldots, n$ the interval J_m has endpoints a_m, e_m, then the *vertices* of J are the points $\mathbf{c} = (c_1, \ldots, c_n)$ such that each c_m is equal to either a_m or e_m.

Certain types of n-intervals can be defined by simple inequalities. Thus the n-box $[a_1, e_1] \times [a_2, e_2] \times \cdots \times [a_n, e_n]$ is the set $\{\mathbf{u} \text{ in } R^n \mid \mathbf{a} \leqslant \mathbf{u} \leqslant \mathbf{e}\}$, where $\mathbf{a} = (a_1, \ldots, a_n)$ and $\mathbf{e} = (e_1, \ldots, e_n)$; this n-box is denoted by $[\mathbf{a}, \mathbf{e}]$. Similarly, the *open n-interval* $\{\mathbf{u} \text{ in } R^n \mid \mathbf{a} < \mathbf{u} < \mathbf{e}\}$ is denoted by $(\mathbf{a}, \mathbf{e})$, the *half-open n-interval* $\{\mathbf{u} \text{ in } R^n \mid \mathbf{a} \leqslant \mathbf{u} < \mathbf{e}\}$ by $[\mathbf{a}, \mathbf{e})$, and the half-open n-interval $\{\mathbf{u} \text{ in } R^n \mid \mathbf{a} < \mathbf{u} \leqslant \mathbf{e}\}$ by $(\mathbf{a}, \mathbf{e}]$. An open or half-open n-interval, if nonempty, has 2^n distinct vertices. The number of distinct vertices of an n-box may be any power of 2 from 1 to 2^n inclusive.

The unit n-cube I^n is the n-box $[\mathbf{0},\mathbf{1}] = [(0,\ldots,0),(1,\ldots,1)]$ and R^n is the n-box $[(-\infty,\ldots,-\infty),(\infty,\ldots,\infty)]$.

Definition 6.1.2. Let B be the n-box $[\mathbf{a},\mathbf{e}]$, where $\mathbf{a} = (a_1,\ldots,a_n)$ and $\mathbf{e} = (e_1,\ldots,e_n)$. Let $\mathbf{c} = (c_1,\ldots,c_n)$ be a vertex of B. If the vertices of B are all distinct (which is equivalent to saying $\mathbf{a} < \mathbf{e}$), then

$$\operatorname{sgn}_B(\mathbf{c}) = \begin{cases} 1, & \text{if } c_m = a_m \text{ for an even number of } m\text{'s}, \\ -1, & \text{if } c_m = a_m \text{ for an odd number of } m\text{'s}. \end{cases} \tag{6.1.1}$$

If the vertices of B are not all distinct, then $\operatorname{sgn}_B(\mathbf{c}) = 0$.

For example, the 2-box $B = [\mathbf{a},\mathbf{e}] = [(a_1,a_2),(e_1,e_2)]$ is the closed rectangle in R^2 whose vertices are the points (a_1,a_2), (e_1,a_2), (e_1,e_2), (a_1,e_2); and if these vertices are distinct, then $\operatorname{sgn}_B(a_1,a_2) = \operatorname{sgn}_B(e_1,e_2) = 1$ and $\operatorname{sgn}_B(e_1,a_2) = \operatorname{sgn}_B(a_1,e_2) = -1$.

Definition 6.1.3. Let $A_1,\ldots,A_n$ be nonempty subsets of R, and let H be a function from $A_1 \times \cdots \times A_n$ into $(-\infty,\infty)$. If B is an n-box whose vertices are all in $\operatorname{Dom} H$, then the *H-volume of B* is the sum

$$V_H(B) = \sum \operatorname{sgn}_B(\mathbf{c}) H(\mathbf{c}), \tag{6.1.2}$$

where the summation is taken over all vertices of B. The function H is *n-increasing* if $V_H(B) \geqslant 0$ for all n-boxes B whose vertices are in $\operatorname{Dom} H$.

For $n = 2$ we therefore have

$$V_H([\mathbf{a},\mathbf{e}]) = H(e_1,e_2) - H(a_1,e_2) - H(e_1,a_2) + H(a_1,a_2). \tag{6.1.3}$$

If $n = 1$, so that $\operatorname{Dom} H$ is a subset of R, then H is 1-increasing if and only if H is nondecreasing. In higher dimensions, on the other hand, the statement "H is n-increasing" neither implies nor is implied by the statement "H is nondecreasing in each place." For example, every vertical and horizontal section of t-conorm Π^* given by $\Pi^*(x,y) = x + y - xy$ is nondecreasing on I, but $V_{\Pi^*}(I^2) = -1$, whence Π^* is not 2-increasing. In the other direction, let H be the function defined on $[-1,1]^2$ by $H(x,y) = xy$. Then H is 2-increasing, but the sections v_x, h_x are decreasing for all x in $[-1,0)$.

There are functions H for which the fact that H is n-increasing implies that H is nondecreasing in each place. Among these are the functions specified by

Definition 6.1.4. Let H be a function from $A_1 \times \cdots \times A_n$ into $(-\infty, \infty)$, where each A_m is a subset of R that contains a least element a_m. Then H is *grounded* if $H(x_1, \ldots, x_n) = 0$ for all $(x_1, \ldots, x_n)$ in Dom H such that $x_m = a_m$ for at least one m.

Lemma 6.1.5. *Let H be n-increasing and grounded. Then H is nondecreasing in each place, in the sense that if $(x_1, \ldots, x_{m-1}, x, x_{m+1}, \ldots, x_n)$ and $(x_1, \ldots, x_{m-1}, y, x_{m+1}, \ldots, x_n)$ are in Dom H and $x < y$, then $H(x_1, \ldots, x, \ldots, x_n) \leqslant H(x_1, \ldots, y, \ldots, x_n)$.*

PROOF. Let $B = [(a_1, \ldots, a_{m-1}, x, a_{m+1}, \ldots, a_n), (x_1, \ldots, x_{m-1}, y, x_{m+1}, \ldots, x_n)]$. Since H is n-increasing, $V_H(B) \geqslant 0$; and since H is grounded, (6.1.2) reduces to

$$V_H(B) = H(x_1, \ldots, y, \ldots, x_n) - H(x_1, \ldots, x, \ldots, x_n),$$

whence the desired conclusion is immediate. □

Definition 6.1.6. Let H be a function from $A_1 \times \cdots \times A_n$ into $(-\infty, \infty)$. Suppose each A_m is a nonempty subset of R and has a maximal element e_m. Then H *has margins*. The *one-dimensional margins* of H are the functions H_m given by

$$\begin{aligned} &\text{Dom}\, H_m = A_m \quad \text{for } m = 1, 2, \ldots, n, \\ &H_m(x) = H(e_1, \ldots, e_{m-1}, x, e_{m+1}, \ldots, e_n) \end{aligned} \tag{6.1.4}$$

for all x in Dom H_m. The higher-dimensional margins are defined similarly, i.e., by fixing fewer places in H.

Combining this definition with Lemma 6.1.5, we obtain

Lemma 6.1.7. *Let H be a function that is n-increasing, grounded, and has margins. Then each one-dimensional margin H_m of H is nondecreasing; and if $(x_1, \ldots, x_{m-1}, x, x_{m+1}, \ldots, x_n)$ is in Dom H, then*

$$0 \leqslant H(x_1, \ldots, x_{m-1}, x, x_{m+1}, \ldots, x_n) \leqslant H_m(x). \tag{6.1.5}$$

A corresponding statement is true for the higher-dimensional margins.

The following result is fundamental:

Lemma 6.1.8. *Suppose H is n-increasing, grounded, and has margins. For any m, $1 \leqslant m \leqslant n$, let $(x_1, \ldots, x_{m-1}, x, x_{m+1}, \ldots, x_n)$, $(x_1, \ldots, x_{m-1}, y, x_{m+1}, \ldots, x_n)$ in* Dom H *be such that $x < y$. Then*

$$\begin{aligned} &H(x_1, \ldots, x_{m-1}, y, x_{m+1}, \ldots, x_n) \\ &\qquad - H(x_1, \ldots, x_{m-1}, x, x_{m+1}, \ldots, x_n) \\ &\qquad \leqslant H_m(y) - H_m(x). \end{aligned} \tag{6.1.6}$$

PROOF. If $n = 2$, then, since H is 2-increasing, (6.1.3) yields

$$H(y, e_2) - H(x, e_2) - H(y, x_2) + H(x, x_2) \geqslant 0,$$

and

$$H(e_1, y) - H(e_1, x) - H(x_1, y) + H(x_1, x) \geqslant 0,$$

which, together, are equivalent to (6.1.6).

If $n > 2$, then we define the n-boxes $B_1, B_2, \ldots, B_{n-1}$ by

$$\begin{aligned} B_1 &= \big[(x_1, a_2, \ldots, a_{m-1}, x, a_{m+1}, \ldots, a_n), \\ &\qquad (e_1, x_2, \ldots, x_{m-1}, y, x_{m+1}, \ldots, x_n)\big], \\ B_2 &= \big[(a_1, x_2, \ldots, a_{m-1}, x, a_{m+1}, \ldots, a_n), \\ &\qquad (e_1, e_2, \ldots, x_{m-1}, y, x_{m+1}, \ldots, x_n)\big], \\ &\;\;\vdots \\ B_{n-1} &= \big[(a_1, a_2, \ldots, a_{m-1}, x, a_{m+1}, \ldots, x_n), \\ &\qquad (e_1, e_2, \ldots, e_{m-1}, y, e_{m+1}, \ldots, e_n)\big], \end{aligned}$$

(with the obvious modifications if x, y are in A_1 or A_n). Since H is n-increasing, $V_H(B_k) \geqslant 0$ for each k, whence

$$\sum_{k=1}^{n-1} V_H(B_k) \geqslant 0. \tag{6.1.7}$$

Now since H is grounded, for each $V_H(B_k)$ the expression (6.1.2) reduces to a four-term sum of the form

$$\begin{aligned} &H(e_1, \ldots, e_k, \ldots, y, \ldots, x_n) - H(e_1, \ldots, e_k, \ldots, x, \ldots, x_n) \\ &\qquad - H(e_1, \ldots, x_k, \ldots, y, \ldots, x_n) + H(e_1, \ldots, x_k, \ldots, x, \ldots, x_n). \end{aligned}$$

When these sums are substituted into the left-hand side of (6.1.7), that side telescopes to

$$\begin{aligned}H(e_1, \ldots, e_{m-1}, y, e_{m+1}, \ldots, e_n)\\ - H(e_1, \ldots, e_{m-1}, x, e_{m+1}, \ldots, e_n)\\ - H(x_1, \ldots, x_{m-1}, y, x_{m+1}, \ldots, x_n)\\ + H(x_1, \ldots, x_{m-1}, x, x_{m+1}, \ldots, x_n),\end{aligned}$$

whence (6.1.7) reduces to (6.1.6). □

Note that by virtue of Lemma 6.1.5, (6.1.6) is equivalent to

$$|H(x_1, \ldots, x_{m-1}, x, x_{m+1}, \ldots, x_n) - H(x_1, \ldots, x_{m-1}, y, x_{m+1}, \ldots, x_n)| \leqslant |H_m(x) - H_m(y)| \quad (6.1.8)$$

for all $(x_1, \ldots, x, \ldots, x_n)$, $(x_1, \ldots, y, \ldots, x_n)$ in $\operatorname{Dom} H$. Applying (6.1.8) n times, we obtain

Lemma 6.1.9. *With H as in Lemma 6.1.8, let $(x_1, \ldots, x_n)$, $(y_1, \ldots, y_n)$ be any points in $\operatorname{Dom} H$. Then*

$$|H(x_1, \ldots, x_n) - H(y_1, \ldots, y_n)| \leqslant \sum_{m=1}^{n} |H_m(x_m) - H_m(y_m)|. \quad (6.1.9)$$

6.2. Joint Distribution Functions, Subcopulas, and Copulas

Definition 6.2.1. Let n be an integer $\geqslant 2$. An *n-dimensional distribution function* (briefly, an *n-d.f.*) is a function H satisfying the following conditions:

i. $\operatorname{Dom} H = R^n$;
ii. H is n-increasing and grounded;
iii. $H(\infty, \infty, \ldots, \infty) = 1$.

A *joint distribution function* (briefly, a *joint d.f.*) is a function H for which there is an integer $n \geqslant 2$ such that H is an n-d.f.

It follows from Lemma 6.1.5 that every one-dimensional margin of a joint d.f. is a distribution function (see Definition 4.1.1). We shall generally denote these one-dimensional margins of an n-d.f. H by $F_1, \ldots, F_n$ instead of $H_1, \ldots, H_n$, and refer to them briefly as "margins."

Definition 6.2.2. Let n be an integer $\geqslant 2$. An *n-dimensional subcopula* (briefly, an *n-subcopula*) is a function C' that satisfies the following conditions:

i. $\operatorname{Dom} C' = A_1 \times A_2 \times \cdots \times A_n$, where each A_m is a subset of I containing 0 and 1;
ii. C' is n-increasing and grounded;
iii. For every positive integer $m \leqslant n$, the margin C'_m of C' satisfies

$$C'_m(x) = x \quad \text{for all } x \text{ in } A_m. \tag{6.2.1}$$

An *n-dimensional copula* (briefly, an *n-copula*) is an n-subcopula C whose domain is the entire unit n-cube I^n.

Lemma 6.2.3. *An n-subcopula C' is uniformly continuous on its domain; and for any* $\mathbf{x}$ *in* $\operatorname{Dom} C'$,

$$0 \leqslant C'(\mathbf{x}) \leqslant \operatorname{Min}(x_1, \ldots, x_n). \tag{6.2.2}$$

PROOF. For $z > 0$, let $\mathbf{x}$ and $\mathbf{y}$ be any points in $\operatorname{Dom} C'$ such that $\sum_{m=1}^{n} |x_m - y_m| < z$. Then (6.1.9) and (6.2.1) yield

$$|C'(\mathbf{x}) - C'(\mathbf{y})| < z.$$

Hence C' is uniformly continuous on its domain. Combining (6.1.5) and (6.2.1) yields (6.2.2). □

The basic connection between the notion of a joint d.f. and that of an n-copula is given by

Theorem 6.2.4. *Let H be an n-d.f. with margins $F_1, \ldots, F_n$. Then there is a unique n-subcopula C', with* $\operatorname{Dom} C' = \operatorname{Ran} F_1 \times \cdots \times \operatorname{Ran} F_n$, *such that*

$$H(x_1, \ldots, x_n) = C'(F_1(x_1), \ldots, F_n(x_n)) \tag{6.2.3}$$

for all $x_1, \ldots, x_n$ in R. The n-subcopula C' is given by

$$C'(y_1, \ldots, y_n) = H(F_1^*(y_1), \ldots, F_n^*(y_n)) \tag{6.2.4}$$

for all $(y_1, \ldots, y_n)$ in $\operatorname{Dom} C'$, *where F_m^* is any function in $Q(F_m)$ (see Definition* 4.4.1). *If the functions $F_1, \ldots, F_n$ are all continuous on R, then C' is an n-copula.*

Conversely, let $F_1, \ldots, F_n$ be distribution functions, let C' be any n-subcopula whose domain contains the set $\operatorname{Ran} F_1 \times \cdots \times \operatorname{Ran} F_n$, *and let H be defined by* (6.2.3). *Then H is an n-d.f. with margins $F_1, \ldots, F_n$.*

PROOF. Suppose H is an n-d.f. with margins $F_1, F_2, \ldots, F_n$. It follows from (6.1.9) that if $F_1(x_1) = F_1(y_1), \ldots, F_n(x_n) = F_n(y_n)$ for $x_1, \ldots, x_n$, $y_1, \ldots, y_n$ in R, then $H(\mathbf{x}) = H(\mathbf{y})$. In other words, for each $\mathbf{x}$, the value $H(\mathbf{x})$ of H depends only on the numbers $F_1(x_1), F_2(x_2), \ldots, F_n(x_n)$. This is the same as saying that there is a unique function C', with $\operatorname{Dom} C' = \operatorname{Ran} F_1 \times \cdots \times \operatorname{Ran} F_n$ such that (6.2.3) holds. That this function C' is an n-subcopula follows directly from the properties of H. In particular, (6.2.1) is established as follows: Since $A_m = \operatorname{Ran} F_m$, for every x in A_m there is an x_m in R such that $F_m(x_m) = x$. Therefore we have

$$\begin{aligned} C'_m(x) &= C'(1, \ldots, 1, x, 1, \ldots, 1) \\ &= C'(F_1(\infty), \ldots, F_{m-1}(\infty), F_m(x_m), F_{m+1}(\infty), \ldots, F_n(\infty)) \\ &= H(\infty, \ldots, \infty, x_m, \infty, \ldots, \infty) = F_m(x_m) = x. \end{aligned}$$

Next, by (4.4.4) we have $F_m(F^*_m(y_m)) = y_m$ for all y_m in $\operatorname{Ran} F_m$ whence, using (6.2.3), we have

$$\begin{aligned} H(F^*_1(y_1), \ldots, F^*_n(y_n)) &= C'(F_1(F^*_1(y_1)), \ldots, F_n(F^*_n(y_n))) \\ &= C'(y_1, \ldots, y_n) \end{aligned}$$

for all $(y_1, \ldots, y_n)$ in $\operatorname{Dom} C'$, which is (6.2.4). If each F_m is continuous, then $\operatorname{Dom} C' = I^n$ and C' is an n-copula.

The converse is a matter of straightforward verification. □

Theorem 6.2.5. *Let H be an n-d.f. with margins $F_1, \ldots, F_n$. Then there is a (generally nonunique) n-copula C such that*

$$H(x_1, \ldots, x_n) = C(F_1(x_1), \ldots, F_n(x_n)) \tag{6.2.5}$$

for all $x_1, \ldots, x_n$ in R.

Theorem 6.2.5 is an immediate consequence of Theorem 6.2.4 by virtue of the following basic extension theorem.

Theorem 6.2.6. *Every subcopula can be extended to a copula; i.e., given any subcopula C', there is a copula C such that*

$$C(\mathbf{y}) = C'(\mathbf{y})$$

for all $\mathbf{y}$ *in* $\operatorname{Dom} C'$.

The proof of Theorem 6.2.6 is too long to be given in detail here. We therefore present a sketch. Given a subcopula C', we first use the uniform continuity of C' to extend C' to a subcopula C'' whose domain is the closure of $\operatorname{Dom} C'$. If $\operatorname{Dom} C''$ is I^n, then we are finished. If not, then since $\operatorname{Dom} C'$ is a Cartesian product, every point in I^n that is not in $\operatorname{Dom} C''$ is in an n-box whose vertices, but no other points, are in $\operatorname{Dom} C''$. For any point $\mathbf{u}$ in such an n-box B, we have

$$\mathbf{u} = \sum \lambda(\mathbf{u}, \mathbf{v})\mathbf{v}$$

where the summation is over the vertices $\mathbf{v}$ of B, each $\lambda(\mathbf{u}, \mathbf{v})$ is a number in I, and $\sum \lambda(\mathbf{u}, \mathbf{v}) = 1$. We then define a function C on B by

$$C(\mathbf{u}) = \sum \lambda(\mathbf{u}, \mathbf{v}) C''(\mathbf{v}).$$

This process yields consistent results for adjacent n-boxes with faces in common, so it can be applied simultaneously on all n-boxes B in I^n. The result is a function C, defined on I^n and satisfying (6.2.5), that can be shown to be a copula.

As a simple illustration of Theorems 6.2.5 and 6.2.6, let $\mathbf{a} = (a_1, \ldots, a_n)$ be a point in R^n, and consider the n-d.f. $\varepsilon_{\mathbf{a}}$ defined by

$$\varepsilon_{\mathbf{a}}(\mathbf{u}) = \begin{cases} 0, & \mathbf{a} \not< \mathbf{u}, \\ 1, & \mathbf{a} < \mathbf{u}. \end{cases} \tag{6.2.6}$$

Evidently, $\varepsilon_{\mathbf{a}}$ is the n-dimensional analog of the one-dimensional d.f. ε_a. Indeed, the margins of $\varepsilon_{\mathbf{a}}$ are the functions $\varepsilon_{a_1}, \ldots, \varepsilon_{a_n}$; conversely, any n-d.f. with margins $\varepsilon_{a_1}, \ldots, \varepsilon_{a_n}$ must coincide with $\varepsilon_{\mathbf{a}}$. With $H = \varepsilon_{\mathbf{a}}$, any n-copula C will satisfy (6.2.5). In this case the construction indicated above yields $C(x_1, x_2, \ldots, x_n) = x_1 x_2 \cdots x_n$.

Although special cases had previously appeared in the literature (see especially [Féron 1956]), copulas in their general form were first introduced in [Sklar 1959] (see also [Sklar 1973]). This was in connection with our work on PM spaces (see Section 1.9) and with the work of Fréchet et al. on joint d.f.'s and their margins (e.g., [Fréchet 1951; Féron 1956; Dall'Aglio 1959, 1972]). The name was chosen to express the fact that a copula embodies the manner in which a joint distribution function is coupled to its one-dimensional margins (see Theorems 6.2.4 and 6.2.5).

A proof of the two-dimensional case of Theorem 6.2.5 appears in [S^2 1974]. An independent proof of the general case was obtained by D. S. Moore and M. C. Spruill [1975]; their proof is probabilistic and makes no reference to copulas.

6.3. Copulas and t-Norms

Every 2-copula is a binary operation on I with identity 1 and null element 0. Some 2-copulas are also t-norms. In fact, comparing the discussion in the preceding section with that in Sections 5.4 and 5.5, we see that a 2-copula is a t-norm if and only if it is associative, and that a t-norm is a 2-copula if and only if it is 2-increasing. Furthermore, we have

Lemma 6.3.1. *The t-norms M, Π, and W are 2-copulas; and for any 2-copula C we have*

$$W \leqslant C \leqslant M. \tag{6.3.1}$$

PROOF. A straightforward computation shows that M, Π, and W are 2-increasing. The inequality $C \leqslant M$ is just (6.2.2). Next, for any x, y in I and any copula C, $V_C([(x, y), (1, 1)]) \geqslant 0$, whence (6.1.3) and (6.2.1) yield $C(x, y) \geqslant x + y - 1$, so that $C \geqslant W$. □

We also have the following two characterization theorems, the first of which is due to R. Moynihan [1978a].

Theorem 6.3.2. *A t-norm T is a copula if and only if it satisfies the Lipschitz condition*

$$T(c,b) - T(a,b) \leqslant c - a \tag{6.3.2}$$

for all a, b, c in I with $a \leqslant c$.

PROOF. If T is a copula, then (6.3.2) follows from (6.1.6). Thus suppose T satisfies (6.3.2) and choose s, t, u, v in I such that $s \leqslant t$ and $u \leqslant v$. By (6.3.2) and (5.4.7), T is continuous. Hence, since $T(0, v) = 0$ and $T(1, v) = v$, there is a w in I such that $T(w, v) = u$. Consequently, since T is associative and commutative, we have

$$\begin{aligned} T(t,u) - T(s,u) &= T(t, T(w,v)) - T(s, T(w,v)) \\ &= T(T(t,v), w) - T(T(s,v), w) \\ &\leqslant T(t,v) - T(s,v), \end{aligned}$$

whence T is 2-increasing. □

Theorem 6.3.3. *Let T be a continuous Archimedean t-norm, additively generated by (f, g). Then T is a copula if and only if either f or g is convex.*

PROOF. Suppose that f is convex. Then for any 2-box $B = [(x_1, x_2), (y_1, y_2)]$ in I^2 either $x_1 = y_1$ and $x_2 = y_2$, in which case $V_T(B) = 0$; or, by Theorem 5.5.2, $g(x_1) - g(y_1) + g(x_2) - g(y_2) > 0$. In the latter case, set

$$\lambda = \frac{g(x_1) - g(y_1)}{g(x_1) - g(y_1) + g(x_2) - g(y_2)}.$$

Then λ is in I and we have

$$\begin{aligned} g(x_1) + g(y_2) &= (1 - \lambda)(g(y_1) + g(y_2)) + \lambda(g(x_1) + g(x_2)), \\ g(x_2) + g(y_1) &= \lambda(g(y_1) + g(y_2)) + (1 - \lambda)(g(x_1) + g(x_2)). \end{aligned}$$

Since f is convex, these yield

$$\begin{aligned} f(g(x_1) + g(y_2)) &\leqslant (1 - \lambda)f(g(y_1) + g(y_2)) + \lambda f(g(x_1) + g(x_2)), \\ f(g(x_2) + g(y_1)) &\leqslant \lambda f(g(y_1) + g(y_2)) + (1 - \lambda)f(g(x_1) + g(x_2)). \end{aligned}$$

Adding the inequalities yields $V_T(B) \geqslant 0$. Hence T is 2-increasing, and therefore a copula.

Conversely, if T is a copula, take x, y in I such that $x < y$. Then $V_T([(x,x),(y,y)]) \geqslant 0$, which yields

$$2f(g(x)+g(y)) \leqslant f(2g(y)) + f(2g(x)),$$

whence f, being continuous, is convex on the interval $[0, 2g(0))$ (see [Boas 1972, §23]). But by (5.4.4) and (5.3.13), this in turn implies that f is convex on all of $[0,\infty)$. The rest of the proof follows immediately from the fact that f is convex iff g is convex. □

In the other direction, the t-norm whose additive generator f is given by

$$f(x) = \begin{cases} \dfrac{3}{2x} - 1, & 0 < x \leqslant \dfrac{1}{2}, \\ \dfrac{2}{x} - 2, & \dfrac{1}{2} \leqslant x \leqslant 1, \end{cases}$$

is strict, stronger than W, but not a copula; and the function C defined on $I \times I$ by

$$C(x,y) = xy + xy(1-x)(1-y)$$

is a commutative copula, but not a t-norm.

Definition 6.3.4. Let T be a binary operation on a set S. Then the *serial iterates* of T (see [S^2 1968]) are the functions T^n, $n = 1, 2, \ldots,$ defined recursively via

$$\operatorname{Dom} T^n = S^{n+1}, \qquad T^1 = T,$$
$$T^{n+1}(x_1, \ldots, x_{n+1}, x_{n+2}) = T(T^n(x_1, \ldots, x_{n+1}), x_{n+2}). \tag{6.3.3}$$

It follows that for $n > 2$ we have

$$M^{n-1}(x_1, \ldots, x_n) = \operatorname{Min}(x_1, \ldots, x_n), \tag{6.3.4}$$

$$\Pi^{n-1}(x_1, \ldots, x_n) = x_1 x_2 \cdots x_n, \tag{6.3.5}$$

$$W^{n-1}(x_1, \ldots, x_n) = \operatorname{Max}(x_1 + \cdots + x_n - n + 1, 0). \tag{6.3.6}$$

Let $\mathbf{u}$ and $\mathbf{v}$ be points in I^n with $\mathbf{u} \leqslant \mathbf{v}$. A simple computation yields

$$V_{\Pi^{n-1}}[\mathbf{u},\mathbf{v}] = (v_1 - u_1) \cdots (v_n - u_n), \tag{6.3.7}$$

while a more elaborate argument yields

$$V_{M^{n-1}}[\mathbf{u},\mathbf{v}] = \operatorname{Max}(0, \operatorname{Min}(v_1, \ldots, v_n) - \operatorname{Max}(u_1, \ldots, u_n)). \tag{6.3.8}$$

As a consequence, we obtain the first part of the following n-dimensional analog of Lemma 6.3.1 (the second part is an easy consequence of Lemmas 6.1.5 and 6.1.9).

Lemma 6.3.5. *The functions M^{n-1} and Π^{n-1} are n-copulas for all $n \geqslant 2$; and if C is any n-copula, then*

$$W^{n-1} \leqslant C \leqslant M^{n-1}. \tag{6.3.9}$$

The function W^{n-1} is not an n-copula for any $n > 2$. Indeed, a straightforward computation yields

$$V_{W^{n-1}}\left[\frac{\mathbf{1}}{\mathbf{2}}, \mathbf{1}\right] = 1 - \frac{n}{2}, \tag{6.3.10}$$

which is negative for all $n > 2$. Nevertheless, the left-hand inequality in (6.3.9) cannot be improved. For, with the aid of the techniques of the proof of Theorem 6.2.6, it can be shown that for any $n > 2$ and any point $\mathbf{x}$ in I^n, there is an n-copula C such that $C(\mathbf{x}) = W^{n-1}(\mathbf{x})$. Indeed, for any $n > 2$ there is an n-copula C and a subset A of I^n such that

$$\lambda_n(A) \geqslant \left(1 - \frac{1}{n}\right)^n + \frac{1}{n^n} \tag{6.3.11}$$

where λ_n denotes n-dimensional Lebesgue measure, and such that

$$C(\mathbf{x}) = W^{n-1}(\mathbf{x}) \quad \text{for every } \mathbf{x} \text{ in } A. \tag{6.3.12}$$

Theorem 6.3.6. *Let T be a continuous Archimedean t-norm. Then T^{n-1} is an n-copula for all $n \geqslant 2$ if and only if T has an outer additive generator f that is completely monotonic, i.e., is real-analytic and satisfies*

$$(-1)^n f^{(n)} \geqslant 0 \quad \textit{for all } n \geqslant 0 \tag{6.3.13}$$

(see [*Widder* 1946, *Ch.* IV]). *If one outer additive generator of T is completely monotonic, then all are, and T satisfies $\Pi \leqslant T < M$.*

C. H. Kimberling [1973b, 1974] has extended Theorem 6.3.6 and used it to characterize joint d.f.'s of sets of exchangeable random variables.

6.4. Dual Copulas

Definition 6.4.1. Let C be a 2-copula. The *dual* of C is the function $\overline{C}$ defined by

$$\overline{C}(x, y) = x + y - C(x, y) \quad \text{for all } x, y \text{ in } I. \tag{6.4.1}$$

The dual of a 2-copula is a *dual copula*.

Note that

$$\overline{M} = M^*, \qquad \overline{\Pi} = \Pi^*, \qquad \overline{W} = W^* \tag{6.4.2}$$

where for any binary operation T on I, T^* is defined by (5.7.1). It is also easy to derive the following:

Lemma 6.4.2. *A dual copula is a binary operation on I, with identity 0 and null element 1, which is continuous and nondecreasing in each place, but not 2-increasing. If $\overline{C}_1$ and $\overline{C}_2$ are dual copulas, then $\overline{C}_1 \leqslant \overline{C}_2$ if and only if $C_2 \leqslant C_1$. Hence every dual copula $\overline{C}$ satisfies*

$$\overline{M} \leqslant \overline{C} \leqslant \overline{W}. \tag{6.4.3}$$

Display (6.4.2) shows that M, Π, and W are each solutions of the functional equation

$$x + y - T(x, y) = 1 - T(1 - x, 1 - y) \quad \text{for all } x, y \text{ in } I. \tag{6.4.4}$$

This equation arises naturally in connection with the study of an important class of triangle functions (see Section 7.4); and M. J. Frank [1979] has shown that in the class of continuous t-norms the only other solutions of (6.4.4) are the functions T_s defined in (5.6.11). All these t-norms are also copulas; this follows from the fact that their additive generators, given in (5.6.10), are convex.

6.5. Copulas and Random Variables

The definition of one-dimensional Borel sets given in Section 2.3 has a natural extension to higher dimensions: Let n be a positive integer and B an n-box. Then there is a unique smallest sigma field $\mathcal{B}$ on B that contains all the n-subintervals of B. This is the *Borel field* on B, and its members are the *Borel sets* in B. Any function from B into R that is measurable with respect to $\mathcal{B}$ is *Borel measurable*.

Just as a function F in Δ defines a Lebesgue–Stieltjes probability measure P_F on R, so an n-d.f. H defines a Lebesgue–Stieltjes probability measure P_H on R^n, and an n-copula C defines a Lebesgue–Stieltjes probability measure P_C on I^n. There are corresponding integrals, for which we use the notation of Section 2.3. Thus, if B is an n-box in I^n and C is an n-copula, then

$$\int_B dC = P_C(B) = V_C(B). \tag{6.5.1}$$

Similarly, if H is an n-d.f. whose margins are in Δ, then

$$\int_{[(-\infty, \ldots, -\infty), \mathbf{u})} dH = P_H[(-\infty, \ldots, -\infty), \mathbf{u}) = H(\mathbf{u}); \tag{6.5.2}$$

while if the margins of H are in Δ^+, then

$$\int_{[\mathbf{0},\mathbf{u})} dH = P_H[\mathbf{0},\mathbf{u}) = H(\mathbf{u}). \tag{6.5.3}$$

An application of Theorem 6.8 in [Kingman and Taylor 1966] yields

Lemma 6.5.1. *Let $H, F_1, \ldots, F_n, C$ be as in Theorem 6.2.5. Let g be a Borel-measurable function from R^n into R and, for $m = 1, \ldots, n$, let F_m^* be in $Q(F_m)$ (see Definition 4.4.1). Then*

$$\int_{R^n} g\,dH = \int_{I^n} g(F_1^*(x_1), \ldots, F_n^*(x_n))\,dC(x_1, \ldots, x_n), \tag{6.5.4}$$

in the sense that if either integral exists, then so does the other and the two are equal. If g is a.s. nonnegative, then both integrals always exist, but may be infinite.

The significance of (6.5.4) lies in the fact that, since C, in contrast to H, is always continuous and since I^n is a bounded region, the integral on the right is often much easier to work with than the one on the left.

Next, if $X_1, \ldots, X_n$ are random variables on a probability space $(\Omega, \mathcal{Q}, P)$, then the function H defined on R^n by

$$H(\mathbf{x}) = P\{\omega \text{ in } \Omega \mid X_1(\omega) < x_1, \ldots, X_n(\omega) < x_n\} \tag{6.5.5}$$

is an n-d.f., the *joint distribution function of $X_1, \ldots, X_n$*; and $df(X_m) = F_m$ for each margin F_m of H. As an immediate consequence of Theorems 6.2.4 and 6.2.5 we thus have

Theorem 6.5.2. *Let $X_1, \ldots, X_n$ be random variables (defined on a common probability space) with respective individual d.f.'s $F_1, \ldots, F_n$, and joint d.f. H. Then there is a unique n-subcopula C' such that (6.2.3) holds; and there are n-copulas C such that (6.2.5) holds. If all the functions F_m are continuous, then there is a unique n-copula C such that (6.2.5) holds.*

The n-subcopula C' of Theorem 6.5.2 will be called the *subcopula of* $X_1, \ldots, X_n$; correspondingly, the n-copulas C will be called *copulas of* $X_1, \ldots, X_n$.

For $n \geqslant 2$ and $1 \leqslant m \leqslant n$, let E_m be the set $\{\omega \text{ in } \Omega \mid X_m(\omega) < x_m\}$. Then, since $P(E_m) = F_m(x_m)$, (6.5.5), Theorem 6.5.2, and (6.3.9) yield

$$W^{n-1}(P(E_1), \ldots, P(E_n)) \leqslant P(E_1 \cap \cdots \cap E_n)$$
$$\leqslant M^{n-1}(P(E_1), \ldots, P(E_n)). \tag{6.5.6}$$

When $n = 2$, (6.5.6) reduces to (2.3.7).

When Theorem 6.5.2 is applied to independent random variables, we obtain

Theorem 6.5.3. *For $n \geqslant 2$, let $X_1, \ldots, X_n$ be random variables with continuous d.f.'s Then $X_1, \ldots, X_n$ are independent if and only if the (unique) copula of $X_1, \ldots, X_n$ is Π^{n-1}.*

The following results illustrate the utility of the copula concept. Theorems 6.5.4 and 6.5.5 are due to Fréchet [1951]. Theorem 6.5.6 is due to Schweizer and Wolff [1976, 1981].

Let X_1, X_2 be random variables on a probability space $(\Omega, \mathscr{Q}, P)$. To say that X_2 is a.s. a function of X_1 means that there is a Borel-measurable function f from R into R and a set N with $P(N) = 0$ such that $X_2(\omega) = f(X_1(\omega))$ for all ω in $\Omega \backslash N$.

Theorem 6.5.4. *Let n and $X_1, \ldots, X_n$ be as in Theorem 6.5.3. Then each of the random variables $X_1, \ldots, X_n$ is a.s. a strictly increasing function of any of the others if and only if the (unique) copula of $X_1, \ldots, X_n$ is M^{n-1}.*

Theorem 6.5.5. *Let X_1, X_2 be random variables with continuous d.f.'s. Then X_1 and X_2 are a.s. strictly decreasing functions of each other if and only if the (unique) copula of X_1, X_2 is W.*

Theorem 6.5.6. *For $n \geqslant 2$ let $X_1, \ldots, X_n$ be random variables with continuous d.f.'s $F_1, \ldots, F_n$, joint d.f. H, and copula C. Let $f_1, \ldots, f_n$ be strictly increasing functions from R into R. Then $f_1(X_1), \ldots, f_n(X_n)$ are random variables (on the same probability space as $X_1, \ldots, X_n$) with continuous d.f.'s and copula C. Thus C is invariant under strictly increasing transformations of $X_1, \ldots, X_n$.*

From (6.2.3), Theorem 6.5.6, and the fact that the d.f. of $f_m(X_m)$ is $F_m \circ f_m^{-1}$, it follows that any functional or "property" of the joint d.f. of $X_1, \ldots, X_n$ that is invariant under strictly increasing transformations of the X_m's is a functional or "property" of their copula (and independent of the individual distribution functions $F_1, \ldots, F_n$). Thus much of the study of nonparametric or order statistics is intimately related to, indeed is often reducible to, the study of copulas. In particular, in view of Theorems 6.5.3–6.5.6 and (5.6.5)—which expresses the fact that the surface for Π (the copula of independence) lies midway between the surfaces for W and M (the copulas of extreme monotone dependence)—it is natural to use any measure of distance between surfaces as a measure of dependence for pairs of random variables. Such measures have been introduced and studied in [Schweizer and Wolff 1976, 1981] and [Johnson and Kotz 1977] and the reader is referred to these references for the details. It turns out that these measures of dependence have many pleasant properties vis-à-vis a set of axioms for such measures introduced by A. Rényi [1959]. Similar measures of dependence can also be defined in higher dimensions [Wolff 1981].

6.6. Margins of Copulas

If $1 \leqslant m < n$ and C is an n-copula, then the m-dimensional margins of C are the functions obtained by setting $n - m$ of the arguments of C equal to 1 (cf. Definition 6.1.6). An n-copula thus has $\binom{n}{m}$ m-dimensional margins; and for $m \geqslant 2$, these margins are themselves m-copulas. In the other direction, however, $\binom{n}{m}$ given m-copulas are generally not the m-dimensional margins of an n-copula; if they are, then the m-copulas are *compatible*.

For example, a 3-copula C has three two-dimensional margins C_{12}, C_{13}, C_{23}, which are given by

$$C_{12}(x, y) = C(x, y, 1), \quad C_{13}(x, z) = C(x, 1, z), \quad C_{23}(y, z) = C(1, y, z); \tag{6.6.1}$$

and if C_1, C_2, C_3 are arbitrary 2-copulas, then C_1, C_2, C_3 are compatible if and only if there is a 3-copula whose two-dimensional margins, in some order, are C_1, C_2, C_3.

G. Dall'Aglio [1959] has established necessary and sufficient conditions for three 2-copulas to be compatible. One of his results admits the following extension:

Theorem 6.6.1. *Let C be an n-copula for some $n \geqslant 3$. If any $\binom{n-1}{2} + 1$ of the $\binom{n}{2}$ two-dimensional margins of C are M, then so are all the remaining ones, and $C = M^{n-1}$. The number $\binom{n-1}{2} + 1$ cannot be decreased.*

Theorem 6.6.2. *Let m, n be integers $\geqslant 2$. Let C_1 be an m-copula and C_2 an n-copula. Let C be the function from I^{m+n} to I defined by*

$$C(x_1, \ldots, x_{m+n}) = M(C_1(x_1, \ldots, x_m), C_2(x_{m+1}, \ldots, x_{m+n})).$$

Then C is an $(m + n)$-copula if and only if $C_1 = M^{m-1}$ and $C_2 = M^{n-1}$.

In contrast, we have

Theorem 6.6.3. *Let m, n, C_1, and C_2 be as in Theorem 6.6.2. Let C', C'', C be the functions defined by*

$$C'(x_1, \ldots, x_{m+1}) = \Pi(C_1(x_1, \ldots, x_m), x_{m+1}),$$
$$C''(x_1, \ldots, x_{n+1}) = \Pi(x_1, C_2(x_2, \ldots, x_{n+1})),$$
$$C(x_1, \ldots, x_{m+n}) = \Pi(C_1(x_1, \ldots, x_m), C_2(x_{m+1}, \ldots, x_{m+n})).$$

Then C' is always an $(m+1)$-copula, C'' an $(n+1)$-copula, and C an $(m+n)$-copula.

6.7. Open Problems

The estimate $(1 - n^{-1})^n + n^{-n}$ in (6.3.11) can certainly be improved. For example, Figure 6.7.1 depicts a 3-copula that coincides with W^2 on $\frac{4}{9}$ of the unit cube. In the figure, the values of the 3-copula are given, within the indicated boxes, by the following expressions:

$$A: \quad (3y + 3z - 2)(3x - 2)/6,$$
$$B: \quad (3z + 3x - 2)(3y - 2)/6,$$
$$C: \quad (3x + 3y - 2)(3z - 2)/6,$$

$$Aa: \quad y(3z - 1)(3x - 2)/2, \qquad Ab: \quad z(3y - 1)(3x - 2)/2,$$
$$Ba: \quad z(3x - 1)(3y - 2)/2, \qquad Bb: \quad x(3z - 1)(3y - 2)/2,$$
$$Ca: \quad x(3y - 1)(3z - 2)/2, \qquad Cb: \quad y(3x - 2)(3z - 2)/2,$$

$$AB: \quad z(3x + 3y - 4)/2,$$
$$BC: \quad x(3y + 3z - 4)/2,$$
$$CA: \quad y(3z + 3x - 4)/2,$$
$$ABC: \quad x + y + z - 2.$$

In the remaining $\frac{11}{27}$ of the unit cube, the value of the 3-copula is 0.

This leads to

Problem 6.7.1. For $n \geqslant 3$, determine the supremum of the numbers $\lambda_n(A)$ for all subsets A of I^n such that there is an n-copula that coincides with W^{n-1} on A. If the supremum is less than 1, is it attained?

Theorem 6.5.4 gives a probabilistic characterization of the functions M^{n-1}. Similarly, Theorems 6.5.3 and 6.5.5 give probabilistic characteriza-

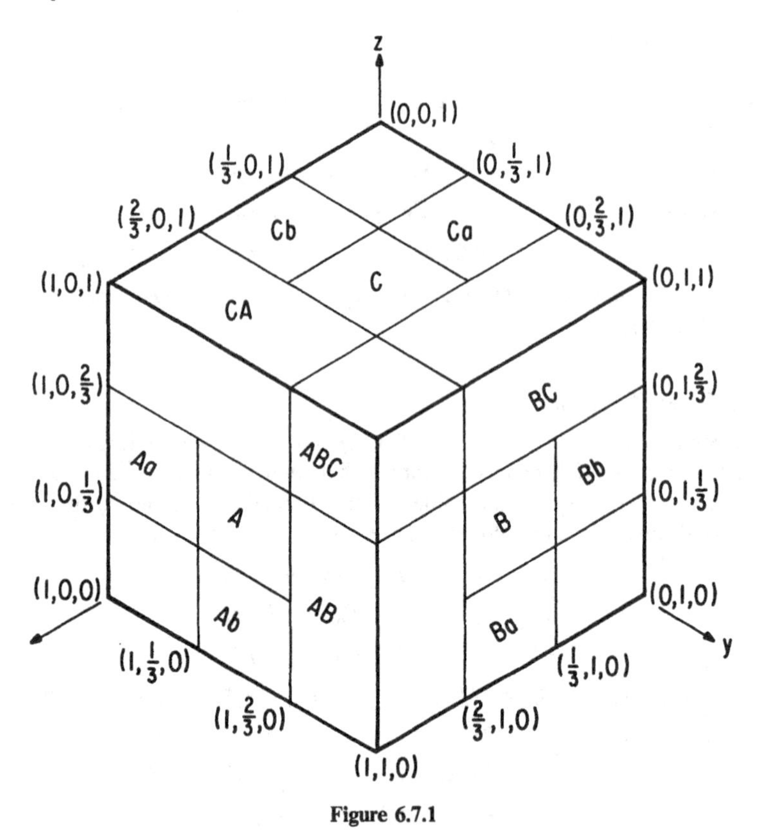

Figure 6.7.1

tions of the functions Π^{n-1} and W, respectively. Now Π and W are continuous Archimedean t-norms, while M enters into any ordinal sum construction (see Definition 5.2.4). This suggests

Problem 6.7.2. Let T be a continuous Archimedean t-norm that is also a 2-copula (see Theorems 6.3.2 and 6.3.3). Give a probabilistic characterization of T; i.e., characterize T in terms of the mutual relations of random variables that can be connected by T. If T^{n-1} is also an n-copula for some $n \geqslant 3$, extend the characterization of T to one of T^{n-1}. If C_1 and C_2 are 2-copulas for which there exist probabilistic characterizations, and C is a 2-copula formed by an ordinal sum construction involving $C_1$and C_2, find a probabilistic characterization of C.

The problem of extending the results of Dall'Aglio [1959, 1972] to higher-dimensional copulas is perhaps the most important open question concerning copulas. The aim here is to find necessary and/or sufficient conditions for the compatibility of any given finite set of copulas. As particular cases, we have

Problem 6.7.3. Let $2 \leqslant m < n$. Let $\{X_1, \ldots, X_n\}$ be a set of random variables, each with a continuous d.f., such that every m-element subset of the set is independent. Find all possible copulas of $X_1, \ldots, X_n$.

Problem 6.7.4. Let $T_1, T_2, \ldots, T_n$ be 2-copulas and let the functions $C_2, C_3, \ldots, C_{n+1}$ be defined via

$$C_2 = T_1, \qquad C_{m+1}(x_1, \ldots, x_{m+1}) = T_m(C_m(x_1, \ldots, x_m), x_{m+1})$$

for $m = 2, 3, \ldots, n$. Find necessary and sufficient conditions on the T_m's for C_m to be an m-copula for all $m \leqslant n$. (Compare Theorem 6.3.6.)

These questions are of course closely related to the problem of determining probability measures with given marginals (see [Jacobs 1978, Appendix B and the references given therein]). Moreover, each of these questions also leads naturally to a corresponding question concerning the possible values of the various measures of dependence for random variables defined in [Schweizer and Wolff 1976, 1981] and [Wolff 1981].

7
Triangle Functions

7.1. Introduction

We recall from Chapter 1 that the need to extend a triangle inequality to a polygonal inequality leads naturally to the consideration of associativity. Thus the Menger triangle inequality leads to the problem of characterizing semigroups on real intervals, while the Šerstnev inequality leads to the corresponding problem on spaces of distribution functions. The results presented in Chapter 5 show that the first problem, or at least that part of it dealing with continuous (i.e., topological) semigroups, can be regarded as solved. In contradistinction, the second is basically untouched. One reason for this is that the problem of characterizing, or even finding, topological semigroups on Δ and its subspaces falls between two chairs. On the one hand, topologists and geometers are not inclined to look upon a space of distribution functions as a geometric object; on the other, probabilists tend to focus their attention on the important special cases that arise in connection with specific properties (most often, independence) of random variables.

In this chapter we define and study several classes of binary operations on Δ^+, principally those that yield triangle functions. We therefore begin by recalling the following definition (cf. Section 1.6).

Definition 7.1.1. A *triangle function* is a binary operation on Δ^+ that is commutative, associative, and nondecreasing in each place, and has ε_0 as identity.

Note that for any triangle function τ and any F in Δ^+, we have

$$\varepsilon_\infty \leqslant \tau(\varepsilon_\infty, F) \leqslant \tau(\varepsilon_\infty, \varepsilon_0) = \varepsilon_\infty,$$

whence

$$\tau(\varepsilon_\infty, F) = \tau(F, \varepsilon_\infty) = \varepsilon_\infty\,; \tag{7.1.1}$$

i.e., ε_∞ is the null element of τ.

Definition 7.1.2. If τ_1 and τ_2 are binary operations on Δ^+, then $\tau_1 \leqslant \tau_2$ means that $\tau_1(F,G) \leqslant \tau_2(F,G)$ for all F, G, which in turn means that for all F, G in Δ^+ and all x in R^+, $\tau_1(F,G)(x) \leqslant \tau_2(F,G)(x)$.

Definition 7.1.3. Let T be a binary operation on I. Then $\mathbf{T}$ is the function on $\Delta^+ \times \Delta^+$ whose value for any F, G in Δ^+ is the function $\mathbf{T}(F,G)$ defined on R^+ by

$$\mathbf{T}(F,G)(x) = T(F(x), G(x)). \tag{7.1.2}$$

If T is a left-continuous t-norm (see Definition 7.1.6), then it is easily verified that $\mathbf{T}$ is a triangle function. Furthermore, we have

Theorem 7.1.4. *If τ is a triangle function, then $\tau \leqslant \mathbf{M}$, i.e., $\mathbf{M}$ is the maximal triangle function.*

PROOF. For any F, G in Δ^+, we have $\tau(F,G) \leqslant \tau(F,\varepsilon_0) = F$ and $\tau(F,G) \leqslant \tau(\varepsilon_0, G) = G$. Thus $\tau(F,G)(x) \leqslant M(F(x), G(x)) = \mathbf{M}(F,G)(x)$ for all x in R^+. □

Definition 7.1.5. Consider the binary operations on I that are nondecreasing in each place. Then $\mathfrak{T}$ is the set of all such binary operations that have 1 as identity; $\mathfrak{T}^*$ is the set of all such binary operations that have 0 as identity; and $\mathfrak{T}_C$ is the set of functions T in $\mathfrak{T}$ that satisfy the Lipschitz condition

$$|T(a,b) - T(c,d)| \leqslant |a-c| + |b-d|.$$

Note that all t-norms are in $\mathfrak{T}$, all t-conorms in $\mathfrak{T}^*$ and all 2-copulas in $\mathfrak{T}_C$. For any binary operation T on I let T^* be defined by (5.7.1). Then a binary operation T' is in $\mathfrak{T}^*$ if and only if there is a T in $\mathfrak{T}$ such that $T' = T^*$.

Definition 7.1.6. Let T be a binary operation on I that is nondecreasing in each place. Then T is *left continuous* if

$$T(x,y) = \sup\{T(u,v) \,|\, 0 < u < x, 0 < v < y\} \tag{7.1.3}$$

for all x, y in $(0,1]$; and T is *right continuous* if

$$T(x,y) = \inf\{T(u,v) \mid x < u < 1, y < v < 1\} \tag{7.1.4}$$

for all x, y in $[0,1)$.

Clearly, for any T in $\mathfrak{T}$, T^* is right continuous if and only if T is left continuous.

Definition 7.1.7. The class $\mathfrak{L}$ is the set of all binary operations L on R^+ that satisfy the following conditions:

i. The range of L is R^+, i.e., L is onto; (7.1.5)
ii. L is nondecreasing in each place; (7.1.6)
iii. L is continuous on $R^+ \times R^+$, except possibly at the points $(0,\infty)$ and $(\infty,0)$. (7.1.7)

It is an immediate consequence of (7.1.5) and (7.1.6) that

$$L(0,0) = 0 \quad \text{and} \quad L(\infty,\infty) = \infty. \tag{7.1.8}$$

The class $\mathfrak{L}$ was introduced in [Frank and Schweizer 1979]. Note that $\mathfrak{L}$ contains all continuous composition laws (see Definition 5.7.5) and the functions K_α given by (5.7.16). In particular, Min, Sum, and Max are in $\mathfrak{L}$.

On occasion we impose additional conditions on functions in $\mathfrak{L}$. These include the following:

iv. If $u_1 < u_2$ and $v_1 < v_2$, then $L(u_1,v_1) < L(u_2,v_2)$. (7.1.9)
v. L has 0 as identity. (7.1.10)
vi. For any x in $[0,\infty)$, the set $A_x = \{(u,v) \mid L(u,v) < x\}$ is bounded, i.e., there is an $\alpha(x)$ in $(0,\infty)$ such that $A_x \subseteq [0,\alpha(x)]^2$. (7.1.11)

The requirement (7.1.9) is sufficient to guarantee that for any x in $(0,\infty)$ the level set $\{(u,v) \mid L(u,v) = x\}$ is a bona fide curve in the first quadrant of the (u,v)-plane. Furthermore, (7.1.9) and (7.1.10) together imply that this curve connects the points $(x,0)$ and $(0,x)$.

Note that (7.1.9) is weaker than the requirement that L be strictly increasing in each place. Thus Sum, Max, and Min all satisfy (7.1.9), but of these only Sum is strictly increasing in each place. Note further that (7.1.10) implies (7.1.11) (with $\alpha(x) = x$), but not conversely, and also that (7.1.10) implies $L \geqslant \text{Max}$. Sum and Max satisfy (7.1.10); Min satisfies neither (7.1.10) nor (7.1.11).

In the sequel (apart from the subscript in d_L), the symbols T and L will be used exclusively to denote functions in $\mathfrak{T}$ and $\mathfrak{L}$, respectively.

7.2. The Operations $\tau_{T,L}$

We begin with a class of operations whose definition is motivated by Šerstnev's reformulation of Menger's triangle inequality as described in Section 1.6.

Definition 7.2.1. For any T in $\mathcal{T}$ and L in $\mathcal{L}$, $\tau_{T,L}$ is the function on $\Delta^+ \times \Delta^+$ whose value, for any F, G in Δ^+, is the function $\tau_{T,L}(F,G)$ defined on R^+ by

$$\tau_{T,L}(F,G)(x) = \sup\{T(F(u),G(v)) \mid L(u,v) = x\}. \qquad (7.2.1)$$

If $L = \text{Sum}$, then we drop the L in $\tau_{T,L}$ and simply write τ_T.

Since both T and L are nondecreasing, with $T(1,1) = 1$ and $L(\infty,\infty) = \infty$, we immediately obtain

Lemma 7.2.2. *The function $\tau_{T,L}(F,G)$ is nondecreasing on R^+ and satisfies*

$$\tau_{T,L}(F,G)(\infty) = 1 \qquad (7.2.2)$$

for any F, G in Δ^+. Moreover, $\tau_{T,L}$ is nondecreasing in each place on $\Delta^+ \times \Delta^+$.

In general, $\tau_{T,L}(F,G)$ need not be left continuous, hence not in Δ^+. However, we do have

Lemma 7.2.3. *If T is left continuous and L satisfies* (7.1.9), *then $\tau_{T,L}$ is a binary operation on Δ^+.*

PROOF. If $u > 0$ and $v > 0$, then (7.1.9) yields $L(u,v) > L(0,0) = 0$. Hence $L(u,v) = 0$ implies $u = 0$ or $v = 0$, which yields

$$\tau_{T,L}(F,G)(0) = \sup\{T(F(u),G(v)) \mid L(u,v) = 0\} = 0.$$

In view of Lemma 7.2.2, it only remains to show that $\tau_{T,L}(F,G)$ is left continuous on $(0,\infty)$; i.e., that for any x in $(0,\infty)$ and any $h > 0$ there is a nonnegative number $y < x$ such that

$$\tau_{T,L}(F,G)(y) > \tau_{T,L}(F,G)(x) - h. \qquad (7.2.3)$$

Given x in $(0,\infty)$, it follows from the continuity of L that there are numbers u, v in $(0,\infty)$ such that $L(u,v) = x$ and

$$T(F(u),G(v)) > \tau_{T,L}(F,G)(x) - h/2. \qquad (7.2.4)$$

By the left-continuity of F, G, and T, there are numbers s, t with $0 \leqslant s < u$,

$0 \leqslant t < v$ such that

$$T(F(s), G(t)) > T(F(u), G(v)) - h/2. \tag{7.2.5}$$

Set $y = L(s,t)$. Then by (7.1.9), $y < x$; and by (7.2.1),

$$\tau_{T,L}(F,G)(y) \geqslant T(F(s), G(t)). \tag{7.2.6}$$

Combining (7.2.4), (7.2.5), and (7.2.6) yields (7.2.3) and completes the proof. □

Left-continuity of T is sufficient but not necessary for $\tau_{T,L}$ to be a binary operation on Δ^+. For example, the t-norm Z defined in (5.6.6) is not left continuous. However, as a bit of calculation shows, for any F, G in Δ^+, we have

$$\tau_Z(F,G)(x) = \mathrm{Max}\left[F\left(\mathrm{Max}(x - G^\wedge(1), 0)\right), G\left(\mathrm{Max}(x - F^\wedge(1), 0)\right)\right], \tag{7.2.7}$$

from which it readily follows that τ_Z is not only a binary operation on Δ^+, but even a triangle function.

In general, to guarantee that $\tau_{L,T}$ is a triangle function we must impose further conditions on T and L.

Theorem 7.2.4. *Let T be a left-continuous t-norm. Let L be commutative and associative and satisfy* (7.1.9) *and* (7.1.10). *Then $\tau_{T,L}$ is a triangle function.*

PROOF. In view of Lemma 7.2.3, it remains to establish commutativity, ε_0 as identity, and associativity. The commutativity of $\tau_{T,L}$ follows immediately from that of T and L. Next, note that from (7.1.9) and (7.1.10), we have

$$\sup\{u \mid \text{there is a } v \text{ such that } L(u,v) = x\} = x$$

for all x in R^+. Therefore, for all F in Δ^+ and x in $(0, \infty)$,

$$\begin{aligned}\tau_{T,L}(F,\varepsilon_0)(x) &= \sup\{T(F(u), \varepsilon_0(v)) \mid L(u,v) = x\}\\ &= \sup\{F(u) \mid L(u,v) = x\}\\ &= F(\sup\{u \mid L(u,v) = x\}) = F(x).\end{aligned}$$

Hence $\tau_{T,L}(F,\varepsilon_0) = F$. Similarly, $\tau_{T,L}(\varepsilon_0, F) = F$, so ε_0 is the identity of $\tau_{T,L}$.

Turning to associativity, we first note that the monotonicity and left-continuity of T imply that if A is a subset of I and $a = \sup A$, then for any b in I,

$$\sup\{T(x,b) \mid x \text{ in } A\} = T(a,b). \tag{7.2.8}$$

From this it follows that for any F, G, H in Δ^+ and x in R^+—abbreviating

$F(u)$ to Fu, etc.—we have

$$\begin{aligned}
&\tau_{T,L}(\tau_{T,L}(F,G),H)x \\
&\quad = \sup\{T(\tau_{T,L}(F,G)s, Hv) \mid L(s,v) = x\} \\
&\quad = \sup\{T(\sup\{T(Ft, Gu) \mid L(t,u) = s\}, Hv) \mid L(s,v) = x\} \\
&\quad = \sup\{\sup\{T(T(Ft, Gu), Hv) \mid L(t,u) = s\} \mid L(s,v) = x\} \\
&\quad = \sup\{T(T(Ft, Gu), Hv) \mid L(t,u) = s, L(s,v) = x\} \\
&\quad = \sup\{T(T(Ft, Gu), Hv) \mid L(L(t,u),v) = x\}.
\end{aligned} \tag{7.2.9}$$

Similarly, we obtain

$$\tau_{T,L}(F,\tau_{T,L}(G,H))x = \sup\{T(Ft, T(Gu, Hv)) \mid L(t, L(u,v)) = x\}. \tag{7.2.10}$$

The associativity of T and L imply that the last term in (7.2.9) and the right-hand side of (7.2.10) are equal, whence $\tau_{T,L}$ is associative, and consequently a triangle function. □

Since Sum and Max are commutative and associative, and satisfy (7.1.9) and (7.1.10), we have the following:

Corollary 7.2.5. *For any left-continuous t-norm T the operations τ_T and $\tau_{T,\text{Max}}$ are triangle functions.*

If we set $L = \text{Max}$ in (7.2.1) and compare with (7.1.2), then we at once obtain

Lemma 7.2.6. *For any T*

$$\tau_{T,\text{Max}} = \mathbf{T}. \tag{7.2.11}$$

Similarly, we obtain

Lemma 7.2.7. *For any T,*

$$\tau_{T,\text{Min}} = \mathbf{M}^*. \tag{7.2.12}$$

Note that the operation $\mathbf{M}^*$ has ε_∞, rather than ε_0, as its identity and so is not a triangle function, although it is clearly commutative, associative, and nondecreasing.

Turning to the question of the continuity of the operations $\tau_{T,L}$, we have

Theorem 7.2.8. *If T is continuous and L satisfies* (7.1.11), *then $\tau_{T,L}$ is uniformly continuous on* (Δ^+, d_L).

PROOF. For brevity, throughout this proof we abbreviate $\tau_{T,L}$ to τ.

Note first that, being continuous, T is uniformly continuous on $I \times I$. Similarly, L is uniformly continuous on any bounded subset of $R^+ \times R^+$ and therefore, by (7.1.11), on each of the sets A_x. Thus, for any η such that $0 < \eta < 1$, there is a $\lambda > 0$ that satisfies the following conditions:

i. $T(\text{Min}(s+\lambda,1),t) < T(s,t) + \eta/4$,

$$T(s,\text{Min}(t+\lambda,1)) < T(s,t) + \eta/4, \quad \text{for all } s, t \text{ in } I; \tag{7.2.13}$$

ii. $$u < 1/\lambda \quad \text{and} \quad v < 1/\lambda \quad \text{whenever} \quad L(u,v) < 2/\eta; \tag{7.2.14}$$

iii. $L(u+\lambda,v) < L(u,v) + \eta/2$,

$$L(u,v+\lambda) < L(u,v) + \eta/2 \quad \text{whenever} \quad L(u,v) < 1/\eta. \tag{7.2.15}$$

Next, note that by (7.2.1), for any F, G in Δ^+ and any x in $(0,2/\eta)$, there exist $u, v > 0$ such that $L(u,v) = x$ and

$$\tau(F,G)(x) < T(F(u),G(v)) + \eta/4. \tag{7.2.16}$$

Now suppose that F' in Δ^+ is such that $d_L(F,F') < \lambda$, so that, by (4.3.1) and (4.3.2),

$$F(u) \leqslant F'(u+\lambda) + \lambda, \quad \text{for } u \text{ in } (0,1/\lambda). \tag{7.2.17}$$

It follows from (7.2.14) that $L(u,v) = x < 2/\eta$ implies $u < 1/\lambda$. Consequently, combining (7.2.16), (7.2.17), and (7.2.13) yields

$$\begin{aligned} \tau(F,G)(x) &< T(\text{Min}(F'(u+\lambda)+\lambda,1),G(v)) + \eta/4 \\ &< T(F'(u+\lambda),G(v)) + \eta/2; \end{aligned} \tag{7.2.18}$$

and now applying (7.2.1) and then (7.2.15), for all x in $(0,2/\eta)$ we obtain

$$\begin{aligned} \tau(F,G)(x) &< \tau(F',G)(L(u+\lambda,v)) + \eta/2 \\ &\leqslant \tau(F',G)(L(u,v)+\eta/2) + \eta/2 \\ &= \tau(F',G)(x+\eta/2) + \eta/2. \end{aligned} \tag{7.2.19}$$

Thus (see (4.3.1)) $[\tau(F',G),\tau(F,G);\eta/2]$ holds. In a similar manner we establish $[\tau(F,G),\tau(F',G);\eta/2]$, whence

$$d_L(\tau(F',G),\tau(F,G)) \leqslant \eta/2.$$

Furthermore, repeating the argument yields the following: If $d_L(G,G') < \lambda$, then

$$d_L(\tau(F',G'),\tau(F',G)) \leqslant \eta/2.$$

Thus, noting that $1/\eta < 2/\eta$, we have: If F, F', G, G' in Δ^+ are such that $d_L(F,F') < \lambda$ and $d_L(G,G') < \lambda$, then

$$\begin{aligned} d_L(\tau(F',G'),\tau(F,G)) & \\ &\leqslant d_L(\tau(F',G'),\tau(F',G)) + d_L(\tau(F',G),\tau(F,G)) \\ &\leqslant \eta/2 + \eta/2 = \eta. \end{aligned}$$

Therefore $\tau_{T,L}$ is uniformly continuous on (Δ^+, d_L) and the theorem is proved. □

Since (7.1.10) implies (7.1.11), we have

Corollary 7.2.9. *If T is a continuous t-norm and L is as in Theorem 7.2.4, then the triangle function $\tau_{T,L}$ is uniformly continuous on (Δ^+, d_L). Thus $(\Delta^+, d_L, \tau_{T,L})$ is a compact topological semigroup.*

Corollary 7.2.10. *Let T be a continuous t-norm. Then τ_T and $\mathbf{T}$ are uniformly continuous on (Δ^+, d_L).*

There are triangle functions that are not continuous on (Δ^+, d_L). One such is τ_Z (see (7.2.7)), as the following example shows.

Let $F_n(x) = 1 - e^{-x/n}$. Then $F_n \xrightarrow{w} \varepsilon_0$, but $\tau_Z(F_n, F_n)$ does not converge weakly to $\tau_Z(\varepsilon_0, \varepsilon_0)$, since $\tau_Z(F_n, F_n) = \varepsilon_\infty$ for all n. Note that the example actually shows more: τ_Z is not continuous on $(\mathcal{D}^+, d_L)$, and in particular not continuous at $(\varepsilon_0, \varepsilon_0)$.

R. Moynihan [1975, 1978ab] has made an extensive study of the semigroups (Δ^+, τ_T). We present a few of his results here in

Theorem 7.2.11. *Let T be a continuous t-norm. Then the following hold:*

i. *If F, G are in Δ^+ and either one is continuous, then $\tau_T(F,G)$ is continuous.*
ii. *The semigroup (Δ^+, τ_T) is not cancellative, i.e., there exist F, G_1, G_2 in Δ^+ with $G_1 \neq G_2$ such that $\tau_T(F,G_1) = \tau_T(F,G_2)$. The semigroup $(\mathcal{D}^+, \tau_M)$, however, is cancellative.*
iii. *Let x be in $(0,\infty)$. Let T^n and τ_T^n denote the serial iterates of T and τ_T, respectively (see Definition 6.3.4). Then*

$$\lim_{n\to\infty} \tau_T^n(F_1, \ldots, F_{n+1})(x) = \sup\left\{ \lim_{n\to\infty} T^n(F_1(x_1), \ldots, F_{n+1}(x_{n+1}))\right\}, \tag{7.2.20}$$

where the supremum in (7.2.20) is taken with respect to all sequences $\{x_n\}$ of positive numbers such that $\sum_{n=1}^{\infty} x_n = x$.

iv. *Let* τ_T^∞ *be defined on sequences* $\{F_n\}$ *in* Δ^+ *by*

$$\tau_T^\infty\{F_n\}(x) = l^-\left(\lim_{n\to\infty}\tau_T^n(F_1,\ldots,F_{n+1})(x)\right). \qquad (7.2.21)$$

Now let $\{F_n\}$ *be a sequence in* $\mathfrak{D}^+$. *Then the supremum on* $(0,\infty)$ *of* $\tau_T^\infty\{F_n\}$ *is either* 0 *or* 1, *whence* $\tau_T^\infty\{F_n\}$ *is either equal to* ε_∞ *or is in* $\mathfrak{D}^+$. *Furthermore,*

$$\tau_\Pi^\infty\{F_n\} = \varepsilon_\infty \quad \text{iff} \quad \tau_W^\infty\{F_n\} = \varepsilon_\infty. \qquad (7.2.22)$$

v. *Let* T_1, T_2 *be continuous and Archimedean t-norms. Then the topological semigroups* $(\Delta^+, d_L, \tau_{T_1})$, $(\Delta^+, d_L, \tau_{T_2})$ *are iseomorphic, i.e., algebraically isomorphic and topologically homeomorphic, if and only if the topological semigroups* (I, T_1), (I, T_2) *are iseomorphic.*

Some of these results undoubtedly extend to the general operations $\tau_{T,L}$.

We conclude this section with several technical results that will be needed in the sequel. First, as an immediate consequence of the definition of $\tau_{T,L}$ we have

Theorem 7.2.12. *Let* T_1, T_2 *and* L_1, L_2 *be such that* $T_1 \leqslant T_2$ *and* $L_1 \leqslant L_2$. *Then*

$$\tau_{T_1,L_2} \leqslant \tau_{T_2,L_1}. \qquad (7.2.23)$$

In particular, if T *is in* $\mathfrak{T}_C$, *then*

$$\tau_{W,L_2} \leqslant \tau_{T,L_2} \leqslant \tau_{T,L_1} \leqslant \tau_{M,L_1}; \qquad (7.2.24)$$

while if L *satisfies* (7.1.10), *then* $L \geqslant \text{Max}$ *and, using* (7.2.11), *for any* T,

$$\tau_{T,L} \leqslant \mathbf{T} \leqslant \mathbf{M}. \qquad (7.2.25)$$

Lemma 7.2.13. *Let* L *satisfy* (7.1.9). *Then for any* T *and all* a, b *in* $(0,\infty)$,

$$\tau_{T,L}(\varepsilon_a, \varepsilon_b) = \varepsilon_{L(a,b)}. \qquad (7.2.26)$$

7.3. The Operations $\tau_{T^*,L}$

We next consider a family of operations that arise naturally in connection with Menger betweenness in PM spaces (see Section 14.3).

Definition 7.3.1. Let T^* be in $\mathfrak{T}^*$ and L in $\mathfrak{L}$ (see Definitions 7.1.5 and 7.1.7). Then $\tau_{T^*,L}$ is defined by

$$\tau_{T^*,L}(F,G)(x) = \inf\{T^*(F(u), G(v)) \mid L(u,v) = x\} \qquad (7.3.1)$$

for any F, G in Δ^+ and any x in R^+. If $L = \text{Sum}$, then we drop the L in $\tau_{T^*,L}$ and simply write τ_{T^*}.

Since the sets $\mathfrak{T}$ and $\mathfrak{T}^*$ are disjoint, no confusion can arise between $\tau_{T^*,L}$ and $\tau_{T,L}$.

The proofs of the following results, except when noted, are similar to the proofs of the corresponding results for $\tau_{T,L}$.

Lemma 7.3.2. *For any F, G in Δ^+ the function $\tau_{T^*,L}(F,G)$ is nondecreasing on R^+ and satisfies $\tau_{T^*,L}(F,G)(0) = 0$ and $\tau_{T^*,L}(F,G)(\infty) \leqslant 1$. Furthermore, the operation $\tau_{T^*,L}$ is nondecreasing in each place on $\Delta^+ \times \Delta^+$.*

Lemma 7.3.3. *If T^* is left continuous and L satisfies* (7.1.9), *then $\tau_{T^*,L}$ is a binary operation on Δ^+.*

The proof that $\tau_{T^*,L}(F,G)(\infty) = 1$ is similar to the proof in Lemma 7.2.3 that $\tau_{T,L}(F,G)(0) = 0$. The proof of left-continuity is considerably more involved than the corresponding one in Lemma 7.2.3, and is an extension of the proof of left-continuity of τ_{T^*} given in Theorem 5.1 of [Schweizer 1975a].

Lemma 7.3.4. *Let T^* be a right-continuous t-conorm. Let L be commutative and associative and satisfy* (7.1.9) *and* (7.1.10). *Then $\tau_{T^*,L}$ is commutative and associative and has ε_0 as identity.*

Combining Lemmas 7.3.2, 7.3.3, and 7.3.4 yields

Theorem 7.3.5. *Let T^* be a continuous t-conorm. Let L be commutative and associative and satisfy* (7.1.9) *and* (7.1.10). *Then $\tau_{T^*,L}$ is a triangle function.*

Corollary 7.3.6. *For any continuous t-conorm T^*, the operation τ_{T^*} is a triangle function.*

Lemma 7.3.7. *For any T^* in $\mathfrak{T}^*$*

$$\tau_{T^*,\text{Max}} = \mathbf{M}, \tag{7.3.2}$$

$$\tau_{T^*,\text{Min}} = \mathbf{T}^*. \tag{7.3.3}$$

Note that $\tau_{T^*,\text{Min}}$ has ε_∞ as identity and thus is not a triangle function.

Theorem 7.3.8. *Let T^* be commutative and continuous. Let L be commutative and satisfy* (7.1.11). *Then $\tau_{T^*,L}$ is uniformly continuous on (Δ^+, d_L).*

Corollary 7.3.9. *Let T^* be a continuous t-conorm. Let L be commutative and associative and satisfy* (7.1.9) *and* (7.1.10). *Then the triangle function $\tau_{T^*,L}$ is uniformly continuous on (Δ^+, d_L). Thus $(\Delta^+, d_L, \tau_{T^*,L})$ is a compact topological semigroup.*

Corollary 7.3.10. *Let T^* be a continuous t-conorm. Then τ_{T^*} is uniformly continuous on (Δ^+, d_L).*

Finally, note that (5.7.6), (7.2.1), and (7.3.1) imply that for all T and all L,

$$\tau_{T,L} \leqslant \tau_{T^*,L}. \tag{7.3.4}$$

7.4. The Operations $\sigma_{C,L}$

For any F, G in Δ^+, the *convolution* $F * G$ of F and G is the function defined on R^+ by $(F * G)(0) = 0$, $(F * G)(\infty) = 1$, and

$$(F * G)(x) = \int_{[0,x)} F(x-t)\,dG(t) \quad \text{for } x \text{ in } (0,\infty). \tag{7.4.1}$$

It is well known (see, e.g., [Loève 1977, § 13.3]) that $F * G$ is in Δ^+ and that the operation of convolution is commutative and associative. Since it is readily seen that convolution is nondecreasing in each place and has ε_0 as identity, it follows that convolution is a triangle function. Furthermore (see [Schweizer 1975a]), convolution is uniformly continuous on (Δ^+, d_L), whence $(\Delta^+, d_L, *)$ is a compact topological semigroup. The algebraic and analytic properties of convolution have been studied in great detail and there is a vast literature devoted to the subject. The rudiments may be found in any advanced probability text; and the books by Linnik [1964] and Lukacs [1970] contain a wealth of further information and extensive bibliographies.

Each of the operations $\tau_{T,L}$ and $\tau_{T^*,L}$ belongs to a family of binary operations on Δ^+. Similarly, convolution is also naturally embedded in such a family. For, as is well known (see, e.g., [Rényi 1970, Ch. IV, §10]), the integral in (7.4.1) can be transformed into

$$\int_{\mathrm{Sum}\{x\}} d\Pi(F(u), G(v)) \tag{7.4.2}$$

where $\mathrm{Sum}\{x\}$ is the region $\{(u,v) \mid u, v \text{ in } R^+, u+v<x\}$; and as noted by Z. Fiedorowicz, Π and Sum in (7.4.2) may be replaced by other functions. This leads to

Definition 7.4.1. Let C be a 2-copula. Then for any L in $\mathfrak{L}$ and any F, G in

Δ^+, $\sigma_{C,L}(F,G)$ is the function defined by

$$\sigma_{C,L}(F,G)(0)=0, \qquad \sigma_{C,L}(F,G)(\infty)=1,$$

and

$$\sigma_{C,L}(F,G)(x)=\int_{L\{x\}} dC(F(u),G(v)) \quad \text{for } x \text{ in } (0,\infty), \tag{7.4.3}$$

where $L\{x\}$ is the region $\{(u,v)\,|\,u,v \text{ in } R^+, L(u,v)<x\}$. If $L=\text{Sum}$, then we drop the L in $\sigma_{C,L}$ and simply write σ_C.

The integral in (7.4.3) is just the Lebesgue–Stieltjes H-measure of the set $L\{x\}$, where

$$H(u,v)=C(F(u),G(v)) \tag{7.4.4}$$

for all u, v in R (see Section 6.5). The standard properties of probability measures (see Section 2.3) then yield the following theorem, which is a straightforward extension of Theorem 5 in [Frank 1975].

Theorem 7.4.2. *For any 2-copula and any L, $\sigma_{C,L}$ is a binary operation on Δ^+ that is nondecreasing in each place, is commutative whenever C and L are commutative, and has ε_0 as identity whenever L satisfies* (7.1.10).

From (6.5.3) we have

$$\begin{aligned}\int_{[0,u)\times[0,v)} dC(F(s),G(t)) &= P_H([0,u)\times[0,v))\\ &= H(u,v)=C(F(u),G(v)).\end{aligned} \tag{7.4.5}$$

Since $\text{Max}\{x\}=[0,x)\times[0,x)$, we therefore obtain

$$\sigma_{C,\text{Max}}=\mathbf{C}. \tag{7.4.6}$$

Next for u, v in $(0,\infty)$ let $\Gamma(u,v)$ be the region defined by

$$\Gamma(u,v)=([0,u)\times R^+)\cup(R^+\times[0,v)). \tag{7.4.7}$$

Then (2.3.6) and (7.4.5) yield

$$\begin{aligned}\int_{\Gamma(u,v)} dC(F(s),G(t)) &= P_H(\Gamma(u,v))\\ &= P_H([0,u)\times R^+)+P_H(R^+\times[0,v))\\ &\quad - P_H([0,u)\times[0,v))\\ &= H(u,\infty)+H(\infty,v)-H(u,v)\\ &= F(u)+G(v)-C(F(u),G(v));\end{aligned}$$

i.e. (see Definition (6.4.1)),

$$P_H(\Gamma(u,v)) = \overline{C}(F(u), G(v)). \tag{7.4.8}$$

In particular, since $\text{Min}\{x\} = \Gamma(x,x)$, we have

$$\sigma_{C,\text{Min}} = \overline{C}. \tag{7.4.9}$$

The operation σ_Π is *convolution*; and if K_α is given by (5.7.16), then σ_{Π,K_α} is the *α-convolution* introduced by K. Urbanik [1964]. If C is a copula, then

$$\sigma_{C,L_2} \leqslant \sigma_{C,L_1} \quad \text{whenever} \quad L_1 \leqslant L_2. \tag{7.4.10}$$

A straightforward calculation yields

$$\sigma_{C,L}(\varepsilon_a, \varepsilon_b) = \varepsilon_{L(a,b)} \tag{7.4.11}$$

for all a, b in R^+. Thus the subset of Δ^+ comprising the unit step-functions ε_a is closed under $\sigma_{C,L}$. This is generally not true for the set of uniform distribution functions U_{ab} defined in (4.1.4). In particular, C. Alsina and E. Bonet [1979] have shown that

i. $\sigma_C(U_{ab}, U_{cd}) = U_{a+c,b+d}$ iff $C = M$; (7.4.12)
ii. $\sigma_C(U_{ab}, U_{cd}) = U_{\text{Min}(a+d,b+c),\text{Max}(a+d,b+c)}$ iff $C = W$. (7.4.13)

On the basis of the corresponding results for $\tau_{T,L}$ and $\tau_{T^*,L}$ one might conjecture that $\sigma_{C,L}$ is associative whenever both C and L are. This is most emphatically not the case, for M. J. Frank [1975, 1979] has proved the following remarkable

Theorem 7.4.3. *Let C be a copula and let L be associative and satisfy* (7.1.11). *Then $\sigma_{C,L}$ is associative if and only if $C = M$, or $C = \Pi$, or (I, C) is an ordinal sum of binary systems (S_α, C_α) such that each S_α is a closed interval $[a_\alpha, e_\alpha]$ and each C_α is either* Min *on S_α, or is given by*

$$C_\alpha(x, y) = a_\alpha + \frac{(x - a_\alpha)(y - a_\alpha)}{e_\alpha - a_\alpha} \quad \textit{for all } x, y \textit{ in } [a_\alpha, e_\alpha].$$

Note that $C_\alpha(x, y) = f_\alpha^{-1}(f(x)f(y))$ where f_α is the linear function given by $f_\alpha(x) = (x - a_\alpha)/(e_\alpha - a_\alpha)$. Thus Theorem 7.4.3 says that $\sigma_{C,L}$ is associative if and only if C is an ordinal sum of M's and Π's.

Corollary 7.4.4. *Let C be a copula and let L be commutative and associative and satisfy* (7.1.9) *and* (7.1.10). *Then $\sigma_{C,L}$ is a triangle function if and only if C has one of the forms specified in Theorem* 7.4.3.

Thus the operations $\sigma_{C,L}$ are not a rich source of triangle functions. Nevertheless, they are important—on the one hand because of their connections with functions of random variables (see Section 7.6), and on the other because of their relationship to other classes of functions on Δ^+.

7.5. The Operations $\rho_{C,L}$

Definition 7.5.1. Let C be in $\mathcal{T}_C$. Then for any L in $\mathfrak{L}$ and any F, G in Δ^+, $\rho_{C,L}(F,G)$ is the function defined by

$$\rho_{C,L}(F,G)(x) = \inf\{\overline{C}(F(u),G(v)) \mid L(u,v) = x\} \tag{7.5.1}$$

for x in R^+, where $\overline{C}(x,y) = x + y - C(x,y)$. If $L = \text{Sum}$, then we simply write ρ_C.

Note that for any C in $\mathcal{T}_C$, the function $\overline{C}$ is a binary operation on I that is continuous and nondecreasing in each place and has 0 as identity and 1 as null element.

Arguments similar to the corresponding ones for $\tau_{T,L}$ (except for left-continuity, which again is proved by an extension of the argument in the proof of Theorem 5.1 in [Schweizer 1975a]) establish the following:

Theorem 7.5.2. *Let C be in $\mathcal{T}_C$ and let L satisfy* (7.1.9). *Then $\rho_{C,L}$ is a binary operation on Δ^+ that is nondecreasing in each place. If, in addition, C and L are commutative, then $\rho_{C,L}$ is commutative; if C and L are associative, then $\rho_{C,L}$ is associative; if L satisfies* (7.1.10), *then $\rho_{C,L}$ has ε_0 as identity.*

Corollary 7.5.3. *Let C in $\mathcal{T}_C$ be commutative and associative. Let L be commutative and associative and satisfy* (7.1.9) *and* (7.1.10). *Then $\rho_{C,L}$ is a triangle function.*

Lemma 7.5.4. *Let C be in $\mathcal{T}_C$. Then $\rho_{C,\text{Max}} = \tau_{C^*,\text{Max}} = \mathbf{M}$ (see* (7.3.2)) *and $\rho_{C,\text{Min}} = \sigma_{C,\text{Min}} = \mathbf{C}$ (see* (7.4.9)).

The operations $\tau_{T,L}$, $\sigma_{C,L}$, and $\rho_{C,L}$ are also related by a number of useful inequalities (see [Moynihan, Schweizer, and Sklar 1978]).

Theorem 7.5.5. *If C is a copula and L satisfies* (7.1.9), *then*

$$\tau_{W,L} \leqslant \tau_{C,L} \leqslant \sigma_{C,L} \leqslant \rho_{C,L} \leqslant \rho_{W,L}. \tag{7.5.2}$$

PROOF. Note first that

$$\tau_{C,L}(F,G)(x) \leqslant \sigma_{C,L}(F,G)(x) \leqslant \rho_{C,L}(F,G)(x)$$

holds automatically for all F, G in Δ^+ when $x = 0$ and when $x = \infty$. Next, by (7.4.3) and (7.4.4), for $0 < x < \infty$ we have

$$\sigma_{C,L}(F,G)(x) = P_H(L\{x\}).$$

Now if u and v in $(0,\infty)$ are such that $L(u,v) = x$, then

$$[0,u) \times [0,v) \subseteq L\{x\} \subseteq \Gamma(u,v),$$

where $\Gamma(u,v)$ is given by (7.4.6). Thus

$$P_H([0,u) \times [0,v)) \leqslant P_H(L\{x\}) \leqslant P_H(\Gamma(u,v)),$$

whence (7.4.5) and (7.4.8) yield

$$C(F(u),G(v)) \leqslant \sigma_{C,L}(F,G)(x) \leqslant \overline{C}(F(u),G(v)).$$

An appeal to (7.2.1), (7.2.24), and (7.5.1) now yields (7.5.2) and completes the proof. □

When $C = M$, then, subject to mild restrictions, some of the inequalities in (7.5.2) can be replaced by equalities. Specifically, we have

Theorem 7.5.6. *If F and G are in $\mathscr{D}^+$, then*

$$\tau_{M,L}(F,G) = \rho_{M,L}(F,G) \tag{7.5.3}$$

for any L. Furthermore, if ∞ is a null element for L and L is continuous on all of $R^+ \times R^+$, then (7.5.3) holds for all F, G in Δ^+.

PROOF. Assume that (7.5.3) is false. Then, in view of (7.5.2), there exist F, G in Δ^+ and x in $(0,\infty)$ such that

$$\tau_{M,L}(F,G)(x) < \rho_{M,L}(F,G)(x). \tag{7.5.4}$$

Let $w = \tau_{M,L}(F,G)(x)$. By (7.5.4), $w < 1$. Now let

$$a = \sup\{t \mid F(t) \leqslant w\}, \qquad b = \sup\{t \mid G(t) \leqslant w\}. \tag{7.5.5}$$

If F and G are in $\mathscr{D}^+$, then a and b are both finite. Since F and G are left continuous, it follows that $F(a) \leqslant w$ and $G(b) \leqslant w$. Now suppose that $L(a,b) < x$. Then by (7.1.5) there exist $a' > a$, $b' > b$ such that $L(a',b') = x$, whence by (7.5.5) we have $F(a') > w$, $G(b') > w$. Consequently,

$$\tau_{M,L}(F,G)(x) \geqslant M(F(a'),G(b')) > w, \tag{7.5.6}$$

which is impossible. Thus $L(a,b) \geqslant x > 0$, and using (7.5.4), Theorem 7.5.2, and (7.5.1), we have

$$\begin{aligned} w < \rho_{M,L}(F,G)(x) &\leqslant \rho_{M,L}(F,G)(L(a,b)) \\ &\leqslant \overline{M}(F(a),G(b)) = \operatorname{Max}(F(a),G(b)) \leqslant w, \end{aligned} \tag{7.5.7}$$

which is a contradiction. Thus (7.5.3) holds if F and G are both in $\mathscr{D}^+$.

If F and G are not both in $\mathscr{D}^+$, then there are several cases to consider. First, if a and b are both finite, then the preceding argument applies. Next, suppose $a = \infty$ and $0 \leqslant b < \infty$. If there is an a'' in $(0, \infty)$ such that $L(a'', b) \geqslant x > 0$, then since $F(a'') \leqslant w$, (7.5.7) again yields a contradiction. Thus we must have $L(a'', b) < x$ for all a'' in $(0, \infty)$. But this too is impossible; for, using the fact that ∞ is a null element for L and the continuity of L, we have the contradiction

$$\infty = L(\infty, b) = \lim_{a'' \to \infty} L(a'', b) \leqslant x. \tag{7.5.8}$$

The same argument holds if $0 \leqslant a < \infty$ and $b = \infty$. Lastly, if $a = b = \infty$, then an argument similar to the foregoing again establishes the contradictions (7.5.7) and (7.5.8). Thus (7.5.4) is false and the proof is complete. □

Combining Theorems 7.5.5 and 7.5.6 yields

Corollary 7.5.7. *If L is continuous on $R^+ \times R^+$, has ∞ as a null element, and satisfies* (7.1.9), *then*

$$\tau_{M,L} = \sigma_{M,L} = \rho_{M,L}. \tag{7.5.9}$$

To see that (7.5.9) does not hold without some restrictions on L, consider the following example: Take $L(a, b) = ab$, with $L(0, \infty) = L(\infty, 0) = \infty$; and for $0 \leqslant \alpha < \beta \leqslant 1$ let $F = \alpha\varepsilon_0 + (1 - \alpha)\varepsilon_\infty$, $G = \beta\varepsilon_0 + (1 - \beta)\varepsilon_\infty$. Then $\tau_{M,L}(F, G) = \sigma_{M,L}(F, G) = F < G = \rho_{M,L}(F, G)$. Note that

$$\tau_{M,L}(F, \varepsilon_0)(x) = l^- F(\infty) \quad \text{for all } x > 0,$$

whence $\tau_{M,L}$ is not a triangle function, even when restricted to $\mathscr{D}^+$. Similarly, in this case, $\sigma_{M,L}$ and $\rho_{M,L}$ are also not triangle functions.

Comparing Definition 7.5.1 with Definition 7.3.1, we see that $\tau_{C^*,L} = \rho_{C,L}$ whenever $C^* = \overline{C}$. Hence we have

Corollary 7.5.8. *For any L,*

$$\tau_{W^*,L} = \rho_{W,L}, \qquad \tau_{\Pi^*,L} = \rho_{\Pi,L}, \qquad \tau_{M^*,L} = \rho_{M,L}. \tag{7.5.10}$$

Hence under the hypotheses of Theorem 7.5.6,

$$\tau_{M^*,L} = \tau_{M,L}. \tag{7.5.11}$$

7.6. Derivability and Nonderivability from Functions of Random Variables

It is well known that convolution of distribution functions corresponds to addition of random variables, in the sense that if F and G are distribution functions (in $\mathscr{D}$), then there exist (independent) random variables X, Y such that F is the distribution function of X, G the distribution function of Y, and their convolution $F * G$ the distribution function of $X+Y$. It is therefore natural to ask: What functions on random variables correspond to the binary operations on Δ^+ discussed in the preceding sections? To make this question precise, we begin with

Definition 7.6.1. Let τ be a binary operation on Δ^+. Then τ is *derivable from a function on random variables* if there exists a Borel-measurable function V from $R^+ \times R^+$ into R^+ satisfying the following condition: For any F, G in Δ^+, there exist random variables X, Y (which may be allowed to assume the value ∞ with positive probability) defined on a common probability space $(\Omega, \mathscr{C}, P)$, such that $df(X) = F$, $df(Y) = G$, and

$$df(V(X,Y)) = \tau(F,G) = \tau(df(X), df(Y)) \tag{7.6.1}$$

where $V(X,Y)$ is the random variable whose value for any ω in Ω is $V(X(\omega), Y(\omega))$.

Lemma 7.6.2. *Let X, Y be random variables defined on a common probability space, with respective distribution functions F, G in Δ^+. Let V be a Borel-measurable function from $R^+ \times R^+$ into R^+. Then $V(X,Y)$ is a random variable whose distribution function $F_{V(X,Y)}$ is in Δ^+ and is given by*

$$F_{V(X,Y)}(x) = \int_{V\{x\}} dC(F(u), G(v)) \tag{7.6.2}$$

for all x in $(0,\infty)$, where $V\{x\} = \{(u,v) | u,v \text{ in } R^+, V(u,v) < x\}$, and C is any copula of X and Y.

PROOF. It is well known (see, e.g., [Rényi 1970, Ch. IV, §10]) that $V(X,Y)$ is a random variable and that

$$F_{V(X,Y)}(x) = \int_{V\{x\}} dH(u,v) \quad \text{for all } x \text{ in } (0,\infty) \tag{7.6.3}$$

where H is the joint d.f. of X and Y. Substituting (6.2.5) into (7.6.3) yields (7.6.2). □

As an immediate consequence of Lemma 7.6.2, we have

Theorem 7.6.3. *Let τ be a binary operation on Δ^+. Then τ is derivable from a function on random variables if and only if there exists a Borel-measurable function V from $R^+ \times R^+$ into R^+ such that for all F, G in Δ^+ there is a*

2-copula C such that

$$\tau(F,G)(x)=\int_{V\{x\}} dC(F(u),G(v)) \quad \textit{for all } x \textit{ in } (0,\infty). \tag{7.6.4}$$

Note that the 2-copula C in (7.6.4) generally depends on F and G. If C is independent of F and G, then τ is a fortiori derivable from a function on random variables. This yields the following:

Corollary 7.6.4. *Let C be a 2-copula and let L satisfy (7.1.9). Then $\sigma_{C,L}$ is derivable from a function (in fact, from the function L) on random variables. Consequently, $\tau_{C,\mathrm{Max}}$, $\tau_{C^*,\mathrm{Max}}$, and $\rho_{C,\mathrm{Max}}$ are derivable from functions on random variables. If, in addition, L satisfies the hypotheses of Corollary 7.5.7, then $\tau_{M,L}$ and $\rho_{M,L}$ are also derivable.*

PROOF. The derivability of $\sigma_{C,L}$ is an immediate consequence of the equality of the integrals in (7.6.4) and (7.4.3). The derivability of $\tau_{M,L}$ and $\rho_{M,L}$ follows from Corollary 7.5.7; that of $\tau_{C,\mathrm{Max}}$ from (7.2.11) and (7.4.6); and that of $\tau_{C^*,\mathrm{Max}}$ and $\rho_{C,\mathrm{Max}}$ from (7.3.2), Lemma 7.5.4, and the fact that, by (7.4.6), both operations are equal to $\sigma_{M,\mathrm{Max}}$. □

There are, however, many operations on Δ^+ that are not derivable from functions on random variables. The principal result of this section is

Theorem 7.6.5. *Let T be any left-continuous function in $\mathcal{T}$ other than M. Then τ_T is not derivable from any function on random variables.*

PROOF. Assume that τ_T is derivable, i.e., that a suitable function V exists. Let X and Y be random variables that are constant almost everywhere, with respective values x and y in R^+. Then $df(X)=\varepsilon_x$ and $df(Y)=\varepsilon_y$, whence by Definition 7.6.1 and Lemma 7.2.13, $df(V(X,Y))=\tau_T(\varepsilon_x,\varepsilon_y)$ $=\varepsilon_{x+y}$, which means that $V(X,Y)$ assumes the value $x+y$ almost everywhere. Therefore $V(x,y)=x+y$, and since x, y are arbitrary, $V=\mathrm{Sum}$. Hence, by (7.6.4) and (7.4.3), for all F, G in Δ^+ we have

$$\tau_T(F,G)=\sigma_C(F,G) \tag{7.6.5}$$

for some 2-copula C which may depend on F and G.

For any a in $(0,1)$, let F_a be defined by

$$F_a(x)=\begin{cases}0, & x=0,\\ a, & x \text{ in } (0,1],\\ 1, & x \text{ in } (1,\infty].\end{cases} \tag{7.6.6}$$

Then a short computation yields

$$\tau_T(F_a, F_b)(x) = \begin{cases} 0, & x = 0, \\ T(a,b), & x \text{ in } (0,1], \\ \mathrm{Max}(a,b), & x \text{ in } (1,2], \\ 1, & x \text{ in } (2,\infty], \end{cases} \tag{7.6.7}$$

and

$$\sigma_C(F_a, F_b)(x) = \begin{cases} 0, & x = 0, \\ C(a,b), & x \text{ in } (0,1], \\ a + b - C(a,b), & x \text{ in } (1,2], \\ 1, & x \text{ in } (2,\infty], \end{cases} \tag{7.6.8}$$

for any t-norm T (including M) and any copula C. In particular,

$$\tau_T(F_a, F_b)(\tfrac{1}{2}) = T(a,b), \qquad \sigma_C(F_a, F_b)(\tfrac{1}{2}) = C(a,b),$$

whence (7.6.5) yields $T(a,b) = C(a,b)$. Similarly, letting $x = \frac{3}{2}$, we obtain $\mathrm{Max}(a,b) = a + b - C(a,b)$, whence $C(a,b) = a + b - \mathrm{Max}(a,b) = M(a,b)$. Thus we have

$$M(a,b) \neq T(a,b) = C(a,b) = M(a,b),$$

which is a contradiction. Hence V cannot exist, and the theorem is proved. □

This result clearly shows that the distinction between working directly with distribution functions (as we generally do in the theory of probabilistic metric spaces) and working with them indirectly, via random variables, is intrinsic and not just a matter of taste. Theorem 7.6.5 thus provides further evidence that there are topics in probability theory that are not encompassed by the standard measure-theoretic model of the theory.

Finally, we remark that the results above all extend to Δ (see [S² 1974]) and to the operations $\tau_{T,L}$.

7.7. Duality

Recall that the operation τ_M is defined by

$$\tau_M(F, G)(x) = \sup\{M(F(u), G(v)) | \mathrm{Sum}(u,v) = x\}. \tag{7.7.1}$$

If F and G are strict d.d.f.'s, then it is easy to see that the supremum on the right-hand side of (7.7.1) is attained precisely when $F(u) = G(v)$. Turning this observation around, we see that for any t in I there exist unique values

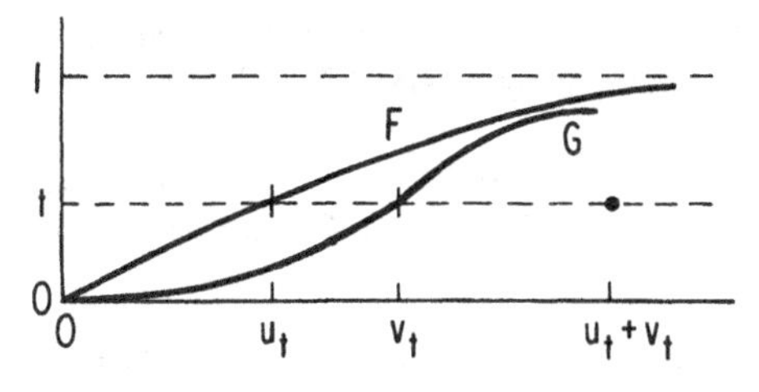

Figure 7.7.1

u_t and v_t such that $F(u_t) = G(v_t) = t$ and $\tau_M(F, G)(u_t + v_t) = t$ (see Figure 7.7.1). Inverting, we have

$$[\tau_M(F,G)]^{-1}(t) = u_t + v_t = F^{-1}(t) + G^{-1}(t),$$

whence

$$[\tau_M(F,G)]^{-1} = F^{-1} + G^{-1}. \tag{7.7.2}$$

Now, with some patience, it can be shown that (7.7.2) remains valid for any F, G in Δ^+, i.e., that

$$[\tau_M(F,G)]^{\wedge} = F^{\wedge} + G^{\wedge}, \tag{7.7.3}$$

from which we at once have

$$\tau_M(F,G) = [F^{\wedge} + G^{\wedge}]^{\wedge} \tag{7.7.4}$$

(cf. [Sherwood and Taylor 1974, Prop. 4]). Display (7.7.4) shows that the operation τ_M on Δ^+ is equivalent to pointwise addition on the space of (left-continuous) quasi-inverses. Since the latter operation is simpler than the former, this is a useful result, frequently applied in the sequel.

Next, since $F^{\wedge}$ and $G^{\wedge}$ are nondecreasing, we may write

$$F^{\wedge}(x) + G^{\wedge}(x) = \inf\{\mathrm{Sum}(F^{\wedge}(u), G^{\wedge}(v)) \mid M(u,v) = x\}. \tag{7.7.5}$$

The expressions on the right-hand sides of (7.7.1) and (7.7.5) are dual in the sense that each may be obtained from the other as follows: Interchange M and Sum, interchange sup and inf, and replace functions by their quasi-inverses. Furthermore, (7.7.3) and (7.7.4) show that each expression is the quasi-inverse of the other. These observations, together with the definition of $\tau_{T,L}$ in (7.2.1), suggest that the foregoing relationships remain valid when M is replaced by any continuous T and Sum by any L. Frank and Schweizer [1979] showed that this is indeed generally the case—and in a much wider context. The results we need are summarized below.

Definition 7.7.1. For any F in Δ^+ let $F^\wedge$ be the left-continuous quasi-inverse of F, as defined by (4.4.5). Then ∇^+ is the set $\{F^\wedge \mid F \text{ in } \Delta^+\}$.

Definition 7.7.2. For any T and any L, $\tau_{T,L}^\wedge$ is the function on $\nabla^+ \times \nabla^+$ whose value for any $F^\wedge$, $G^\wedge$ in ∇^+ is the function $\tau_{T,L}^\wedge(F^\wedge, G^\wedge)$ defined on I by

$$\tau_{T,L}^\wedge(F^\wedge, G^\wedge)(x) = \inf\{L(F^\wedge(u), G^\wedge(v)) \mid T(u,v) = x\}. \quad (7.7.6)$$

If $L = \text{Sum}$, then we write $\tau_T^\wedge$.

Theorem 7.7.3. *Suppose that L has ∞ as a null element and is continuous on all of $R^+ \times R^+$, and that T is continuous. Then for any F, G in Δ^+ we have*

$$\tau_{T,L}(F,G) = [\tau_{T,L}^\wedge(F^\wedge, G^\wedge)]^\wedge \quad (7.7.7)$$

and

$$\tau_{T,L}^\wedge(F^\wedge, G^\wedge) = [\tau_{T,L}(F,G)]^\wedge. \quad (7.7.8)$$

Thus $\tau_{T,L}^\wedge$ is a binary operation on ∇^+ that is nondecreasing in each place and has $\varepsilon_0^\wedge$ as identity.

If we let $T = M$ in (7.7.6), then we obtain

$$\tau_{M,L}^\wedge(F^\wedge, G^\wedge) = L(F^\wedge, G^\wedge), \quad (7.7.9)$$

whence (7.7.7) yields

$$\tau_{M,L}(F,G) = [L(F^\wedge, G^\wedge)]^\wedge, \quad (7.7.10)$$

and letting $L = \text{Sum}$ in (7.7.10) brings us back to (7.7.4).

Finally, we remark that the duality theorems of [Frank and Schweizer 1979] can also be applied to the operations $\rho_{C,L}$ defined by (7.5.1). Thus, if we define

$$\rho_{C,L}^\wedge(F^\wedge, G^\wedge)(x) = \sup\{L(F^\wedge(u), G^\wedge(v)) \mid \overline{C}(u,v) = x\}, \quad (7.7.11)$$

then (7.7.7) and (7.7.8) remain valid when τ and T are replaced by ρ and C, respectively. In particular, letting $C = M$ and using (7.7.9), we have

$$\rho_{M,L}^\wedge(F^\wedge, G^\wedge) = L(F^\wedge, G^\wedge) = \tau_{M,L}^\wedge(F^\wedge, G^\wedge), \quad (7.7.12)$$

which, on taking quasi-inverses, yields $\rho_{M,L} = \tau_{M,L}$ and an immediate proof of Theorem 7.5.6.

7.8. The Conjugate Transform

There are functions Φ from Δ^+ into sets of real- or complex-valued functions that have the property

$$\Phi(F * G) = \Phi(F)\Phi(G) \quad \text{for all } F, G \text{ in } \Delta^+ \tag{7.8.1}$$

where $*$, as usual, denotes convolution. Among these are the Fourier-Stieltjes transform or characteristic function (see e.g., [Loève 1977, §§13–15]), and the Laplace-Stieltjes transform or moment-generating function (see, e.g., [Widder 1946, Ch. 2, §1]). These transforms are indispensable tools in the theory of probability and in other branches of mathematics. It is thus natural to ask whether one can find similar transforms for other triangle functions. This question has been considered by R. Moynihan. In his doctoral dissertation of 1975 and in [Moynihan 1980ab] he introduced and studied a function c_T on (Δ^+, τ_T) that is an analog of the Laplace-Stieltjes transform on $(\Delta^+, *)$. The salient features of his work are summarized in this section.

In what follows, it is convenient to extend the exponential function to all of R by setting $e^{-\infty} = 0$ and $e^{\infty} = \infty$.

Definition 7.8.1. Let T be a continuous Archimedean t-norm and let k be a fixed inner multiplicative generator[1] of T (see Definition 5.5.6). Then for any F in Δ^+ the *T-conjugate transform* of F is the function $c_T F$ defined on R^+ by

$$\begin{aligned} c_T F(s) &= \sup\{e^{-sx}k(F(x)) \mid x \text{ in } (0, \infty)\} \quad \text{for } 0 \leqslant s < \infty, \\ c_T F(\infty) &= \inf\{c_T F(s) \mid s < \infty\}. \end{aligned} \tag{7.8.2}$$

Note that in view of Theorem 5.5.5, we have

$$c_T F = c_\Pi(k \circ F). \tag{7.8.3}$$

The Π-conjugate transform c_Π is the maximum transform of R. Bellman and W. Karush [1961, 1962, 1963] applied to the semigroup (Δ^+, τ_Π). This maximum transform is in turn closely related to the conjugate transform of convex analysis (see, e.g., [Fenchel 1953]); and many of the properties of c_Π (e.g., (7.8.12)) are either equivalent to, or simple consequences of, previously established results of Fenchel and of Bellman and Karush. However, just as characteristic functions have many properties not shared by Fourier–Stieltjes transforms in general, so the fact that the domain of c_T is

[1] The particular choice of k does not affect the validity of the results.

a set of bounded nondecreasing functions yields a rich theory with a structure of its own.

Theorem 7.8.2. *Let T and k be as in Definition 7.8.1. Then for any F in Δ^+ other than ε_∞ the function $c_T F$ has the following properties.*

i. $c_T F$ *is nonincreasing on* R^+. (7.8.4)
ii. $c_T F$ *is positive on* $[0, \infty)$ *and* $\geqslant k(0)$ *on* R^+. (7.8.5)
iii. $c_T F$ *is continuous on* R^+. (7.8.6)
iv. $\log c_T F$ *is convex on* $[0, \infty)$. (7.8.7)

Furthermore,

v. $c_T \varepsilon_\infty(s) = k(0)$ *for all* s *in* R^+. (7.8.8)
vi. $c_T \varepsilon_a(s) = \text{Max}(e^{-sa}, k(0))$, (7.8.9)
for a *in* $[0, \infty)$, s *in* $[0, \infty)$;
vii. $c_T F(s) = 1$ *for some* s *in* $(0, \infty)$ *iff* $F = \varepsilon_0$. (7.8.10)

Conversely, if f is function on R^+ that satisfies (7.8.4)–(7.8.7), then there is an F in Δ^+ other than ε_∞ such that $c_T F = f$.

Corollary 7.8.3. Ran c_T *is the set of all functions f from R^+ to I that satisfy (7.8.4)–(7.8.7), together with the function defined in (7.8.8).*

Theorem 7.8.4. *The transform c_T is order preserving on Δ^+; i.e., if $F \leqslant G$, then $c_T F \leqslant c_T G$.*

The raison d'être of T-conjugate transforms is brought out in

Theorem 7.8.5. *Let T and k be as in Definition 7.8.1. Then for any F, G in Δ^+ we have*

$$c_T \tau_T(F, G) = \text{Max}(c_T F \cdot c_T G, k(0)). \tag{7.8.11}$$

In particular, if T is strict, then

$$c_T \tau_T(F, G) = c_T F \cdot c_T G. \tag{7.8.12}$$

In order to use (7.8.11) effectively, we need to know the extent to which we can pass from d.d.f's to their T-conjugate transforms and back again, i.e., we need an inverse transform.

Definition 7.8.6. Let T and k be as in Definition 7.8.1 and let h be the outer multiplicative generator of T associated with k. Let f be in Ran c_T (see Corollary 7.8.3). Then the *inverse T-conjugate transform* of f is the function $c_T^* f$ defined on R^+ by

$$c_T^* f(0) = 0, \qquad c_T^* f(\infty) = 1 \tag{7.8.13a}$$

and

$$c_T^* f(x) = l^- G(x), \quad \text{for } 0 < x < \infty, \tag{7.8.13b}$$

where

$$G(x) = \inf\left\{ h\left(e^{xt} f(t)\right) \middle| t \text{ in } R^+ \right\}. \tag{7.8.14}$$

Definition 7.8.7. Let T and k be as in Definition 7.8.1. Then F in Δ^+ is *T-log-concave* if $\log(k \circ F)$ is concave on (b_F, ∞), where $b_F = \sup\{x | k(F(x)) = 0\}$ if $k(0) = 0$, and $b_F = 0$ otherwise. The set of all T-log-concave functions in Δ^+ is denoted by Δ_T^+.

The set Δ_T^+ depends only on T and not on the particular choice of k.

Theorem 7.8.8. *The T-conjugate transform c_T is a one-to-one mapping from Δ_T^+ onto* Ran c_T *with inverse c_T^*.*

Corollary 7.8.9. *The mappings c_T and c_T^* are quasi-inverses.*

It follows that for any f in Ran c_T, $c_T^* f$ is T-log-concave. Thus if we set

$$F_T = c_T^* c_T F \tag{7.8.15}$$

for any F in Δ^+, then F_T is in Δ_T^+ and

$$c_T F_T = c_T c_T^* c_T F = c_T F. \tag{7.8.16}$$

Furthermore, $F = F_T$ if and only if F is T-log-concave. In addition, we have

$$c_T^* (c_T F \cdot c_T G) = \tau_T(F_T, G_T), \tag{7.8.17}$$

whence

$$c_T \tau_T(F_T, G_T) = c_T F \cdot c_T G \tag{7.8.18}$$

if $c_T F \cdot c_T G \geqslant k(0)$ on R^+. These observations yield

Theorem 7.8.10. *For any F, G, H in Δ^+ such that $H \neq \varepsilon_\infty$,*

$$H_T = \tau_T(F_T, G_T) \tag{7.8.19}$$

if and only if

$$c_T H = c_T F \cdot c_T G. \tag{7.8.20}$$

Theorem 7.8.11. *Let T be a strict t-norm with multiplicative generator k. Then $\tau_T(F, G)$ is T-log-concave whenever F and G are. Thus (Δ_T^+, τ_T) is a subsemigroup of (Δ^+, τ_T). Moreover, the cancellation law holds in (Δ_T^+, τ_T). Lastly, if F is in Δ_T^+ and if, for any $\alpha \geqslant 0$, $F^{[\alpha]}$ is defined by*

$$F^{[\alpha]}(x) = k^{-1}\left[k(F(x/\alpha))\right]^{\alpha}, \tag{7.8.21}$$

then $F^{[\alpha]}$ is in Δ_T^+ and

$$\tau_T(F^{[\alpha]}, F^{[\beta]}) = F^{[\alpha+\beta]}. \tag{7.8.22}$$

Thus, every F in Δ_T^+ can be embedded in a continuous iteration semigroup and, in particular, has roots of all positive orders, i.e., is "infinitely divisible."

Theorem 7.8.12. *Let T be an Archimedean t-norm and let $\{F_n\}$ be a sequence in Δ^+. If $F_n \xrightarrow{w} F$, then $c_T F_n(s) \to c_T F(s)$ for all $s > 0$. Furthermore, for each s in $(0, \infty)$, $c_T F_n(s) \to 1$ if and only if $F_n \xrightarrow{w} \varepsilon_0$.*

Thus c_T is a continuous map from (Δ^+, τ_T) to $\operatorname{Ran} c_T$. Moreover, when $\operatorname{Ran} c_T$ is endowed with a suitable metric, then c_T induces a homeomorphism between Δ_T^+ and $\operatorname{Ran} c_T$; and each of these spaces is compact and complete.

7.9. Open Problems

As indicated in Section 7.1, the focal question of this chapter is the characterization of continuous triangle functions. Here, in spite of considerable progress, the basic problems remain unsolved. They are

Problem 7.9.1. Determine all distinct, continuous, nondecreasing solutions of the functional equation of associativity (or equivalently, all non-isomorphic, partially ordered, topological semigroups) on the space (Δ, d_L) and its various subspaces. In particular, determine all continuous triangle functions (see Definition 7.1.1) and, if possible, find a representation corresponding to the one given in Theorems 5.3.8 and 5.4.1.

Problem 7.9.2. Determine the arithmetic structure of each of the distinct solutions obtained in Problem 7.9.1, i.e., for any such solution, do what has been done in detail for the convolution semigroup of d.f.'s (see, e.g., [Linnik 1964]) and in part for the triangle functions τ_T (see [Moynihan 1980ab]).

Problems 7.9.1 and 7.9.2 obviously present an ambitious long-term program. Some specific questions are given in

Problem 7.9.3. Given a (continuous) triangle function τ, determine the following:

i. The subsets of Δ^+ closed under τ. In particular, is $\mathcal{D}^+$ closed under τ? Is the set of d.f.'s $\{a\varepsilon_b + (1-a)\varepsilon_\infty | a \text{ in } I \text{ and } b \text{ in } R^+\}$ closed under τ? When is the set of unimodal d.f.'s closed under τ? (Compare [Feller 1966, p. 164; Gnedenko and Kolmogorov 1954, Appendix 2; Ibragimov 1956; Hengartner and Theodorescu 1979]).
ii. The idempotent elements of τ.
iii. The nilpotent elements of τ, i.e., those functions F in Δ^+ such that $F^n = \varepsilon_\infty$ for some $n > 0$, where F^n is the nth τ-power of F (see Definition 5.1.3).
iv. The infinitely divisible elements of τ, i.e., those F's in Δ^+ that have the property that for every $n > 0$ there is a G in Δ^+ such that $G^n = F$. More generally, find the elements of τ that can be embedded in a continuous iteration semigroup (cf. Theorem 7.8.11).
v. The indecomposable elements of τ. Here, F in Δ^+ is *decomposable* if there exist F_1, F_2, neither of which is of the form ε_a for some $a \geqslant 0$, such that $F = \tau(F_1, F_2)$ (see Moynihan [1975]).
vi. The subsets of Δ^+ on which τ is cancellative (see Part ii of Theorem 7.2.11).
vii. The relationship of the set of discontinuities of $\tau(F, G)$ to the set of discontinuities of F and G. (For convolution the set of discontinuities of $\tau(F, G)$ is the direct sum of the set of discontinuities of F and the set of discontinuities of G (see [Lukacs 1970, §3.3]); for τ_T it is a subset of this direct sum (see [Moynihan 1975]).)
viii. The means, medians, etc., of $\tau(F, G)$ in terms of the means, medians, etc., of F and G.

Turning to the various triangle functions studied in this chapter, we have

Problem 7.9.4. Let T be a continuous t-norm and suppose that F and G in Δ^+ are absolutely continuous (see [Kingman and Taylor 1966, §9.2]). Is $\tau_{T,L}(F, G)$ absolutely continuous? (This problem is due to Moynihan [1978a].) An affirmative answer would imply that $\tau_{T,L}$ induces a corresponding operation on probability density functions.

Problem 7.9.5. Suppose that T is a continuous t-norm. To what extent does the structure of T determine the structure of $\tau_{T,L}$ (cf. Part v of Theorem 7.2.11)? In particular, if T is an ordinal sum, is $\tau_{T,L}$ an ordinal sum?

Problem 7.9.6. Find conditions on T and L that are both necessary and sufficient (rather than merely sufficient) for $\tau_{T,L}$ to be a triangle function. Similarly, find necessary and sufficient conditions on T and L for $\tau_{T,L}$ to be continuous.

Clearly, the same problem can be posed for $\tau_{T^*,L}$ and for $\rho_{C,L}$.

Problem 7.9.7. Is the set of triangle functions of the form $\tau_{T,L}$ dense in the set of all triangle functions? This question can be taken in either a pointwise or a uniform sense. Pointwise: Given a triangle function τ, a number h in $(0,1)$, and a pair (F,G) of d.d.f.'s, does there exist a pair (T,L) such that $d_L(\tau(F,G),\tau_{T,L}(F,G)) < h$? Uniform: Given a triangle function τ and a number h in $(0,1)$, does there exist a pair (T,L) such that $d_L(\tau(F,G),\tau_{T,L}(F,G)) < h$ for all pairs (F,G) of d.d.f's? If the answer to the first question is "yes," and to the second question "no," does there exist a topological structure on the set of all triangle functions relative to which the answer to the appropriately modified version of the second question is "yes"?

Problem 7.9.8. Find necessary and sufficient conditions on C and L for $\sigma_{C,L}$ to be continuous.

Problem 7.9.9. As remarked in Section 7.4, the operations $\sigma_{C,L}$ include some of the generalized convolutions introduced by Urbanik [1964] (see also [Bingham 1971] and [Gilewski 1972]). What about the rest of them? What is the exact nature of the relationship between the class of operations $\sigma_{C,L}$ and the class of all generalized convolutions?

Problem 7.9.10. Extend the theory of conjugate transforms to all the semigroups $(\Delta^+,\tau_{T,L})$ where T is continuous and Archimedean and L is associative and commutative and satisfies (7.1.9) and possibly (7.1.10).

Problem 7.9.11. Study the interconnections between these various operations (e.g., distributive laws, modular laws, inequalities).

For any F in Δ the *concentration function* of F is the function Q_F defined on R^+ by

$$Q_F(t) = \sup_x[F(x+t) - F(x)].$$

It is easy to show that Q_F is in Δ^+.

Concentration functions were introduced by P. Lévy [1937], who used them to establish a number of important limit theorems for sums of independent random variables. Since then they have continued to play an important role in probability theory (see [Hengartner and Theodorescu 1973, 1979]). An important property is

$$Q_{F*G} \leqslant \mathbf{M}(Q_F, Q_G), \tag{7.9.1}$$

or equivalently, if X and Y are independent random variables, then

$$Q_{df(X+Y)} \leqslant \mathbf{M}(Q_{df(X)}, Q_{df(Y)}). \tag{7.9.2}$$

This motivates

Problem 7.9.12. Establish inequalities similar to (7.9.1) for the operations τ_T, σ_C, $\tau_{T,L}$, $\sigma_{C,L}$, etc., and use these inequalities to obtain corresponding limit theorems.

We say that a triangle function τ is *causal* if it satisfies the following condition: For any $a > 0$, if $F_1(x) \leqslant F_2(x)$ for $x \leqslant a$ and $G_1(x) \leqslant G_2(x)$ for $x \leqslant a$, then $\tau(F_1, G_1)(x) \leqslant \tau(F_2, G_2)(x)$ for $x \leqslant a$.

Causality is clearly a strengthening of (1.6.2). Moreover, all the triangle functions introduced in this chapter are causal and we have no example of a triangle function that is not causal. Thus we pose

Problem 7.9.13. Is every triangle function causal?

These problems are all important in one way or another in the theory of PM spaces. Most of them are also of intrinsic interest. The entire program of study may also be justified on philosophical grounds, for upon looking at the development of measurement processes during the past century, one soon observes that with increasing frequency the raw data are distribution functions (or frequency functions) rather than real numbers. This is so in the physical sciences, and in the biological and social sciences it is the rule rather than the exception. Thus one may convincingly argue that probability distribution functions are the "numbers" of the future. This being so, the program outlined here represents a first step toward an understanding of their arithmetic.

8
Probabilistic Metric Spaces

8.1. Probabilistic Metric Spaces in General

Having developed the necessary machinery, we now turn to our main topic. As indicated in Section 1.6, our definition of a probabilistic metric space is based on that of A. N. Šerstnev.

Definition 8.1.1. A *probabilistic metric space* (briefly, a *PM space*) is a triple $(S, \mathfrak{F}, \tau)$ where S is a nonempty set (whose elements are the *points* of the space), $\mathfrak{F}$ is a function from $S \times S$ into Δ^+, τ is a triangle function, and the following conditions are satisfied for all p, q, r in S:

i. $\mathfrak{F}(p,p) = \varepsilon_0$. (8.1.1)
ii. $\mathfrak{F}(p,q) \neq \varepsilon_0$ if $p \neq q$. (8.1.2)
iii. $\mathfrak{F}(p,q) = \mathfrak{F}(q,p)$. (8.1.3)
iv. $\mathfrak{F}(p,r) \geqslant \tau(\mathfrak{F}(p,q), \mathfrak{F}(q,r))$. (8.1.4)

If $(S, \mathfrak{F}, \tau)$ is a PM space, then we also say that $(S, \mathfrak{F})$ is a *PM space under* τ. If (8.1.1) and (8.1.3) are satisfied, then $(S, \mathfrak{F})$ is a *probabilistic premetric space* (briefly, a *pre-PM space*). If (8.1.1), (8.1.2), and (8.1.3) are satisfied, then $(S, \mathfrak{F})$ is a *probabilistic semimetric space* (briefly, a *PSM space*). If (8.1.1), (8.1.3), and (8.1.4) are satisfied, then $(S, \mathfrak{F}, \tau)$ is a *probabilistic pseudometric space* (briefly, a *PPM space*), and $(S, \mathfrak{F})$ is a *PPM space under* τ.

In the sequel we shall frequently denote the distribution function $\mathfrak{F}(p,q)$ by F_{pq} and its value at x by $F_{pq}(x)$. Hence, for example, (8.1.4) can also be written in the form

$$F_{pr} \geqslant \tau(F_{pq}, F_{qr}). \tag{8.1.5}$$

Definition 8.1.2. Let $(S_1, \mathcal{F}_1)$, $(S_2, \mathcal{F}_2)$ be pre-PM spaces. Then $(S_1, \mathcal{F}_1)$ and $(S_2, \mathcal{F}_2)$ are *isometric* if and only if there exists a one-to-one function φ from S_1 onto S_2 such that

$$\mathcal{F}_1(p,q) = \mathcal{F}_2(\varphi(p), \varphi(q)) \quad \text{for all } p, q \text{ in } S; \tag{8.1.6}$$

φ and its inverse φ^{-1} are *isometries* between $(S_1, \mathcal{F}_1)$ and $(S_2, \mathcal{F}_2)$.

If $(S, \mathcal{F}, \tau)$ is a PPM space, then the relation defined on S by

$$p \sim q \quad \text{if and only if} \quad \mathcal{F}(p,q) = \varepsilon_0 \tag{8.1.7}$$

is an equivalence relation. If $p_1 \sim p_2$ and $q_1 \sim q_2$, then we have

$$\begin{aligned} \mathcal{F}(p_1, q_1) &\geqslant \tau(\mathcal{F}(p_1, q_2), \mathcal{F}(q_2, q_1)) = \tau(\mathcal{F}(p_1, q_2), \varepsilon_0) = \mathcal{F}(p_1, q_2) \\ &\geqslant \tau(\mathcal{F}(p_1, p_2), \mathcal{F}(p_2, q_2)) = \tau(\varepsilon_0, \mathcal{F}(p_2, q_2)) = \mathcal{F}(p_2, q_2). \end{aligned}$$

Similarly, $\mathcal{F}(p_2, q_2) \geqslant \mathcal{F}(p_1, q_1)$, whence $\mathcal{F}(p_1, q_1) = \mathcal{F}(p_2, q_2)$. Since $\mathcal{F}(p,q) \neq \varepsilon_0$ if $p \not\sim q$, we therefore obtain

Lemma 8.1.3. *Let $(S, \mathcal{F}, \tau)$ be a* PPM *space. For any p in S, let p^* denote the equivalence class containing p and let S^* denote the set of these equivalence classes. Then the expression*

$$\mathcal{F}^*(p^*, q^*) = \mathcal{F}(p,q) \quad \textit{for any } p \textit{ in } p^*, \ q \textit{ in } q^* \tag{8.1.8}$$

defines a function $\mathcal{F}^*$ from $S^* \times S^*$ into Δ^+ and the triple $(S^*, \mathcal{F}^*, \tau)$ is a PM *space, the quotient space of $(S, \mathcal{F}, \tau)$.*

Probabilistic metric spaces can be classified in many different ways. We begin with a partial classification based on the properties of the triangle function τ.

Definition 8.1.4. Let $(S, \mathcal{F}, \tau)$ be a PM space. Then $(S, \mathcal{F}, \tau)$ is *proper* if

$$\tau(\varepsilon_a, \varepsilon_b) \geqslant \varepsilon_{a+b} \quad \text{for all } a, b \text{ in } R^+. \tag{8.1.9}$$

If $\tau = \tau_T$ for some t-norm T, then $(S, \mathcal{F}, \tau_T)$ is a *Menger space,* or equivalently, $(S, \mathcal{F})$ is a *Menger space under T*. If τ is convolution, then $(S, \mathcal{F})$ is a *Wald space.*

The condition $F \geqslant \varepsilon_a$ is equivalent to $F(a+) = 1$ but not to $F(a) = 1$. Thus the class of proper PM spaces is slightly larger than the class of weak PM spaces defined by (1.4.2). On the other hand, the foregoing definition of a Menger space is equivalent to the earlier definition given in Section 1.4. For if (1.2.1) holds and if x is in R^+, then for all u, v in R^+ such that $u + v = x$, we have

$$F_{pr}(x) \geqslant T(F_{pq}(u), F_{qr}(v)), \tag{8.1.10}$$

whence

$$F_{pr}(x) \geqslant \sup\{T(F_{pq}(u), F_{qr}(v)) \mid u + v = x\} = \tau_T(F_{pq}, F_{qr}), \tag{8.1.11}$$

so that (8.1.4) holds with τ_T. In the other direction, (8.1.4) with τ_T obviously implies (1.2.1).

If (S, d) is a metric space, $\mathfrak{F}$ is defined on $S \times S$ by

$$\mathfrak{F}(p, q) = \varepsilon_{d(p,q)} \quad \text{for all } p, q \text{ in } S, \tag{8.1.12}$$

and τ is a triangle function such that

$$\tau(\varepsilon_a, \varepsilon_b) = \varepsilon_{a+b} \quad \text{for all } a, b \text{ in } R^+ \tag{8.1.13}$$

(e.g., if $\tau = \tau_T$ or $\tau = \sigma_C$—see (7.2.26) and (7.4.11)), then $(S, \mathfrak{F}, \tau)$ is a proper PM space. In the other direction, if $(S, \mathfrak{F}, \tau)$ is a proper PM space, and there is a function d from $S \times S$ into R^+ such that (8.1.12) holds, then (S, d) is a metric space. Moreover, for any τ satisfying (8.1.13) the semigroups $(R^+, +)$ and $(\{\varepsilon_a \mid a \text{ in } R^+\}, \tau)$ are isomorphic. Consequently, the class of metric spaces may be viewed as a subclass of the class of proper PM spaces. Since we generally want a given class of PM spaces to include all metric spaces, we often need to know that all the PM spaces in this class are proper. The following theorem, which is an immediate consequence of Theorem 7.2.12 and Lemma 7.2.13, provides a useful criterion.

Theorem 8.1.5. *If $(S, \mathfrak{F}, \tau_{T,L})$ is a* PM *space where T is a t-norm, L satisfies* (7.1.9), *and $L \leqslant$* Sum, *then $(S, \mathfrak{F}, \tau_{T,L})$ is proper.*

Corollary 8.1.6. *Every Menger space and every* PM *space of the form $(S, \mathfrak{F}, \mathbf{T})$ is proper.*

Inequalities among triangle functions also lead to corresponding relations among PM spaces. Thus an immediate consequence of Definition 8.1.1 is

Lemma 8.1.7. *Let τ_1 and τ_2 be triangle functions such that $\tau_1 \leqslant \tau_2$. If $(S, \mathfrak{F}, \tau_2)$ is a* PM *space, then so is $(S, \mathfrak{F}, \tau_1)$.*

Lemma 8.1.7 can be interpreted as follows: Every property of $(S, \mathfrak{F}, \tau_1)$ that depends only on the triangle inequality also holds for $(S, \mathfrak{F}, \tau_2)$. Thus, since (7.2.25) yields $\mathbf{T} \geqslant \tau_T$ for any t-norm T, it follows at once that any such property of a Menger space $(S, \mathfrak{F}, \tau_T)$ holds a fortiori in a so-called *non-Archimedean* PM *space* $(S, \mathfrak{F}, \mathbf{T})$ (see [Istrăţescu 1976]).

Now (7.2.23) also yields $\tau_T \leqslant \tau_M$ for all T in $\mathfrak{T}$; and (7.5.2), (7.4.10), (7.2.11), and (7.4.6) yield $\tau_\Pi \leqslant \sigma_\Pi \leqslant \sigma_{\Pi,\mathrm{Max}} = \mathbf{\Pi}$. Recalling that σ_Π is convolution, we therefore obtain

Theorem 8.1.8.

i. *If $(S, \mathfrak{F})$ is a Menger space under M, then $(S, \mathfrak{F})$ is a Menger space under any t-norm T for which τ_T is a triangle function.*
ii. *If $(S, \mathfrak{F}, \mathbf{\Pi})$ is a* PM *space, then $(S, \mathfrak{F})$ is a Wald space.*
iii. *Every Wald space is a Menger space under* Π.

Corollary 8.1.9. *Every Wald space is proper.*

In addition, J. B. Brown [1973] has shown that Part iii of Theorem 8.1.8 is best-possible, in the sense that if T is a t-norm stronger then Π, then there exist Wald spaces that are not Menger spaces under T.

In Section 8.4 we exhibit a PM space that is a Menger space under M, but not a Wald space. It follows that for no t-norm T does the fact that $(S, \mathfrak{F})$ is a Menger space under T imply that $(S, \mathfrak{F})$ is a Wald space.

8.2. Transformed Triangle Inequalities and Derived Metrics

From this point on we make the convention that *unless explicitly stated otherwise, "T" means "continuous t-norm T" and "L" means "function L in $\mathfrak{L}$ that is commutative and associative and satisfies* (7.1.9) *and* (7.1.10)" (cf. p. 98). With this convention $\tau_{T,L}$ is always a continuous triangle function (see Theorems 7.2.4 and 7.2.8).

In addition, the phrases "metric space" and "pseudometric space" are to be taken in the extended sense, i.e., distance functions are allowed to assume the value ∞.

The duality theorems of Section 7.7 often enable us to transform the Šerstnev inequality (8.1.5) into a form that is more perspicuous and easier to work with. Thus we have

Theorem 8.2.1. *If $(S, \mathfrak{F}, \tau_{T,L})$ is a* PM *space and if $\tau_{T,L}^{\wedge}$ is the operation dual to $\tau_{T,L}$ given by* (7.7.6), *then*

$$F_{pr}^{\wedge} \leqslant \tau_{T,L}^{\wedge}\left(F_{pq}^{\wedge}, F_{qr}^{\wedge}\right) \tag{8.2.1}$$

for all p, q, r in S. Conversely, if $(S, \mathfrak{F})$ is a PSM *space in which* (8.2.1) *holds for all p, q, r in S, then $(S, \mathfrak{F}, \tau_{T,L})$ is a* PM *space.*

PROOF. For any F, G in Δ^+, it follows from Corollary 4.4.3 that $F \geqslant G$ if and only if $F^\wedge \leqslant G^\wedge$. Hence (8.1.5) yields

$$F_{pr}^\wedge \leqslant \left[\tau_{T,L}(F_{pq}, F_{qr})\right]^\wedge,$$

which, by virtue of (7.7.8), is equivalent to (8.2.1). The converse is immediate. □

An appeal to (7.7.9) and its special case (7.7.4) then yields

Corollary 8.2.2. *If $T = M$ in Theorem 8.2.1, then*

$$F_{pr}^\wedge \leqslant L\left(F_{pq}^\wedge, F_{qr}^\wedge\right) \tag{8.2.2}$$

for all p, q, r in S. If, in addition, $L = \text{Sum}$ (so that $(S, \mathfrak{F})$ is a Menger space under M), then

$$F_{pr}^\wedge \leqslant F_{pq}^\wedge + F_{qr}^\wedge \tag{8.2.3}$$

for all p, q, r in S. Conversely, if $(S, \mathfrak{F})$ is a PSM space in which (8.2.3) *holds for all p, q, r in S, then $(S, \mathfrak{F})$ is a Menger space under M.*

Inequalities (8.2.2) and (8.2.3) are reminiscent of, and frequently lead directly to, the ordinary triangle inequality. Thus for any c in I, (8.2.3) yields

$$F_{pr}^\wedge(c) \leqslant F_{pq}^\wedge(c) + F_{qr}^\wedge(c) \tag{8.2.4}$$

for all p, q, r in S, which proves

Theorem 8.2.3. *Let $(S, \mathfrak{F})$ be a Menger space under M. For any c in I, define d_c on $S \times S$ by*

$$d_c(p,q) = F_{pq}^\wedge(c). \tag{8.2.5}$$

Then (S, d_c) is a pseudometric space; and (S, d_c) is a metric space if and only if $F_{pq}(0+) < c$ for all pairs p, q of distinct points of S.

Note that the function d_c is well defined in any pre-PM space and satisfies (3.1.1) and (3.1.3).

The number $F^\wedge(\frac{1}{2})$ is the *median* of F, the numbers $F^\wedge(\frac{1}{4})$ and $F^\wedge(\frac{3}{4})$ are the *quartiles* of F, and for any c in I, the number $F^\wedge(c)$ is the *c-tile* of F. Thus, put another way, Theorem 8.2.3 says that in any Menger space under M, the c-tiles of the d.d.f's are either metrics or pseudometrics. Theorem 8.2.3 is an important special case of a more general situation. This is a consequence of the following lemma, which is itself a direct consequence of (7.7.6).

Lemma 8.2.4. *If c in I is an idempotent of the t-norm T, i.e., if $T(c,c) = c$, then for all $F^\wedge$, $G^\wedge$ in ∇^+ we have*

$$\tau^\wedge_{T,L}(F^\wedge, G^\wedge)(c) \leqslant L\big(F^\wedge(c), G^\wedge(c)\big). \tag{8.2.6}$$

Theorem 8.2.5. *Let $(S, \mathfrak{F}, \tau_{T,L})$ be a PM space. If $L \leqslant$ Sum, and $T(c,c) = c$ for some c in I, then the function d_c defined by (8.2.5) is a pseudometric on S. For $c > 0$, d_c is a metric if and only if $F_{pq}(0+) < c$ whenever $p \neq q$.*

PROOF. For any p, q, r in S, (8.2.1) and (8.2.6) yield

$$F^\wedge_{pr}(c) \leqslant \tau^\wedge_{T,L}(F^\wedge_{pq}, F^\wedge_{qr})(c) \leqslant L\big(F^\wedge_{pq}(c), F^\wedge_{qr}(c)\big) \leqslant F^\wedge_{pq}(c) + F^\wedge_{qr}(c).$$

The rest is immediate. □

In some cases the means or other moments of the d.d.f.'s F_{pq} also give rise to metrics on S.

Theorem 8.2.6. *Let $(S, \mathfrak{F}, \tau_{M,L})$ be a PM space in which $L = K_\alpha$ for some $\alpha \geqslant 1$ (see (5.7.16)). Let $d_{(\beta)}$ be defined on $S \times S$ by*

$$d_{(\beta)}(p,q) = \begin{cases} m^{(\alpha\beta)}F_{pq}, & \beta \text{ in } (0,1], \\ \left(m^{(\alpha\beta)}F_{pq}\right)^{1/\beta}, & \beta \text{ in } (1,\infty), \end{cases} \tag{8.2.7}$$

where $m^{(\alpha\beta)}F_{pq}$ is the moment defined by (4.4.20). Then $(S, d_{(\beta)})$ is a metric space.

PROOF. For any p, q, r in S, and any t in I, (8.2.2) yields

$$\left(F^\wedge_{pr}(t)\right)^\alpha \leqslant \left(F^\wedge_{pq}(t)\right)^\alpha + \left(F^\wedge_{qr}(t)\right)^\alpha, \tag{8.2.8}$$

whence it follows from (4.4.20) that $d_{(1)}$ is a pseudometric. Next, since for $0 < \beta < 1$, $f(x) = x^\beta$ is a metric transform (see Section 3.2), it follows that $d_{(\beta)}$ is a pseudometric when $0 < \beta \leqslant 1$. When $\beta > 1$, using (8.2.8) and the Minkowski inequality (see [Kingman and Taylor 1966, Th. 7.8]), we obtain

$$\begin{aligned} m^{(\alpha\beta)}F_{pr} &\leqslant \int_0^1 \left[\left(F^\wedge_{pq}(t)\right)^\alpha + \left(F^\wedge_{qr}(t)\right)^\alpha\right]^\beta dt \\ &\leqslant \left[\left(m^{(\alpha\beta)}F_{pq}\right)^{1/\beta} + \left(m^{(\alpha\beta)}F_{qr}\right)^{1/\beta}\right]^\beta \end{aligned}$$

whence it follows that $d_{(\beta)}$ is again a pseudometric. Lastly, it follows from (4.4.20) that $m^{(\beta)}F = 0$ if and only if $F = \varepsilon_0$. Hence (8.1.1) and (8.1.2) imply that $d_{(\beta)}(p,q) = 0$ if and only if $p = q$, which means that $(S, d_{(\beta)})$ is a metric space for all β in $(0, \infty)$ and concludes the proof. □

Similar arguments involving (4.4.21) and (4.4.22) in place of (4.4.20) yield

Theorem 8.2.7. *Let $(S, \mathfrak{F}, \tau)$ be a* PM *space. Let $d_{(0)}$ and $d_{(\infty)}$ be defined on $S \times S$ by*

$$d_{(0)}(p,q) = m^{(0)}F_{pq} \tag{8.2.9}$$

and

$$d_{(\infty)}(p,q) = \lim_{\beta\to\infty} \left(m^{(\beta)}F_{pq}\right)^{1/\beta}. \tag{8.2.10}$$

If $\tau \geqslant \tau_W$, then $(S, d_{(0)})$ is a metric space. If $\tau = \tau_{T,L}$ where $L \leqslant$ Sum, *then $(S, d_{(\infty)})$ is a metric space.*

By using the results of Section 7.8, it is possible to get inequalities similar to (8.2.3) for Wald spaces and for Menger spaces under Archimedean t-norms. Thus, in (7.8.1) let Φ denote the Laplace-Stieltjes transform, which transforms functions in Δ^+ into nonnegative functions and is nondecreasing on Δ^+. Let $(S, \mathfrak{F})$ be a Wald space. Then (8.1.5) and (7.8.1) with τ equal to convolution yield

$$\Phi F_{pr} \geqslant \Phi(F_{pq} * F_{qr}) = (\Phi F_{pq})(\Phi F_{qr}).$$

Hence

$$-\log \Phi F_{pr} \leqslant (-\log \Phi F_{pq}) + (-\log \Phi F_{qr}). \tag{8.2.11}$$

The functions $-\log \Phi F_{pq}$ are defined, nonnegative, continuous, and nondecreasing on R^+. Thus we have the following result, which is due to J. F. C. Kingman [1964].

Theorem 8.2.8. *Let $(S, \mathfrak{F})$ be a Wald space. For any x in $(0, \infty)$, let d_{*x} be defined on $S \times S$ by*

$$d_{*x}(p,q) = (-\log \Phi F_{pq})(x). \tag{8.2.12}$$

Then (S, d_{*x}) is a pseudometric space.

For Menger spaces under Archimedean t-norms, we apply the T-conjugate transform c_T of Section 7.8 to obtain

Theorem 8.2.9. *Let $(S, \mathfrak{F})$ be a Menger space under an Archimedean t-norm T. Then*

$$-\log c_T F_{pr} \leqslant (-\log c_T F_{pq}) + (-\log c_T F_{qr}). \tag{8.2.13}$$

Furthermore, for all p, q in S and any x in $(0, \infty)$, $(-\log c_T F_{pq})(x) = 0$ if and only if $F_{pq} = \varepsilon_0$. Thus if $d_{T,x}$ is defined on $S \times S$ by

$$d_{T,x}(p,q) = (-\log c_T F_{pq})(x), \tag{8.2.14}$$

then $(S, d_{T,x})$ is a metric space.

8.3. Equilateral Spaces

We now turn to the consideration of various particular classes of PM spaces. Here it is useful to begin with

Definition 8.3.1. Let $\mathcal{C}$ be a class of PSM spaces and let τ be a triangle function. Then τ is *universal for* $\mathcal{C}$ if every PSM space in $\mathcal{C}$ is a PM space under τ.

Note that if τ is universal for $\mathcal{C}$ and $\tau_1 \leqslant \tau$, then τ_1 is universal for $\mathcal{C}$.

The simplest metric spaces are the equilateral spaces, i.e., those in which the distance between any two distinct points is a fixed positive number. Similarly, the simplest PM spaces are defined by

Definition 8.3.2. A PSM space $(S, \mathfrak{F})$ is *equilateral* if there is a d.d.f. F distinct from ε_0 and ε_∞ such that

$$\mathfrak{F}(p,q) = F \quad \text{for all } p, q \text{ in } S \text{ with } p \neq q. \tag{8.3.1}$$

Equilateral spaces, trivial as they may seem, nevertheless arise naturally as the spaces generated by the action of mixing transformations on metric spaces (see Section 11.3).

Theorem 8.3.3. *Every equilateral* PSM *space $(S, \mathfrak{F})$ is a* PM *space under* **M**.

PROOF. Let F be as in (8.3.1). Then for any p, q in S, either $F_{pq} = F$ or $F_{pq} = \varepsilon_0$, whence the conclusion follows from the fact that $\varepsilon_0 \geqslant F = \mathbf{M}(F, \varepsilon_0) \geqslant \mathbf{M}(F, F)$. □

Since **M** is the maximal triangle function (see Theorem 7.1.4), it follows that every triangle function is universal for the class of equilateral spaces.

Furthermore, note that if $(S, \mathcal{F})$ is equilateral, then each of the metric spaces derived from $(S, \mathcal{F})$ by any of the methods presented in the preceding section is also equilateral.

8.4. Simple Spaces

Definition 8.4.1. Let (S,d) be a metric space and G a d.d.f. distinct from ε_0 and ε_∞. Then (S,d,G), *the simple space generated by* (S,d) *and* G, is the PSM space $(S,\mathcal{F})$ in which $\mathcal{F}$ is defined on $S \times S$ by

$$\mathcal{F}(p,q)(x) = G(x/d(p,q)) \quad \text{for all } x \text{ in } R^+. \tag{8.4.1}$$

Here we make the convention that $G(x/0) = G(\infty) = 1$ for $x > 0$, and $G(0/0) = G(0) = 0$. (Note that the function $\mathcal{F}$ defined by (8.4.1) automatically satisfies (8.1.1), (8.1.2), and (8.1.3).)

If $G = \varepsilon_a$ for some a in $(0,\infty)$, then $F_{pq} = \varepsilon_{ad(p,q)}$, whence (S,d,ε_a) determines a metric space that is homothetic to (S,d). Thus metric spaces are special cases of simple spaces. If G does not coincide with any of the functions ε_a, then (S,d,G) can be regarded as a metric space whose distances are "smeared out" or randomized by the distribution function G.

If (S,d) is an equilateral metric space, then the PSM space (S,d,G) is equilateral; and any equilateral PSM space can be obtained in this way. Thus every equilateral space is a simple space.

Theorem 8.4.2. *Every simple space is a Menger space under M, i.e., τ_M is universal for the class of simple spaces.*

PROOF. It follows from (8.4.1) that

$$F_{pq}^{\wedge} = d(p,q)G^{\wedge} \quad \text{for all } p, q \text{ in } S. \tag{8.4.2}$$

Since d is a metric on S, we therefore have

$$F_{pr}^{\wedge} = d(p,r)G^{\wedge} \leqslant (d(p,q) + d(q,r))G^{\wedge} = F_{pq}^{\wedge} + F_{qr}^{\wedge} \tag{8.4.3}$$

for all p, q, r in S. Hence, using (7.7.10) with $L = \text{Sum}$, we obtain

$$F_{pr} \geqslant (F_{pq}^{\wedge} + F_{qr}^{\wedge})^{\wedge} = \tau_M(F_{pq}, F_{qr}) \tag{8.4.4}$$

for all p, q, r in S, and (8.1.4) holds with $\tau = \tau_M$. □

Corollary 8.4.3. *Let (S,d,G) be a simple space. Then the means, the appropriate roots of moments, and the c-tiles, for $c > G(0+)$, of the d.d.f's each form a metric space homothetic to (S,d).*

Theorem 8.4.4. *If $\tau_{T,L}$ is universal for the class of all simple spaces, then $\tau_{T,L} \leqslant \tau_M$. In other words, in the set of triangle functions of the form $\tau_{T,L}$, τ_M is the strongest one that is universal for the class of simple spaces.*

PROOF. Take (S,d) to be the Euclidean line, and G to be ε_1. Then in the simple space (S,d,ε_1) we have $F_{pq} = \varepsilon_{|p-q|}$ for all p, q in $(-\infty,\infty)$. Given a, b in $(0,\infty)$, let $p=0$, $q=a$, and $r=a+b$. Then (7.2.24) yields

$$F_{pr} = \varepsilon_{a+b} \geqslant \tau_{T,L}(F_{pq}, F_{qr}) = \tau_{T,L}(\varepsilon_a, \varepsilon_b) = \varepsilon_{L(a,b)},$$

whence $a+b \leqslant L(a,b)$. Since a and b are arbitrary, this yields Sum $\leqslant L$; and since $T \leqslant M$ for any t-norm T, an appeal to Theorem 7.2.12 completes the proof. □

Thus, within the class of triangle functions $\tau_{T,L}$, the result of Theorem 8.4.2 is best-possible.

While every simple space is a Menger space under M, there are Menger spaces under M that are not simple. For example, let $S = I$ and define $\mathfrak{F}$ via

$$F_{pq}^{\wedge}(c) = \begin{cases} 0, & c \leqslant \operatorname{Min}(p,q), \\ 1, & c > \operatorname{Min}(p,q). \end{cases} \tag{8.4.5}$$

It is readily checked that $F_{pr}^{\wedge} \leqslant F_{pq}^{\wedge} + F_{qr}^{\wedge}$ for any p, q, r in S, whence by Corollary 8.2.2, $(S,\mathfrak{F})$ is a Menger space under M. But since the d.d.f.'s given by (8.4.5) are not all positive multiples of a single quasi-inverse of some G in Δ^+, (8.4.2) fails and $(S,\mathfrak{F})$ is not a simple space.

There are also simple spaces that are not Wald spaces. For example, take (S,d) to be the Euclidean line and $G(x) = 1 - e^{-x}$. Then, with $p=0$, $q=1$, $r=2$, we have $F_{pq} = F_{qr} = G$ and $F_{pr}(x) = G(x/2)$. Hence $(F_{pq} * F_{qr})(x) = (G * G)(x) = 1-(1+x)e^{-x}$, while $F_{pr}(x) = 1 - e^{-x/2}$. A little computation now shows that if $x \geqslant 2.513$, then $e^{x/2} > 1 + x$, i.e., $F_{pr}(x) < (F_{pq} * F_{qr})(x)$, which contradicts the Wald inequality.

8.5. Ellipse m-Metrics and Hysteresis

In [Erber, Schweizer, and Sklar 1971], we developed a PM space model to describe the macroscopic behavior of certain physical systems under hysteresis. The relevant PM spaces are constructed as follows.

Let (S,d) be a metric space. For any p, q in S and any $x \geqslant 0$, the *elliptical region* $E(p,q;x)$ is the set $\{r \mid d(p,r) + d(q,r) < x\}$. Note that $E(p,q;x)$ is empty for $x \leqslant d(p,q)$. We say that (S,d) is *admissible* if

i. (S,d) admits segments joining any two of its points (see Section 3.3);

ii. (S,d) admits a measure μ such that every elliptical region is μ-measurable and

$$0 < \mu(E(p,q;x)) < \infty \quad \text{for } d(p,q) < x < \infty. \tag{8.5.1}$$

For any admissible metric space (S,d) and any m in $(1,\infty)$ the *ellipse m-metric* on (S,d) is the function $\mathscr{F}$ from $S \times S$ into Δ^+ defined via

$$F_{pq}(x) = \mathrm{Min}\left(\frac{\mu(E(p,q;x))}{\mu(E(p,q;md(p,q)))}, 1 \right) \tag{8.5.2}$$

when $p \neq q$ and by $\mathscr{F}(p,q) = \varepsilon_0$ when $p = q$. Since $\mathscr{F}(p,q) \leqslant \varepsilon_{d(p,q)}$ for all p, q in S, $(S,\mathscr{F})$ is a PSM space.

When (S,d_2) is the Euclidean plane (i.e., $\alpha = 2$ in (3.4.3)), then $E(p,q;x)$, when nonempty, is a region bounded by an ordinary ellipse with major axis x, minor axis $(x^2 - (d_2(p,q))^2)^{1/2}$, and eccentricity $d_2(p,q)/x$. In this case, for any $m > 1$, $(S,\mathscr{F})$ is the simple space generated by (S,d_2,G_2), where

$$G_2(x) = \begin{cases} 0, & x \text{ in } [0,1], \\ \dfrac{x}{m}\left(\dfrac{x^2-1}{m^2-1}\right)^{1/2}, & x \text{ in } [1,m], \\ 1, & x \text{ in } [m,\infty]. \end{cases} \tag{8.5.3}$$

A similar construction can be carried out in certain Minkowski planes (see Section 3.4) provided the line segment joining p and q is parallel to one of the coordinate axes (the case of physical interest). Thus if we take $\alpha = 1$ in (3.4.3), then $E(p,q;x)$, for p and q in "standard position" and $x > d_1(p,q)$, is the hexagonal region with vertices at the points

$$\left(\pm\frac{x}{2}, 0\right), \quad \left(\frac{-d_1(p,q)}{2}, \pm\frac{x-d_1(p,q)}{2}\right), \quad \left(\frac{d_1(p,q)}{2}, \pm\frac{x-d_1(p,q)}{2}\right)$$

(see Figure 8.5.1). In this case $(S,\mathscr{F})$ is the simple space (S,d_1,G_1), where

$$G_1(x) = \begin{cases} 0, & x \text{ in } [0,1], \\ \dfrac{x^2-1}{m^2-1}, & x \text{ in } [1,m], \\ 1, & x \text{ in } [m,\infty]. \end{cases} \tag{8.5.4}$$

If we take the metric d_∞ of (3.4.4), then $E(p,q;x)$, for p and q in standard position and $x > d_\infty(p,q)$, is the octagonal region with vertices at the points

$$\left(\pm\frac{x}{2}, \pm\frac{x-d_\infty(p,q)}{2}\right), \quad \left(\pm\frac{x-d_\infty(p,q)}{2}, \pm\frac{x}{2}\right)$$

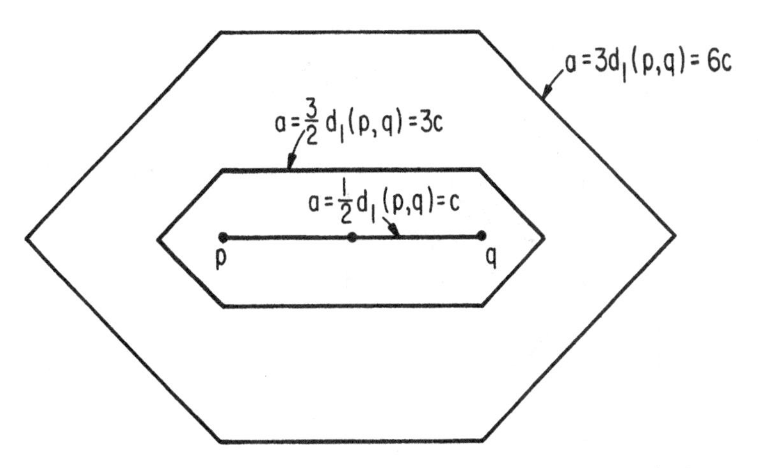

Figure 8.5.1. Elliptical regions in the Minkowski plane with metric d_1.

(see Figure 8.5.2). As x decreases to $d_\infty(p,q)$, the octagon reduces to a square. Here $(S, \mathfrak{F})$ is the simple space generated by (S, d_∞, G_∞), where

$$G_\infty(x) = \begin{cases} 0, & x \text{ in } [0,1], \\ \dfrac{2x^2-1}{2m^2-1}, & x \text{ in } (1,m], \\ 1, & x \text{ in } [m,\infty]. \end{cases} \tag{8.5.5}$$

Note that G_∞ has a jump of height $1/(2m^2-1)$ at 1.

These ellipse m-metrics arise naturally in the probabilistic description of the evolution of large-scale physical systems subjected to external forces. In general, the state of such a system can be associated with a point in a phase space of very high dimension; for instance, in statistical mechanics the dimensionality is typically scaled by Avogadro's number. When such a system is acted on by an external force, its response can be visualized as a motion of the associated point in the phase space. A system is said to exhibit hysteresis if its response to the external forces is not single valued; in particular this will occur if the phase space trajectory of the system does not return to its initial state after a complete cycle of the external force. For certain model systems (e.g., arrays of small magnets exposed to varying external magnetic fields) it can be shown rigorously that this irreversible behavior arises from singularities—bifurcations and jump discontinuities—in the phase space trajectories. The macroscopic behavior of such a complex system can often be adequately described by two parameters: a *hysteresis coordinate* v, which indicates the variation of the external force;

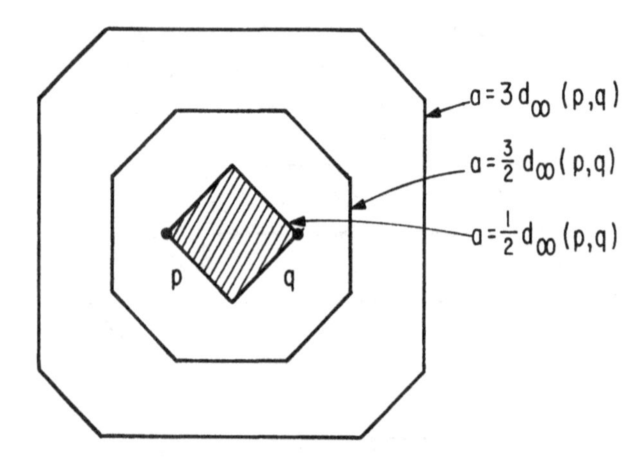

Figure 8.5.2. Elliptical regions in the Minkowski plane with metric d_∞.

and a *configuration coordinate* u, which characterizes the state of the system. For instance, in magnetic systems u corresponds to the induced magnetization, and in the deformation of solids u represents the strain. When this is the case, we say that we have a *simple hysteresis system*. It follows that a simple hysteresis system may, at any given instant, be visualized as a point in the (u, v)-plane. The existence of hysteresis is manifested by the fact that if v is returned to its initial value, then the corresponding final value of u differs from the original one. A *hysteresis cycle* is therefore a path in the (u, v)-plane from a point p to a point q such that $v(p) = v(q)$ but $u(p) \neq u(q)$.

Given this description, we then make certain assumptions—principally on physical grounds, but partly for the sake of mathematical simplicity—which enable us to construct a mathematical model. The physical assumptions are based on results given in [Erber, Harmon, and Latal 1971] and [Erber, Guralnick, and Latal 1972], and have received independent experimental confirmation in the work of J. L. Porteseil [1978]. The model predicts that the average energy loss per cycle, for all hysteresis cycles between p and q whose length is less than x, is proportional to

$$\int_{[0,x)} t\, dF_{pq}(t) = \int_0^{F_{pq}(x)} F_{pq}^{\wedge}(s)\, ds \tag{8.5.6}$$

where F_{pq} is given by (8.5.2). It is this average energy loss as a function H of the maximum excursion of the hysteresis coordinate v that is of practical interest. The latter excursion is the minor axis y of the ellipse bounding $E(p, q; x)$ (see Figure 8.5.3). In the cases discussed above, it is easy to

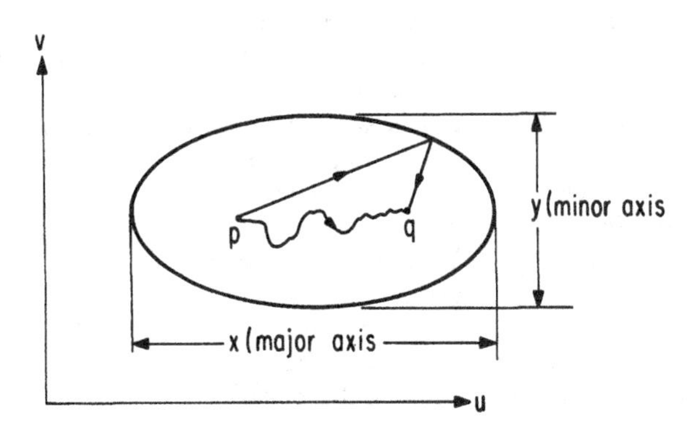

Figure 8.5.3. Elliptical region in the hysteresis model.

obtain explicit expressions for $H(y)$. Thus in the Minkowski case $\alpha = 1$, we have $x = y + d_1(p,q)$. If we set $c = d_1(p,q)$, then a short computation using (8.5.4) yields

$$H_1(y) = \begin{cases} \dfrac{2}{3}\,\dfrac{y^3 + 3cy^2 + 3c^2y}{(m^2-1)c^2}, & y \leqslant (m-1)c, \\[2ex] \dfrac{2}{3}mc\left(1 + \dfrac{1}{m^2+m}\right), & y \geqslant (m-1)c. \end{cases} \tag{8.5.7}$$

In the Euclidean case, we have $x = (y^2 + (d_2(p,q))^2)^{1/2}$. We set $c = d_2(p,q)$ and using (8.5.3), we obtain

$$H_2(y) = \begin{cases} \dfrac{2}{3}\,\dfrac{y^3 + (3/2)c^2y}{m(m^2-1)^{1/2}c^2}, & y \leqslant (m^2-1)^{1/2}c, \\[2ex] \dfrac{2}{3}mc\left(1 + \dfrac{1}{m^2}\right), & y \geqslant (m^2-1)^{1/2}c. \end{cases} \tag{8.5.8}$$

In the other Minkowski case, we have $x = y$. We set $c = d_\infty(p,q)$, and using (8.5.5) we obtain

$$H_\infty(y) = \begin{cases} 0, & y \leqslant c, \\[1ex] \dfrac{2}{3}\,\dfrac{y^3 - c^3/4}{(m^2-1/2)c^2}, & c < y \leqslant mc, \\[2ex] \dfrac{2}{3}mc\left(1 + \dfrac{2m-1}{2m(2m^2-1)}\right), & mc \leqslant y. \end{cases} \tag{8.5.9}$$

Note the jump of height $c/(2m^2-1)$ of H_∞ at c.

What is most striking about these expressions is their overall similarity in spite of the differences in the shapes of the regions $E(p,q;x)$. In particular, for $m \gg 1$ and y comparable in magnitude to mc, all three expressions yield virtually identical cubic dependences. Such cubic dependence is familiar in ferromagnetic hysteresis as Rayleigh's law and in stress–strain hysteresis as Dorey's rule. When y is small, the dominant term in (8.5.7) and (8.5.8) is linear. A transition from cubic to linear dependence has also been observed experimentally in magnetic cooperative systems and structural "shake-down," and even the jump discontinuity exhibited in (8.5.9) corresponds to something observed in nature: the so-called Portevin-LeChatelier jumps in the constituent materials of structures. Fuller discussions of these points may be found in the papers cited in this section.

8.6. α-Simple Spaces

Definition 8.6.1. Let (S,d) be a metric space, α a number in $[0,\infty)$, and G a d.d.f. distinct from ε_0 and ε_∞. Then $(S,d,G;\alpha)$, the *α-simple space generated by (S,d) and G*, is the PSM space $(S,\mathfrak{F})$ in which $\mathfrak{F}$ is defined on $S \times S$ by

$$\mathfrak{F}(p,q)(x) = G\big(x/(d(p,q))^{\alpha}\big) \tag{8.6.1}$$

for all x in R^+. Here we make the same conventions as in Definition 8.4.1 and, in addition, set $(d(p,p))^0 = 0$.

A 0-simple space is equilateral; a 1-simple space is simple; and for α in $(0,1)$ the function d^α is itself a metric on S, whence the α-simple space generated by (S,d) and G is identical to the simple space generated by (S,d^α) and G. The situation changes when $\alpha > 1$, for then d^α may not be a metric on S and consequently there may be triangle functions τ for which (8.1.4) fails (see the example following (1.5.4)). In fact, we have

Theorem 8.6.2. *Let τ be a triangle function such that*

$$\sup\{\tau(\varepsilon_a,\varepsilon_a) \mid a > 0\} = \varepsilon_0. \tag{8.6.2}$$

Then τ is not universal for the class of all α-simple spaces.

PROOF. Let $G = \varepsilon_1$ and let (S,d) be the Euclidean line. Take $p = 0$, $q = \frac{1}{2}$, and $r = 1$. Then (8.6.1) yields $F_{pq} = F_{qr} = \varepsilon_{2^{-\alpha}}$, and $F_{pr} = \varepsilon_1$, whence (8.1.5) takes the form

$$\varepsilon_1 \geqslant \tau(\varepsilon_{2^{-\alpha}},\varepsilon_{2^{-\alpha}}). \tag{8.6.3}$$

But since $\tau(\varepsilon_{2^{-\alpha}},\varepsilon_{2^{-\alpha}})$ increases with α, it follows from (8.6.2) that for sufficiently large α,

$$d_L(\tau(\varepsilon_{2^{-\alpha}}, \varepsilon_{2^{-\alpha}}), \varepsilon_0) < 1 = d_L(\varepsilon_1, \varepsilon_0),$$

which by Lemma 4.3.4 contradicts (8.6.3). □

Corollary 8.6.3. *No continuous triangle function is universal for the class of all α-simple spaces.*

On the positive side, we have the following results from [S² 1963b].

Lemma 8.6.4. *Let α be in $(1, \infty)$, let G be a strict d.d.f., and let g be the function defined on I by*

$$g(x) = \left(G^{-1}(x)\right)^{1/(1-\alpha)}. \tag{8.6.4}$$

Then g is the additive generator of the strict t-norm T_G given by

$$T_G(u, v) = G\left\{\left(\left[G^{-1}(u)\right]^{1/1-\alpha} + \left[G^{-1}(v)\right]^{1/1-\alpha}\right)^{1-\alpha}\right\}. \tag{8.6.5}$$

In the other direction, if g is the additive generator of a strict t-norm and G is the function defined on R^+ by

$$G(x) = g^{-1}(x^{1/(1-\alpha)}), \tag{8.6.6}$$

then G is a strict d.d.f. and g is given in terms of G by (8.6.4).

Theorem 8.6.5. *Let (S, d) be a metric space and let α be in $(1, \infty)$. Let G and g be as in Lemma* 8.6.4, *and let T_G be the strict t-norm given by* (8.6.5). *Then the α-simple space $(S, d, G; \alpha)$ is a Menger space under T_G.*

Next, it follows after some calculation that for any p, q, r in S and all x in R^+

$$\tau_{T_G}(F_{pq}, F_{qr})(x) = G\left(\frac{x}{(d(p,q) + d(p,r))^{\alpha}}\right) \tag{8.6.7}$$

where F_{pq} and F_{qr} are given by (8.6.1); and (8.6.7) leads at once to

Theorem 8.6.6. *The result in Theorem* 8.6.5 *is best-possible in the following sense: Let α, G, g, and T_G be as above. Then there exist metric spaces (S, d) such that T_G is the strongest t-norm under which the α-simple space $(S, d, G; \alpha)$ is a Menger space.*

Finally, we note that in any α-simple space $(S, d, G; \alpha)$, for $\beta \geqslant 1$ the $\alpha\beta$th roots of the moments of order β of the d.d.f.'s form a metric space homothetic to (S,d); and similarly for the αth roots of the c-tiles whenever $\alpha \geqslant 1$ and $G^\wedge(c) > 0$.

8.7. Best-Possible Triangle Inequalities

Since a stronger triangle function provides more information about the probabilistic distances of a PM space than a weaker one, it is of interest to consider the problem of finding best-possible triangle functions for a given PM space (see Theorems 8.4.4 and 8.6.6). This motivates

Definition 8.7.1. Let $\mathcal{Q}$ be a set of triangle functions and $(S,\mathcal{F})$ a PSM space. Suppose that τ is a triangle function in $\mathcal{Q}$ and that $(S,\mathcal{F},\tau)$ is a PM space. Then τ is a *maximal* (resp., *the best-possible*) *triangle function for* $(S,\mathcal{F})$ *within* $\mathcal{Q}$ if for every τ' in $\mathcal{Q}$ such that $(S,\mathcal{F},\tau')$ is a PM space τ' is not stronger than τ (resp., $\tau' \leqslant \tau$).

The following result is due to E. O. Thorp [1960].

Theorem 8.7.2. *Let* $(S,\mathcal{F},\tau_{T_0})$ *be a Menger space. Then there exists a maximal triangle function for* $(S,\mathcal{F})$ *within the set of all triangle functions of the form* τ_T.

Thorp's result was extended by Šerstnev [1964a, 1965], who proved

Theorem 8.7.3. *Let* $(S,\mathcal{F},\tau_0)$ *be a* PM *space with* τ_0 *continuous. Then there exists a maximal triangle function for* $(S,\mathcal{F})$ *within the set of all continuous triangle functions.*

The proofs of Theorems 8.7.2 and 8.7.3 involve appeals to Zorn's lemma, hence are nonconstructive. Thus these theorems cannot be applied to the problem of actually finding best-possible triangle functions.

Thorp [1960] also considered the converse problem of finding a space for which a given triangle function is best-possible. He showed, by means of an explicit construction, that if T_0 is a given t-norm, then there is a Menger space for which τ_{T_0} is best-possible within the set of all triangle functions of the form τ_T. Since this construction does not make use of the associativity of T, and since the maximum of two t-norms need not be a t-norm, Thorp was also able to exhibit a Menger space for which there is no best-possible triangle function of the form τ_T. Subsequently, Šerstnev [1964a, 1965] extended Thorp's construction and showed that if τ is any triangle function, then there is a PM space for which τ is best-possible within the set of all triangle functions. This completely settles the converse problem.

8.8. Open Problems

First and foremost, there is the long-term project of the further development of the simple space model for hysteresis. This includes extending its realm of applicability and gathering additional experimental data to determine its limitations. Since a system undergoing hysteresis may be viewed as one experiencing a large number of catastrophes per cycle, connections between our model and the average behavior of an ensemble of catastrophes merit exploration.

In addition, we have

Problem 8.8.1. Prove or disprove the conjecture that τ_M is the strongest triangle function that is universal for the class of simple spaces.

Problem 8.8.2. Let (S,d) be a metric space. Determine all pairs (G,α), where G is a d.d.f. and α a number in $(1,\infty)$, such that the α-simple space $(S,d,G;\alpha)$ is a PM space. Similarly, given a d.d.f. G distinct from ε_0 and ε_∞, characterize those metric spaces (S,d) and α's in $(0,\infty)$ such that $(S,d,G;\alpha)$ is a PM space.

9

Random Metric Spaces

9.1. *E*-Spaces

In this and the following two chapters we turn our attention to PM spaces derived from, or related to, families of random variables. We begin with a simple illustrative example. Let (I, λ) be the unit interval I endowed with Lebesgue measure λ, and let $L_1(I)$ be the set of all a.e. finite Lebesgue-measurable functions on I. As usual, functions that are equal a.e. are identified. Many different metrics can be defined on $L_1(I)$, each of which, of course, associates exactly one number with every pair of functions in $L_1(I)$. Given such a pair of functions p, q in $L_1(I)$, we can, however, make a much more refined measurement of the "distance" between them. Namely, for any $x \geqslant 0$ we can determine the size (λ-measure) of the set of all points t in I at which the difference $|p(t) - q(t)|$ is less than x. Accordingly, we can define the function F_{pq} whose value for any $x \geqslant 0$ is given by

$$F_{pq}(x) = \lambda\{t \text{ in } I \mid |p(t) - q(t)| < x\}. \tag{9.1.1}$$

It is readily checked that F_{pq} is in Δ^+ (indeed in $\mathfrak{D}^+$) and that the mapping $\mathfrak{F}$ from $L_1(I) \times L_1(I)$ into Δ^+ defined by $\mathfrak{F}(p, q) = F_{pq}$ satisfies conditions (8.1.1)–(8.1.4) with $\tau = \tau_W$ (for the proof see Theorem 9.1.2). Thus, in this natural fashion, the pair $(L_1(I), \mathfrak{F})$ is a PM space, which we designate by $\mathcal{E}(I)$.

The foregoing construction carries over to a more abstract setting; and this leads us to the family of E-spaces, which were independently defined by H. Sherwood in his doctoral dissertation of 1966 (see [Sherwood 1969]), and by D. H. Muštari and A. N. Šerstnev [1966]. (See also [Drossos 1977].)

Definition 9.1.1. Let $(\Omega, \mathcal{A}, P)$ be a probability space, (M,d) be a metric space, and S a set of functions from Ω into M. Let $\mathcal{F}$ be a mapping from $S \times S$ into Δ^+. Then $(S, \mathcal{F})$ is an *E-space*, with *base* $(\Omega, \mathcal{A}, P)$ and *target* (M,d) if

i. For all p, q in S and all x in R^+ the set
$$\{\omega \text{ in } \Omega \mid d(p(\omega), q(\omega)) < x\}$$
belongs to $\mathcal{A}$; i.e., the composite function $d(p,q)$ from Ω into R^+ is P-measurable and therefore in $L_1^+(\Omega)$.

ii. For all p, q in S, $\mathcal{F}(p,q) = F_{pq} = df(d(p,q))$, i.e.,
$$F_{pq}(x) = P\{\omega \text{ in } \Omega \mid d(p(\omega), q(\omega)) < x\}. \tag{9.1.2}$$

It follows immediately from (9.1.2) that $\mathcal{F}$ satisfies (8.1.1) and (8.1.3). If $\mathcal{F}$ also satisfies (8.1.2), then $(S, \mathcal{F})$ is a *canonical* E-space.

Theorem 9.1.2. *Let* $(S, \mathcal{F})$ *be an E-space. Then* $(S, \mathcal{F})$ *is a* PPM *space under* τ_W. *If* $(S, \mathcal{F})$ *is canonical, then it is a Menger space under* W.

PROOF. We need only establish (8.1.4). For any p, q, r in S and any x in R^+, let u, v in R^+ be such that $u + v = x$. Define the sets A, B, C by

$$A = \{\omega \text{ in } \Omega \mid d(p(\omega), q(\omega)) < u\},$$
$$B = \{\omega \text{ in } \Omega \mid d(q(\omega), r(\omega)) < v\},$$
$$C = \{\omega \text{ in } \Omega \mid d(p(\omega), r(\omega)) < x\}.$$

Since d satisfies the triangle inequality, it follows that $A \cap B \subseteq C$. Hence (2.3.7) yields
$$P(C) \geqslant P(A \cap B) \geqslant W(P(A), P(B)).$$
But by (9.1.2), $P(A) = F_{pq}(u)$, $P(B) = F_{qr}(v)$ and $P(C) = F_{pr}(x)$. Thus
$$F_{pr}(x) \geqslant W(F_{pq}(u), F_{qr}(v)), \tag{9.1.3}$$
from which (8.1.4) with τ_W follows via (8.1.10) and (8.1.11). □

The proof of Theorem 9.1.2 is due to R. R. Stevens [1968]. Within the class of Menger spaces, the result is best-possible in the sense that there exists an E-space that is not a Menger space under any t-norm stronger than W (see Theorem 9.2.5).

Because the functions $d(p,q)$ are random variables, the conclusion of Theorem 8.2.6 is valid in any E-space even though the hypotheses generally are not. This is a consequence of

Lemma 9.1.3. *Let* $(S,\mathfrak{F})$ *be an E-space with base* $(\Omega,\mathcal{Q},P)$ *and target* (M,d). *Then for any* p, q *in* S *and any* β *in* $(0,\infty)$ *we have*

$$E\big(d(p,q)^{\beta}\big)=\int_{\Omega}\big[d(p,q)\big]^{\beta}dP=\int_{I}\big(F_{pq}^{\wedge}(x)\big)^{\beta}dx. \tag{9.1.4}$$

PROOF. By (4.4.16) and (9.1.2), we have

$$P\{\omega \text{ in } \Omega \mid d(p(\omega),q(\omega))<x\}=\lambda\{y \mid F_{pq}^{\wedge}(y)<x\}.$$

Hence $d(p,q)$ and $F_{pq}^{\wedge}$ are equimeasurable (see [Hardy, Littlewood, and Pólya 1952, §10.12]), from which (9.1.4) follows immediately.

Theorem 9.1.4. *Let* $(S,\mathfrak{F})$ *be an E-space. For* β *in* $(0,\infty)$, *let* $d_{(\beta)}$ *be defined on* $S\times S$ *by* (8.2.7) (*with* $\alpha=1$). *Then* $(S,d_{(\beta)})$ *is a pseudometric space; and* $(S,d_{(\beta)})$ *is a metric space if and only if* $(S,\mathfrak{F})$ *is canonical.*

PROOF. Let $(\Omega,\mathcal{Q},P)$ and (M,d) be as in Lemma 9.1.3. For any p, q, r in S, we have

$$d(p,r)(\omega)\leqslant d(p,q)(\omega)+d(q,r)(\omega) \quad \text{for all } \omega \text{ in } \Omega,$$

whence, using (9.1.4) and arguing as in Theorem 8.2.6, the conclusion follows. □

Corollary 9.1.5. *Let* $(S,\mathfrak{F})$ *be a canonical E-space and let* $d_{(0)}$ *and* $d_{(\infty)}$ *be as defined in Theorem* 8.2.7. *Then both* $(S,d_{(0)})$ *and* $(S,d_{(\infty)})$ *are metric spaces.*

Theorem 9.1.4 and Corollary 9.1.5 are due to Fréchet ([1948, pp. 229–232] and [1956], respectively) who, however, explicitly assumed [1948, p. 227] that the individual functions $p,q,\ldots,$ rather than the composite functions $d(p,q)$, are P-measurable. This assumption is unnecessary. In fact, there are E-spaces in which $p,q,\ldots$ are not P-measurable. For example, let $(\Omega,\mathcal{Q},P)$ be (I,λ) and (M,d) be the Euclidean line. Let S be the set of all functions p_c on I given by

$$p_c(x)=p_0(x)+c$$

where c is in $(-\infty,\infty)$ and p_0 is a nonmeasurable function on I. Let $\mathfrak{F}$ be defined by (9.1.2). Then $(S,\mathfrak{F})$ is an E-space, indeed a canonical E-space, but none of the functions in S are λ-measurable.

On the other hand, when the functions $p, q, \ldots$ are themselves P-measurable, then the P-measurability of $d(p,q)$ is a conclusion rather than a hypothesis (see [Neveu 1965, pp. 36–37]). In this case we say that the E-space in question is *random variable generated*. Note further that in these circumstances any indexed collection of points $\{p_a \mid a \text{ in } A\}$ of a random variable generated E-space is a stochastic process with parameter set A.

One of the consequences of Theorem 9.1.4 is that an entire family of function spaces can be obtained from a single PM space (see [S² 1980]). For example, the space $\mathfrak{S}(I)$ discussed at the beginning of this section is a canonical E-space and therefore a Menger space under W. In this case, for any β in $[1, \infty)$ the space $(S, d_{(\beta)})$ of Theorem 9.1.4 is precisely the classical function space $L_\beta(I)$. Hence, not only is each of the spaces $L_\beta(I)$ obtainable from the single PM space $\mathfrak{S}(I)$, but the space $\mathfrak{S}(I)$ contains at least as much information about the set of a.e. finite Lebesgue-measurable functions on I as all the spaces $L_\beta(I)$ taken together.

Finally, it should be observed that the basic idea behind the definition of an E-space extends to other settings, in the sense that the metric space (M, d) in Definition 9.1.1 may be replaced, e.g., by a normed linear space. In the same manner, any suitable collection of functions from a probability space into a space with a specified deterministic structure induces a space in which this structure is probabilistic. Further details are given in Chapter 15.

9.2. Pseudometrically Generated Spaces, Sherwood's Theorem

If $(S, \mathfrak{F})$ is an E-space with base $(\Omega, \mathfrak{A}, P)$ and target (M, d), then for every ω in Ω the function d_ω from $S \times S$ into R^+ defined by

$$d_\omega(p,q) = d(p(\omega), q(\omega)) \tag{9.2.1}$$

is a pseudometric on S. Since we may have $p(\omega) = q(\omega)$ for a particular ω even when p and q are distinct functions, d_ω need not be a metric; nevertheless, if p and q are distinct, there is at least one ω in Ω such that $p(\omega) \neq q(\omega)$, which means that there is at least one pseudometric d_ω such that $d_\omega(p,q) > 0$. Thus, noting that

$$P\{\omega \text{ in } \Omega \mid d(p(\omega), q(\omega)) < x\} = P\{\omega \text{ in } \Omega \mid d_\omega(p,q) < x\},$$

we have the fact that *any E-space $(S, \mathfrak{F})$ is pseudometrically generated*, i.e., satisfies the conditions of the following:

Definition 9.2.1. Let S be a set and $\mathfrak{F}$ a function from $S \times S$ into Δ^+. Then $(S, \mathfrak{F})$ is a *pseudometrically generated space* if there is a probability space

$(\Omega, \mathcal{A}, P)$, called the *base*, such that

i. for each ω in Ω there is a pseudometric d_ω on S;
ii. for any distinct p and q in S there is an ω in Ω such that $d_\omega(p,q) > 0$;
iii. for all p, q in S and all x in R^+, the set $\{\omega \text{ in } \Omega \mid d_\omega(p,q) < x\}$ belongs to $\mathcal{A}$, i.e., is P-measurable;
iv. for all p, q in S, $\mathcal{F}(p,q) = F_{pq}$, where

$$F_{pq}(x) = P\{\omega \textit{ in } \Omega \mid d_\omega(p,q) < x\}. \tag{9.2.2}$$

If, in addition, each d_ω is a metric on S, then $(S, \mathcal{F})$ is *metrically generated.*

For example, $\mathcal{E}(I)$ is a pseudometrically generated space whose base is (I, λ) and in which $F_{pq}(x)$ is given by (9.1.1). Since there exist distinct functions p, q in $L_1(I)$ for which $d_t(p,q) = |p(t) - q(t)| = 0$ for some t in I, $\mathcal{E}(I)$ is not metrically generated.

In a pseudometrically generated space $(S, \mathcal{F})$, (8.1.1) and (8.1.3) are automatically satisfied. If (8.1.2) is also satisfied, so that $(S, \mathcal{F})$ is a PSM space, then $(S, \mathcal{F})$ is a *separated* pseudometrically generated space. Clearly, any canonical E-space is a separated pseudometrically generated space.

At first glance, it might seem that the class of pseudometrically generated spaces is larger than the class of E-spaces, i.e., that whereas every E-space is pseudometrically generated, there are pseudometrically generated spaces that are not E-spaces. This is not the case: H. Sherwood [1969] has established the remarkable fact that, up to isometry, the class of E-spaces and the class of pseudometrically generated spaces coincide. Since we have already observed that every E-space is pseudometrically generated, Sherwood's result will follow from

Theorem 9.2.2. *Every pseudometrically generated space is isometric to an E-space.*

PROOF. Suppose that $(S, \mathcal{F})$ is a pseudometrically generated space with base $(\Omega, \mathcal{A}, P)$. The space $(\Omega, \mathcal{A}, P)$ will also be the base for the E-space to be constructed. Let $M = S \times \Omega$, fix a point p_0 in S, and define a function ρ from $M \times M$ into R^+ by

$$\rho((p,\omega_1),(q,\omega_2)) = \begin{cases} d_{\omega_1}(p, p_0) + d_{\omega_2}(p_0, q) + 1, & \omega_1 \neq \omega_2, \\ d_{\omega_2}(p,q), & \omega_1 = \omega_2. \end{cases} \tag{9.2.3}$$

It is easily verified that ρ is a pseudometric on M. Let (M^*, ρ^*) be the quotient space of (M, ρ) (see (3.1.7)).

Now for any p in S let π_p denote the function from Ω into M^* given by $\pi_p(\omega) = (p,\omega)^*$ for all ω in Ω and let $S' = \{\pi_p \mid p \text{ in } S\}$. Notice that for any π_p, π_q in S' and any x in R^+, we have

$$\begin{aligned}\{\omega \text{ in } \Omega \mid \rho^*(\pi_p(\omega),\pi_q(\omega)) < x\} &= \{\omega \text{ in } \Omega \mid \rho^*((p,\omega)^*,(q,\omega)^*) < x\} \\ &= \{\omega \text{ in } \Omega \mid \rho((p,\omega),(q,\omega)) < x\} \\ &= \{\omega \text{ in } \Omega \mid d_\omega(p,q) < x\}. \qquad (9.2.4)\end{aligned}$$

The last set in (9.2.4) is P-measurable. Thus, if $\mathfrak{F}'$ is the function from $S' \times S'$ to Δ^+ defined via

$$\mathfrak{F}'(\pi_p,\pi_q)(x) = P\{\omega \text{ in } \Omega \mid \rho^*(\pi_p(\omega),\pi_q(\omega)) < x\},$$

then $(S',\mathfrak{F}')$ is an E-space, and furthermore $\mathfrak{F}'(\pi_p,\pi_q) = \mathfrak{F}(p,q)$ for all p, q in S.

It remains to show that the correspondence between S and S' is one to one, i.e., that if $p \neq q$, then $\pi_p \neq \pi_q$. Now if $p \neq q$, then there is an ω in Ω such that $d_\omega(p,q) > 0$. Consequently (p,ω) is not equivalent to (q,ω), whence $(p,\omega)^* \neq (q,\omega)^*$. But this in turn means that $\pi_p(\omega) \neq \pi_q(\omega)$, whence $\pi_p \neq \pi_q$ and the proof is complete. □

Sherwood's theorem shows that E-spaces and pseudometrically generated spaces are two sides of the same coin. At times it is more convenient to work with one, at times with the other.

Corollary 9.2.3. *Every separated pseudometrically generated space is isometric to a canonical E-space.*

Corollary 9.2.4. *Let $(S,\mathfrak{F})$ be a pseudometrically generated space. Then $(S,\mathfrak{F})$ is a* PPM *space under τ_W. If $(S,\mathfrak{F})$ is separated, then it is a Menger space under W.*

In the other direction we have

Theorem 9.2.5. *There exists a metrically generated space, and hence an E-space, that is not a Menger space under any t-norm stronger than W. Thus, within the class of Menger spaces, Theorem* 9.1.2 *and Corollary* 9.2.4 *are best-possible.*

The space in question consists of a collection of triangles, one for each interior point of the unit square. If $\{p,q,r\}$ is the triangle corresponding to the point (a, b), then

$$F_{pq}(x) = \begin{cases} 0, & x \leqslant \frac{1}{2}, \\ a, & \frac{1}{2} < x \leqslant 1, \\ 1, & 1 < x, \end{cases} \qquad F_{qr}(x) = \begin{cases} 0, & x \leqslant \frac{1}{2}, \\ b, & \frac{1}{2} < x \leqslant 1, \\ 1, & 1 < x, \end{cases}$$

$$F_{pr}(x) = \begin{cases} 0, & x \leqslant 1, \\ W(a,b), & 1 < x \leqslant \frac{3}{2}, \\ 1, & \frac{3}{2} < x; \end{cases}$$

and if p and p' belong to different triangles, then $F_{pp'} = \varepsilon_1$. Thorp [1960] showed that W is the strongest t-norm under which this space is a Menger space; Brown [1972] showed that it is metrically generated. Theorem 9.2.5 is an improvement of a result of Stevens [1968], who had shown that if T is stronger than W, then there exists a metrically generated space that is not a Menger space under T.

We conclude this section with several additional results due to Stevens.

Theorem 9.2.6. *Let $(S, \mathfrak{F})$ be a Menger space under M. Then $(S, \mathfrak{F})$ is pseudometrically generated by the family of pseudometrics d_c given by*

$$d_c(p,q) = F_{pq}^{\wedge}(c) \quad \textit{for } c \textit{ in } (0,1]. \tag{9.2.5}$$

If F_{pq} is continuous at 0 whenever $p \neq q$, then each d_c is a metric and $(S, \mathfrak{F})$ is metrically generated. In the other direction, if $(S, \mathfrak{F})$ is pseudometrically generated, and the set $\{d_\omega \mid \omega \text{ in } \Omega\}$ of generating pseudometrics is linearly ordered by

$$d_\omega \leqslant d_\nu \quad \textit{if and only if} \quad d_\omega(p,q) \leqslant d_\nu(p,q)$$

for all p, q in S, then $(S, \mathfrak{F})$ is a Menger space under M.

Corollary 9.2.7. *Let $(S, \mathfrak{F})$ be the simple space (S, d, G). Then $(S, \mathfrak{F})$ is pseudometrically generated by the pseudometrics d_c defined in (9.2.5), which in this case are given by*

$$d_c(p,q) = d(p,q)G^{\wedge}(c). \tag{9.2.6}$$

Each d_c is a metric, and $(S, \mathfrak{F})$ is metrically generated, if and only if G is continuous at 0.

Note that each of the pseudometrics d_c in (9.2.6) is a multiple of the generating metric d. Hence for any c such that $G^{\wedge}(c) > 0$, the metric spaces (S, d_c) and (S, d) are homothetic.

9.3. Random Metric Spaces

Suppose that $(S, \mathcal{F})$ is a pseudometrically generated space with base $(\Omega, \mathcal{C}, P)$. For every pair of points p, q in S, let X_{pq} be the function from Ω into R^+ defined by

$$X_{pq}(\omega) = d_\omega(p, q). \tag{9.3.1}$$

Then it follows from Definition 9.2.1 that X_{pq} is in $L_1^+(\Omega)$, and furthermore, that

i. $X_{pq}(\omega) = 0$ for all ω in Ω iff $p = q$; (9.3.2)
ii. $X_{pq} = X_{qp}$ for all p, q in S; (9.3.3)
iii. $X_{pr} \leqslant X_{pq} + X_{qr}$ for all p, q, r in S. (9.3.4)

In the other direction, let S be a set, $(\Omega, \mathcal{C}, P)$ a probability space, and $\mathcal{X}$ a function from $S \times S$ into $L_1^+(\Omega)$. Denote the random variable $\mathcal{X}(p, q)$ by X_{pq} and let $\mathcal{F}$ be the function from $S \times S$ into Δ^+ defined by

$$\mathcal{F}(p, q) = df(X_{pq}). \tag{9.3.5}$$

Suppose that (9.3.2), (9.3.3), and (9.3.4) are satisfied. Then for each ω in Ω, the function d_ω defined by (9.3.1) is a pseudometric on S and $(S, \mathcal{F})$ is a pseudometrically generated space with base $(\Omega, \mathcal{C}, P)$.

It follows that the notion of a pseudometrically generated space is equivalent to the notion of a space of nonnegative random variables satisfying (9.3.2), (9.3.3), and (9.3.4). Such spaces were first considered by A. Špaček [1956, 1960], and following him are called "random metric spaces." The next definition extends Špaček's notion by allowing exceptional sets of measure 0.

Definition 9.3.1. Let S be a set and $(\Omega, \mathcal{C}, P)$ a probability space. Let $\mathcal{X}$ be a function from $S \times S$ into $L_1^+(\Omega)$. Then $(S, \mathcal{X})$ is a *random premetric space* (briefly, a *pre-RM space*) with *base* $(\Omega, \mathcal{C}, P)$ if, for any p, q in S,

i. $X_{pq}(\omega) = 0$ a.s. if $p = q$, (9.3.6)
ii. $X_{pq}(\omega) = X_{qp}(\omega)$ a.s. (9.3.7)

where $X_{pq} = \mathcal{X}(p, q)$. If $(S, \mathcal{X})$ is a pre-RM space and if, for all pairs of distinct points p, q in S,

iii. $P\{\omega \text{ in } \Omega \mid X_{pq}(\omega) > 0\} > 0$, (9.3.8)

then $(S, \mathcal{X})$ is a *random semimetric space* (briefly an *RSM space*). If $(S, \mathcal{X})$ is a pre-RM space and if, for any p, q, r in S,

iv. $X_{pr}(\omega) \leqslant X_{pq}(\omega) + X_{qr}(\omega)$ a.s., (9.3.9) then $(S, \mathscr{X})$ is a *random pseudometric space* (briefly, an *RPM space*). If $(S, \mathscr{X})$ is both an RSM and an RPM space, then $(S, \mathscr{X})$ is a *random metric space* (briefly, an *RM space*). Finally, if there is a single set N such that $P(N) = 0$ and such that (9.3.6), (9.3.7), and (9.3.9) hold for all points in $\Omega \backslash N$, then $(S, \mathscr{X})$ is a *uniform* RPM space.

In a random metric space, the random variables X_{pq} are all defined on a common probability space. Therefore, all their joint distribution functions exist and are completely determined. In contradistinction, in an arbitrary PM space, joint distributions are not even defined. Thus by virtue of their very definition, RM spaces are endowed with considerably more structure than PM spaces. For example, since distinct random variables can be identically distributed, the assertion (9.3.7) that $X_{pq} = X_{qp}$ a.s. is much stronger than the corresponding assertion (8.1.3) that $F_{pq} = F_{qp}$.

If $(S, \mathscr{X})$ is a pre-RM space and $\mathscr{F}$ is defined by (9.3.5), then $(S, \mathscr{X})$ is a pre-PM space. In this case, we say that $(S, \mathscr{F})$ is *determined* by $(S, \mathscr{X})$. It is clear that $(S, \mathscr{F})$ is uniquely determined by $(S, \mathscr{X})$, and that if $(S, \mathscr{X})$ is an RSM space, then $(S, \mathscr{F})$ is a PSM space. As regards the triangle inequality in $(S, \mathscr{F})$, we have

Lemma 9.3.2. *Suppose that* $(S, \mathscr{X})$ *is an* RPM *space and that* $(S, \mathscr{F})$ *is determined by* $(S, \mathscr{X})$. *Then for any* p, q, r *in* S, *we have*

$$\mathscr{F}(p,r) \geqslant \sigma_{C_{pqr}}(\mathscr{F}(p,q), \mathscr{F}(q,r)), \tag{9.3.10}$$

where C_{pqr} *is a copula of* X_{pq} *and* X_{qr}.

PROOF. From (9.3.9) and (9.3.5) it follows that

$$\mathscr{F}(p,r) \geqslant df(X_{pq} + X_{qr}). \tag{9.3.11}$$

But by (7.6.2) and (7.4.3), $df(X_{pq} + X_{qr}) = \sigma_{C_{pqr}}(\mathscr{F}(p,q), \mathscr{F}(q,r))$, where C_{pqr} is any copula of X_{pq} and X_{qr}. □

Theorem 9.3.3. (i) *If* $(S, \mathscr{X})$ *is an* RPM *space and if* $(S, \mathscr{F})$ *is determined by* $(S, \mathscr{X})$, *then* $(S, \mathscr{F}, \tau_W)$ *is a* PPM *space.* (ii) *If* $(S, \mathscr{X})$ *is an* RM *space, then* $(S, \mathscr{F})$ *is a Menger space under* W. (iii) *If the random variables* X_{pq}, X_{qr}, *and* X_{pr} *are pairwise independent whenever* p, q, r *are distinct, then* $(S, \mathscr{F})$ *is a Wald space.*

PROOF. Parts (i) and (ii) follow from (9.3.10) and the fact that, by (7.5.2), $\sigma_{C_{pqr}} \geqslant \tau_{C_{pqr}} \geqslant \tau_W$. Part (iii) follows from the fact that if X_{pq}, X_{qr}, and X_{pr} are pairwise independent, then $C_{pqr} = \Pi$ and $\sigma_{C_{pqr}}$ is convolution. □

The definitions and the discussion at the beginning of this section, together with Definition 9.2.1, yield

Theorem 9.3.4. *A* PM *space is a separated pseudometrically generated space if and only if it is determined by a uniform* RM *space. A* PPM *space is pseudometrically generated if and only if it is determined by a uniform* RPM *space having the property that, for any distinct p, q in S, there is an ω in $\Omega \backslash N$ such that $X_{pq}(\omega) > 0$.*

Combining Theorem 9.3.4 with Theorem 9.2.6 and Corollary 9.2.7, we obtain the following result, which is due to A. N. Šerstnev [1967].

Corollary 9.3.5. *Every Menger space under M, and in particular every simple space, is determined by a uniform* RM *space.*

Furthermore, it follows from (9.2.6) that the simple space (S, d, G) is determined by the RM space whose base is (I, λ) and for which $X_{pq} = d(p,q)G^{\wedge}$. In the other direction suppose that (S, d) is a metric space, ξ a nonnegative random variable (not a.s. equal to 0 or ∞) defined on the probability space $(\Omega, \mathcal{Q}, P)$, and $\mathfrak{X}$ the mapping from $S \times S$ into $L_1^+(\Omega)$ defined by

$$X_{pq}(\omega) = d(p,q)\xi(\omega). \tag{9.3.12}$$

Then $(S, \mathfrak{X})$ is a random metric space and the PM space determined by $(S, \mathfrak{X})$ is the simple space $(S, d, df(\xi))$.

Theorem 9.3.4 and the results of the preceding section show that the notions "canonical E-space," "separated pseudometrically generated space," and "uniform RM space" are all equivalent. However, when the RM space is not uniform this relationship with pseudometrically generated spaces and E-spaces can fail with a vengeance. For example, let $S = I$, let $(\Omega, \mathcal{Q}, P)$ be (I, λ), and let X_{pq} be given by

$$X_{pq}(\omega) = \begin{cases} |p - q|, & \omega \neq p, q, \\ q, & \omega = p, \\ pq, & \omega = q, q \neq p. \end{cases}$$

Then $(S, \mathfrak{X})$ is an RM space. Now, as in (9.3.1), let $d_\omega(p,q) = X_{pq}(\omega)$. Then for any distinct q and ω in $(0,1)$, we have

$$d_\omega(\omega, q) = q \neq \omega q = d_\omega(q, \omega),$$

whence for all ω in $(0,1)$, d_ω fails to satisfy (3.1.3). Similarly, for any ω in $(0,1)$, and for any h in $(0, \frac{1}{2})$ distinct from ω or $1 - \omega$, let $p = 1 - h$ and

$r = h$. Then $p > r$, and the triangle inequality

$$d_\omega(p,r) \leqslant d_\omega(p,\omega) + d_\omega(\omega,r)$$

is equivalent to $1 - 2h \leqslant (1-h)\omega + h$, which fails for h sufficiently small. Thus, for all ω in $(0,1)$, d_ω also fails to satisfy (3.1.4).

In the class of Menger spaces, the result of Corollary 9.3.5 is best-possible, for [Brown 1972, Th. 3] has demonstrated the following:

Theorem 9.3.6. *Let T be a t-norm weaker than M. Then there is a Menger space $(S,\mathcal{F},\tau_T)$ that is not determined by an* RM *space—indeed, not determined by any* RSM *space in which* (9.3.10) *is satisfied.*

Theorem 9.3.6 is an extension of a result of [Stevens 1968, Th. 11], and an improvement of a result of [Šerstnev 1967]. It also follows at once from Theorems 9.2.5 and 9.3.6 that Parts i and ii of Theorem 9.3.3 are best-possible in the sense that W cannot be replaced by any stronger t-norm.

9.4. The Probability of the Triangle Inequality

The results of the preceding section indicate both the scope and the limitations of the notion of a random metric space. Thus while all canonical E-spaces and all Menger spaces under M are determined by RM spaces, most PM spaces are not. However, every PM space $(S,\mathcal{F})$ is determined by a pre-RM space $(S,\mathcal{X})$—namely, the one whose base is (I,λ) and for which $X_{pq} = F_{pq}^{\wedge}$. Furthermore, for any p, q, r in S, the set $J_{pqr} = \{t \text{ in } I \mid X_{pr}(t) \leqslant X_{pq}(t) + X_{qr}(t)\}$ is an event in (I,λ). Thus the probability $\lambda(J_{pqr})$ is well defined and $\lambda(J_{pqr}) = 1$ for all p, q, r in S if and only if $(S,\mathcal{X})$ is an RPM space. These observations motivate the following questions: If a PM space $(S,\mathcal{F})$ is determined by a pre-RM space $(S,\mathcal{X})$, then what can one say about the probabilities of the events $\{X_{pr} \leqslant X_{pq} + X_{qr}\}$? In particular, what restrictions does the triangle inequality in $(S,\mathcal{F})$ impose on these probabilities? This section is devoted to some partial answers.

Definition 9.4.1. Let $(S,\mathcal{X})$ be a pre-RM space with base $(\Omega,\mathcal{Q},P)$. For any fixed ordered triple (p,q,r) of points in S, let $X_1 = X_{pq}$, $X_2 = X_{qr}$, $X_3 = X_{rp}$. With (ijk) representing any cyclic permutation of (123), let

$$J_i = \{\omega \text{ in } \Omega \mid X_i(\omega) \leqslant X_j(\omega) + X_k(\omega)\}, \tag{9.4.1}$$

$$J_{ij} = J_i \cap J_j, \tag{9.4.2}$$

$$J = J_1 \cap J_2 \cap J_3. \tag{9.4.3}$$

Thus $P(J)$ is the probability that the triangle inequality holds for the triangle with vertices p, q, r. Clearly, a necessary and sufficient condition for $(S, \mathfrak{X})$ to be an RPM space is that $P(J) = 1$ for every triangle in S.

Lemma 9.4.2. *For any ordered triple (p, q, r) of points in a pre-RM space $(S, \mathfrak{X})$ we have*

$$P(J_{ij}) = P(J_i) + P(J_j) - 1, \tag{9.4.4}$$

$$P(J) = P(J_1) + P(J_2) + P(J_3) - 2, \tag{9.4.5}$$

$$2P(J) = P(J_{12}) + P(J_{23}) + P(J_{31}) - 1, \tag{9.4.6}$$

$$\mathrm{Max}(P(J_1), P(J_2), P(J_3)) \geqslant \tfrac{2}{3}, \tag{9.4.7}$$

$$\mathrm{Max}(P(J_{12}), P(J_{23}), P(J_{31})) \geqslant \tfrac{1}{3}. \tag{9.4.8}$$

Furthermore, if the events J_1, J_2, J_3 are pairwise independent, then at least two of the numbers $P(J_1)$, $P(J_2)$, $P(J_3)$ must equal 1.

PROOF. Both (9.4.4) and (9.4.5) follow directly from the fact that the complements of the sets J_i are pairwise disjoint; and the rest is immediate. □

Lemma 9.4.2 shows that if p, q, and r are the vertices of a triangle in a pre-RM space, then:

i. The probability that the triangle inequality holds for at least one side is $\geqslant \frac{2}{3}$.
ii. The probability that it holds for at least two sides is $\geqslant \frac{1}{3}$.

Nevertheless, these inequalities impose almost no restrictions on $P(J)$. This is brought out in striking fashion by the following examples, which are due to P. Calabrese [1978].

Example 9.4.3. There is a three-point RSM space $(\{p, q, r\}, \mathfrak{X})$ that determines a Menger space under Π and in which $P(J_1) = 0$.

Example 9.4.4. There is a three-point RSM space $(\{p, q, r\}, \mathfrak{X})$ that determines a Menger space under W and is such that X_1, X_2, X_3 each a.s. assume exactly three values, and in which $P(J_1) = 0$.

Example 9.4.5. Let $(S, \mathfrak{X})$ be an RSM space that determines an equilateral PM space (recall that this means that for distinct p, q, r in S, the random variables X_{pq}, X_{qr}, X_{pr} are identically distributed, although in general not identical). Then for any p, q, r in S, $P(J_i) > 0$ for $i = 1, 2, 3$. This result is best-possible: for any $\eta > 0$, there is a three-point RSM space $(\{p, q, r\}, \mathfrak{X})$ that determines an equilateral PM space and is such that $P(J_1) < \eta$.

Example 9.4.6. If $(S, \mathfrak{X})$ is an RSM space that determines a Menger space under a positive t-norm T (see Definition 5.5.10), and if for some p, q, r in S the random variables X_1, X_2 a.s. assume only a finite number of values, then $P(J_3) > 0$. If each of X_1, X_2, X_3 a.s. assumes at most two values, then $\mathrm{Min}(P(J_1), P(J_2), P(J_3)) \geqslant \frac{1}{3}$. If T is stronger than W, but not necessarily positive, and each X_1, X_2, X_3 a.s. assumes at most two values, then $\mathrm{Min}(P(J_1), P(J_2), P(J_3)) \geqslant \frac{1}{4}$. The bounds $\frac{1}{3}$ and $\frac{1}{4}$ are sharp.

The following far-reaching examples are due to M. S. Matveičuk (see [Šerstnev 1968]).

Example 9.4.7. There is a three-point PM space $(\{p, q, r\}, \mathfrak{F}, \tau_W)$ such that $P(J) = 0$ in any RSM space that determines it.

Example 9.4.8. For any $\eta > 0$, there is a three-point Wald space $(\{p, q, r\}, \mathfrak{F}, *)$ such that $P(J) < \eta$ in any RSM space that determines it.

An immediate consequence of Example 9.4.8 is

Theorem 9.4.9. *There are Wald spaces that are not determined by any* RM *space, and hence are not pseudometrically generated.*

In the construction of Example 9.4.7, Matveičuk defines three integer-valued sequences ξ_1, ξ_2, ξ_3 by

$$\xi_1(1) = \xi_2(1) = 0, \qquad \xi_3(1) = 1, \tag{9.4.9}$$

$$\left.\begin{aligned} \xi_1(n+1) &= \xi_2(n) + \xi_3(n) + 1, \\ \xi_2(n+1) &= \xi_2(n) + 2\xi_3(n) + 2, \\ \xi_3(n+1) &= 2\xi_2(n) + 3\xi_3(n) + 4, \end{aligned}\right\} \quad \text{for } n \geqslant 1, \tag{9.4.10}$$

and then defines the distribution functions $F_1 = F_{pq}$, $F_2 = F_{qr}$, and $F_3 = F_{pr}$ by

$$F_i(x) = \begin{cases} 0, & x \text{ in } \left[0, \xi_i(1)\right], \\ 1 - \dfrac{1}{2^n}, & x \text{ in } \left[\xi_i(n), \xi_i(n+1)\right]. \end{cases} \tag{9.4.11}$$

In Example 9.4.8, Matveičuk again employs the sequences defined by (9.4.10), but with the initial conditions (9.4.9) replaced by

$$\xi_1(1) = \xi_3(1) = 1, \qquad \xi_2(1) = 2. \tag{9.4.12}$$

Then given $\eta > 0$, he chooses the integer $N > 1/(2\eta)$, and defines F_i for

$i = 1, 2, 3$ by

$$F_i(x) = \begin{cases} 0, & x \text{ in } [0, \xi_i(1)], \\ 1 - \frac{n}{N}, & x \text{ in } (\xi_i(n), \xi_i(n+1)], \quad n \leqslant N, \\ \frac{1}{2}\left(\frac{2}{3}\right)^{n-N}, & x \text{ in } (\xi_i(n), \xi_i(n+1)], \quad n > N. \end{cases} \quad (9.4.13)$$

The rest is then a matter of straightforward, but nontrivial, verification.

9.5. W-Spaces

J. B. Brown [1972] introduced and studied the following class of spaces:

Definition 9.5.1. An RSM space $(S, \mathcal{X})$ is a *W-space* if for all p, q, r in S and x in R

$$P\{\omega \text{ in } \Omega \mid X_{pr}(\omega) < x\} \geqslant P\{\omega \text{ in } \Omega \mid X_{pq}(\omega) + X_{qr}(\omega) < x\}. \quad (9.5.1)$$

Since (9.3.9) implies (9.5.1), it follows that all RM spaces, E-spaces, pseudometrically generated spaces, and Menger spaces under M are W-spaces. Furthermore, since (9.5.1) and (9.3.10) are equivalent, the proof of Lemma 9.3.2 yields

Theorem 9.5.2. *If $(S, \mathcal{X})$ is a W-space, then the space $(S, \mathcal{F})$ determined by $(S, \mathcal{X})$ is a Menger space under W. If the* r.v.'s *X_{pq}, X_{qr}, and X_{pr} are independent for any distinct p, q, r in S, then $(S, \mathcal{F})$ is a Wald space.*

In addition, Brown also established

Theorem 9.5.3. *A* PM *space $(S, \mathcal{F})$ is a Wald space if and only if it is determined by a W-space with the property that the* r.v.'s *X_{pq} and X_{rs} are independent whenever the unordered pairs $\{p, q\}$ and $\{r, s\}$ are distinct.*

(Note that in view of (9.3.7), if X_{pq} is not constant a.e., then X_{pq} and X_{qp} are dependent with copula M.)

It follows at once from Corollary 9.3.5 that every Menger space under M is determined by a W-space, and from Theorem 9.3.6 that for any t-norm T weaker than M, there is a PM space $(S, \mathcal{F}, \tau_T)$ that is not determined by any W-space. This last result merits comparison with the fact that if $T \neq M$, then τ_T is not derivable from any function on random variables (see Section 7.6).

9.6. Open Problems

The outstanding open problem for this chapter may aptly be called the *Špaček problem.*

Problem 9.6.1. Find necessary and sufficient conditions for a PM space to be determined by an RM space.

The difficulty of this problem stems from the strength of the requirement (9.3.9). Relaxing this requirement should therefore yield an easier problem. For example, replacing (9.3.9) by (9.5.1) leads to

Problem 9.6.2. Find necessary and sufficient conditions for a PM space to be determined by a W-space.

In an RM space we have $P(J) = 1$ for every triangle in S (see Definition 9.4.1). What about a W-space? Thus:

Problem 9.6.3. What can be said about the probabilities $P(J_i)$, $P(J_{ij})$, $P(J)$ in a W-space? In particular, can the inequalities (9.4.7) and (9.4.8) be improved, and do we have $P(J) > 0$ for every triangle in S?

10

Distribution-Generated Spaces

10.1. Introduction

In this chapter we turn to a class of spaces that we first studied in depth in [S² 1962].

Let S be a set. With each point p of S associate an n-dimensional distribution function G_p whose margins are in $\mathcal{D}$, and with each pair (p, q) of distinct points of S associate a $2n$-dimensional distribution function H_{pq} such that

$$H_{pq}(\mathbf{u}, (\infty, \ldots, \infty)) = G_p(\mathbf{u})$$

and (10.1.1)

$$H_{pq}((\infty, \ldots, \infty), \mathbf{v}) = G_q(\mathbf{v})$$

for any $\mathbf{u} = (u_1, \ldots, u_n)$ and $\mathbf{v} = (v_1, \ldots, v_n)$ in R^n. For any $x \geqslant 0$, let $Z(x)$ be the cylinder in R^{2n} given by

$$Z(x) = \{(\mathbf{u}, \mathbf{v}) \text{ in } R^{2n} \mid |\mathbf{u} - \mathbf{v}| < x\} \tag{10.1.2}$$

and define F_{pq} in $\mathcal{D}^+$ via

$$F_{pq}(x) = \int_{Z(x)} dH_{pq} = P_{H_{pq}}(Z(x)). \tag{10.1.3}$$

If we think of the elements of S as "particles," then for any Borel set A in R^n the integral $\int_A dG_p$ is naturally interpreted as the probability that the particle p is in the set A and $F_{pq}(x)$ as the probability that the distance between the particles p and q is less than x.

In order to obtain a PM space by means of the foregoing construction, the definition of F_{pq} given by (10.1.3) must be extended to the case in which

$p = q$ in such a way that (10.1.1) holds and $F_{pp} = \varepsilon_0$. We achieve this with the aid of

Lemma 10.1.1. *Let G be an n-d.f. all of whose margins are in $\mathfrak{D}$, and let H be the $2n$-d.f. given by*

$$H(\mathbf{u}, \mathbf{v}) = G(\operatorname{Min}(u_1, v_1), \ldots, \operatorname{Min}(u_n, v_n)) \tag{10.1.4}$$

for all $\mathbf{u} = (u_1, \ldots, u_n)$ and $\mathbf{v} = (v_1, \ldots, v_n)$ in R^n. Then

i. $H(\mathbf{u}, (\infty, \ldots, \infty)) = H((\infty, \ldots, \infty), \mathbf{u}) = G(\mathbf{u})$ *for all $\mathbf{u}$ in R^n.* (10.1.5)

ii. *For any positive integer $m \leqslant n$*

$$H(u_1, \ldots, u_m, \ldots, u_n, v_1, \ldots, v_m, \ldots, v_n)$$
$$= H(u_1, \ldots, v_m, \ldots, u_n, v_1, \ldots, u_m, \ldots, v_n). \tag{10.1.6}$$

iii. $P_H(Z(x)) = 1$ *for all $x > 0$.* (10.1.7)

PROOF. Suppose that H is given by (10.1.4). Then (10.1.5) and (10.1.6) are immediate. As for (10.1.7), let $\mathbf{k}$, $\mathbf{l}$ be arbitrary lattice points (i.e., points with integer coordinates) in R^n and consider the $2n$-boxes $B(\mathbf{k}, \mathbf{l}) = [(\mathbf{k}, \mathbf{l}), (\mathbf{k}+\mathbf{1}, \mathbf{l}+\mathbf{1})]$. These boxes partition $(-\infty, \infty)^{2n}$, i.e., every finite point in R^{2n} is in at least one box and no two distinct boxes overlap. Hence the sum of the H-volumes of these boxes is $P_H((-\infty, \infty)^{2n}) = P_H(R^{2n}) = 1$, since all the margins of H are in $\mathfrak{D}$.

If $\mathbf{k} \neq \mathbf{l}$, then these points differ in at least one coordinate. Suppose, for example, that $l_1 < k_1$. Then, since l_1 and k_1 are integers, $l_1 + 1 \leqslant k_1$. Now the 2^{2n} vertices of $B(\mathbf{k}, \mathbf{l})$ can be grouped into 2^{2n-1} pairs, the vertices in each pair differing only in their first coordinates. Let $\mathbf{a} = (k_1, k_2, \ldots, k_{2n})$ and $\mathbf{b} = (k_1 + 1, k_2, \ldots, k_{2n})$ be such a pair. Then by (10.1.4), we have $H(\mathbf{a}) = H(\mathbf{b})$; and by (6.1.1), we have $\operatorname{sgn}_{B(\mathbf{k},\mathbf{l})}(\mathbf{a}) = -\operatorname{sgn}_{B(\mathbf{k},\mathbf{l})}(\mathbf{b})$. Hence (6.1.2) yields $P_H(B(\mathbf{k}, \mathbf{l})) = 0$. The same conclusion follows when $k_1 < l_1$ and whenever $k_m \neq l_m$ for any m. Thus $P_H(B(\mathbf{k}, \mathbf{l})) = 0$ whenever $\mathbf{k} \neq \mathbf{l}$; and consequently $\sum P_H(B(\mathbf{l}, \mathbf{l})) = 1$, where the summation is over all lattice points $\mathbf{l}$ in R^n. But for any $\mathbf{l}$ in R^n, the box $B(\mathbf{l}, \mathbf{l})$ is contained in the cylinder $Z(\sqrt{2n} + 1)$, whence $P_H(Z(\sqrt{2n} + 1)) = 1$. Finally, let $\alpha > 0$ be given. Then repeating the above argument using the $2n$-boxes $[\alpha(\mathbf{k}, \mathbf{l}), \alpha(\mathbf{k}+\mathbf{1}, \mathbf{l}+\mathbf{1})]$ yields $P_H(Z(\alpha\sqrt{2n} + \alpha)) = 1$, whence, since α is arbitrary, we obtain (10.1.7). □

The converse of Lemma 10.1.1 is also valid, so that the conditions (10.1.5), (10.1.6), (10.1.7) completely characterize all $2n$-d.f.'s H of the form (10.1.4). Since this fact is not needed in the sequel, we omit the proof.

It follows from Lemma 10.1.1 that if H_{pp} is defined by (10.1.4) with

$G = G_p$, and F_{pp} by (10.1.3), then $F_{pp} = \varepsilon_0$. Furthermore, if $H_{pq}(\mathbf{u},\mathbf{v}) = H_{qp}(\mathbf{v},\mathbf{u})$ then $F_{pq} = F_{qp}$, which brings us to

Definition 10.1.2. Let S be a set, $\mathfrak{F}$ a function from $S \times S$ into Δ^+, and n a positive integer. Then $(S, \mathfrak{F})$ is a *distribution-generated space* (*over* R^n) if the following conditions are satisfied:

i. For every p in S there is an n-d.f. G_p whose margins $G_{p,1}, \ldots, G_{p,n}$ are in $\mathfrak{D}$.
ii. For every p, q in S there is a $2n$-d.f. H_{pq} which satisfies (10.1.1), is such that $H_{pq}(\mathbf{u},\mathbf{v}) = H_{qp}(\mathbf{v},\mathbf{u})$ for all $\mathbf{u}$, $\mathbf{v}$ in R^n, and is given by (10.1.4) with $G = G_p$ when $p = q$.
iii. For every p, q in S, F_{pq} is given by (10.1.3).

The preceding discussion at once yields

Theorem 10.1.3. *Every distribution-generated space is a pre-*PM *space.*

Since all of the margins of each H_{pq} are in $\mathfrak{D}$, it follows easily from (10.1.3) that each distance distribution function F_{pq} is in $\mathfrak{D}^+$.

As regards the triangle inequality, virtually nothing can be said at this level of generality. To illustrate this, consider the distribution-generated space $(S, \mathfrak{F})$ determined by

$$S = \{p, q, r\},$$

$$G_p = G_r = \tfrac{1}{2}\varepsilon_0 + \tfrac{1}{2}\varepsilon_{100}, \qquad G_q = \tfrac{1}{2}\varepsilon_1 + \tfrac{1}{2}\varepsilon_{101}, \tag{10.1.8}$$

$$H_{pq} = \tfrac{1}{2}\varepsilon_{(0,1)} + \tfrac{1}{2}\varepsilon_{(100,101)},$$
$$H_{qr} = \tfrac{1}{2}\varepsilon_{(1,0)} + \tfrac{1}{2}\varepsilon_{(101,100)}, \tag{10.1.9}$$
$$H_{pr} = \tfrac{1}{2}\varepsilon_{(0,100)} + \tfrac{1}{2}\varepsilon_{(100,0)}.$$

Thus H_{pq}, H_{qr}, and H_{pr} are the 2-d.f.'s determined, respectively, by masses of $\frac{1}{2}$ at the points (0, 1) and (100, 101), (1, 0) and (101, 100), and (0, 100) and (100, 0). Now (10.1.3) yields

$$F_{pq} = F_{qr} = \varepsilon_1, \quad \text{but} \quad F_{pr} = \varepsilon_{100},$$

whence it follows that there is no triangle function under which $(S, \mathfrak{F})$ is a proper PM space.

10.2. Consistency, Triangle Inequalities

Let C_{pq}, C_{qr}, and C_{pr} be the subcopulas associated with the d.f.'s defined by (10.1.8) and (10.1.9). Then

$$C_{pq}(\tfrac{1}{2}, \tfrac{1}{2}) = C_{qr}(\tfrac{1}{2}, \tfrac{1}{2}) = \tfrac{1}{2} \quad \text{and} \quad C_{pr}(\tfrac{1}{2}, \tfrac{1}{2}) = 0.$$

These subcopulas are incompatible (see Section 6.6). For if there were a 3-copula C_{pqr} such that

$$C_{pqr}(\tfrac{1}{2},\tfrac{1}{2},1) = C_{pq}(\tfrac{1}{2},\tfrac{1}{2}) = \tfrac{1}{2} = C_{qr}(\tfrac{1}{2},\tfrac{1}{2}) = C_{pqr}(1,\tfrac{1}{2},\tfrac{1}{2})$$

and

$$C_{pqr}(\tfrac{1}{2},1,\tfrac{1}{2}) = C_{pr}(\tfrac{1}{2},\tfrac{1}{2}) = 0,$$

then we would have $C_{pqr}(\frac{1}{2},\frac{1}{2},\frac{1}{2}) = 0$, whence (6.1.2) would yield

$$V_{C_{pqr}}\left[(\tfrac{1}{2},0,\tfrac{1}{2}),(1,\tfrac{1}{2},1)\right] = -\tfrac{1}{2} < 0,$$

a contradiction. It is this incompatibility that lies at the root of the counterexample of the preceding section.

Theorem 10.2.1. *Let $(S,\mathcal{F})$ be a distribution-generated space over R^n. For any p, q in S let C_{pq} be a $2n$-copula of G_p, G_q, and H_{pq}, so that*

$$H_{pq}(\mathbf{u},\mathbf{v}) = C_{pq}(G_{p,1}(u_1),\ldots,G_{p,n}(u_n),G_{q,1}(v_1),\ldots,G_{q,n}(v_n)) \quad (10.2.1)$$

for all $\mathbf{u}$, $\mathbf{v}$ in R^n. Suppose that for any p, q, r in S the copulas C_{pq}, C_{qr}, and C_{pr} are compatible, i.e., that there is a $3n$-copula C_{pqr} such that

$$\begin{aligned} C_{pq}(\mathbf{a},\mathbf{b}) &= C_{pqr}(\mathbf{a},\mathbf{b},\mathbf{1}), \qquad C_{qr}(\mathbf{b},\mathbf{c}) = C_{pqr}(\mathbf{1},\mathbf{b},\mathbf{c}),\\ C_{pr}(\mathbf{a},\mathbf{c}) &= C_{pqr}(\mathbf{a},\mathbf{1},\mathbf{c}), \end{aligned} \quad (10.2.2)$$

for all $\mathbf{a}$, $\mathbf{b}$, $\mathbf{c}$ in I^n. Then

$$F_{pr} \geqslant \tau_W(F_{pq},F_{qr}). \quad (10.2.3)$$

PROOF. Let H_{pqr} be the $3n$-d.f. determined by C_{pqr}, G_p, G_q, and G_r. Let x, y, z in R^+ be such that $x = y + z$ and let A, B, C be the hyperslabs in R^{3n} given by $A = \{(\mathbf{u},\mathbf{v},\mathbf{w}) \mid |\mathbf{u}-\mathbf{v}| < y\}$, $B = \{(\mathbf{u},\mathbf{v},\mathbf{w}) \mid |\mathbf{v}-\mathbf{w}| < z\}$ and $C = \{(\mathbf{u},\mathbf{v},\mathbf{w}) \mid |\mathbf{u}-\mathbf{w}| < x\}$. Then, since $H_{pqr}(\mathbf{u},\mathbf{v},\mathbf{1}) = H_{pq}(\mathbf{u},\mathbf{v})$, we have

$$F_{pq}(y) = \int_{Z(y)} dH_{pq} = \int_A dH_{pqr} = P(A),$$

where P is the probability measure in R^{3n} determined by H_{pqr}. Similarly,

$$F_{pr}(z) = P(B) \quad \text{and} \quad F_{pr}(x) = P(C).$$

In view of the triangle inequality in R^{3n}, $A \cap B \subseteq C$, whence, arguing as in the proof of Theorem 9.1.2, we have (10.2.3). □

Definition 10.2.2. Let $(S,\mathcal{F})$ be a distribution-generated space over R^n and k an integer greater than 1. Then $(S,\mathcal{F})$ is *k-consistent* if, for every m-tuple $(p_1,\ldots,p_m)$ of elements of S, $2 \leqslant m \leqslant k$, there is an mn-copula

$C_{p_1,\ldots,p_m}$ such that, for any $\mathbf{x}_1, \ldots, \mathbf{x}_m$ in I^n,

$$C_{p_1,\ldots,p_m}(\mathbf{x}_1, \ldots, \mathbf{x}_m) = C_{p_{\pi(1)},\ldots,p_{\pi(m)}}(\mathbf{x}_{\pi(1)}, \ldots, \mathbf{x}_{\pi(m)}), \quad (10.2.4)$$

for any permutation π of the set $\{1, \ldots, m\}$, and

$$C_{p_1,\ldots,p_{m-1}}(\mathbf{x}_1, \ldots, \mathbf{x}_{m-1}) = C_{p_1,\ldots,p_m}(\mathbf{x}_1, \ldots, \mathbf{x}_{m-1}, \mathbf{1}). \quad (10.2.5)$$

The space $(S, \mathfrak{F})$ is *ω-consistent* if it is k-consistent for every $k \geq 2$.

By definition, every distribution-generated space is 2-consistent. And as an immediate consequence of Theorem 10.2.1, we have

Theorem 10.2.3. *A 3-consistent distribution-generated space $(S, \mathfrak{F})$ is a* PPM *space under τ_W; if in addition, $p \neq q$ implies $F_{pq} \neq \varepsilon_0$, then $(S, \mathfrak{F})$ is a Menger space under W.*

The concept of k-consistency may equally well be couched in the language of joint d.f.'s—replace "C" and "I" in Definition 10.2.2 by "H" and "R," respectively. In this form, it is due to R. M. Dudley, who also established Theorem 10.2.3. The next result, which is a consequence of the Kolmogorov consistency theorem, is due to H. Sherwood [1969].

Theorem 10.2.4. *The space $(S, \mathfrak{F})$ is an ω-consistent distribution-generated space if and only if it is isometric to a random variable generated E-space (i.e., an E-space whose points are random n-vectors—see Section* 9.1).

Theorems 10.2.3 and 10.2.4 are direct consequences of assumptions that certain copulas (or equivalently, certain joint d.f.'s) exist. As such, they clearly show how strong such assumptions are. Indeed, saying that a distribution-generated space is 3-consistent is tantamount to saying that, for all points p, q, r, the distances $d(p,q)$, $d(q,r)$, and $d(p,r)$ can be measured simultaneously. Similar remarks apply to k-consistency and ω-consistency.

10.3. C-Spaces

Definition 10.3.1. A function g from R^n into R^+ is an (n-dimensional) *density* if the function G defined on R^n by

$$G(\mathbf{u}) = \int_{[(-\infty, \ldots, -\infty), \mathbf{u})} g(\mathbf{v})\, d\mathbf{v} \quad (10.3.1)$$

is an n-d.f. If G is an n-d.f., and if there is an n-dimensional density g such that (10.3.1) holds, then G is *absolutely continuous* and g is a *density of* G.

A consequence of (10.3.1) is

Lemma 10.3.2. *Let G be an absolutely continuous n-d.f. and g a density of G. If B is any Borel subset of R^n, then*

$$P_G(B) = \int_B g(\mathbf{v})\, d\mathbf{v}; \tag{10.3.2}$$

and if f is any Borel-measurable function from R^n into R, then

$$\int_{R^n} f\, dG = \int_{R^n} f(\mathbf{v}) g(\mathbf{v})\, d\mathbf{v}, \tag{10.3.3}$$

in the sense that if either integral exists, then so does the other and the two are equal.

If p is a point in a distribution-generated space over R^n such that G_p is absolutely continuous, then any density g_p of G_p may be visualized as a "cloud" in R^n—a cloud whose "density" at any point of R^n measures the relative likelihood of finding the particle p in the vicinity of that point.

Definition 10.3.3. Let g be an n-dimensional density and $\mathbf{c}$ a point in $(-\infty, \infty)^n$. Then

i. g is *spherically symmetric*, with *center* $\mathbf{c}$, if

$$g(\mathbf{c} + \mathbf{u}) = g(\mathbf{c} + \mathbf{v}) \quad \text{whenever} \quad |\mathbf{u}| = |\mathbf{v}|. \tag{10.3.4}$$

ii. g is *unimodal*, with *mode* $\mathbf{c}$, if

$$g(\mathbf{c} + \mathbf{u}) \geqslant g(\mathbf{c} + \mathbf{v}) \quad \text{whenever} \quad |\mathbf{u}| \leqslant |\mathbf{v}|, \tag{10.3.5a}$$

and

$$g(\mathbf{c}) > g(\mathbf{c} + \mathbf{v}) \quad \text{whenever} \quad |\mathbf{v}| > 0. \tag{10.3.5b}$$

We shall also need the concept of a *singular point*.

Definition 10.3.4. A point p in a distribution-generated space over R^n is *singular* if there is a point $\mathbf{c}_p$ in R^n such that

$$G_p = \varepsilon_{\mathbf{c}_p}. \tag{10.3.6}$$

A singular point p behaves, for all practical purposes, like the ordinary Euclidean point $\mathbf{c}_p$ and may be identified with it; looked at in another way, the particle p is fixed at the point $\mathbf{c}_p$. Thus a distribution-generated space all of whose points are singular is in essence indistinguishable from the subset of R^n consisting of all the points $\mathbf{c}_p$. Indeed, one of the reasons for including singular points in the discussion is to have ordinary Euclidean spaces appear as special cases of distribution-generated spaces.

Definition 10.3.5. A distribution-generated space over R^n is *independent* if

$$C_{pq} = \Pi^{2n-1} \quad \text{whenever} \quad p \neq q. \tag{10.3.7}$$

It follows from (10.1.1), (10.2.1), and (10.3.7) that for any point p in an independent distribution-generated space, G_p is given by

$$G_p(\mathbf{u}) = G_{p,1}(u_1) \cdot \cdots \cdot G_{p,n}(u_n). \tag{10.3.8}$$

The most significant consequence of (10.3.7) is

Theorem 10.3.6. *An independent distribution-generated space over R^n is ω-consistent.*

PROOF. We define the appropriate kn-copulas recursively. First, for $k = 2$, $C_{p_1p_2}$ is given by (10.3.7) if $p_1 \neq p_2$; and if $p_1 = p_2$, then in view of (10.1.4), (10.2.1), (10.3.8), and the fact that $G(M(u,v)) = M(G(u), G(v))$ for any d.f. G, we have

$$C_{p_1p_2}(\mathbf{x}, \mathbf{y}) = M(x_1, y_1) \cdot \cdots \cdot M(x_n, y_n). \tag{10.3.9}$$

For $k > 2$ and $C_{p_1, \ldots, p_{k-1}}$ given, we define $C_{p_1, \ldots, p_k}$ by

$$C_{p_1, \ldots, p_k}(x_1, \ldots, x_{kn}) = C_{p_1, \ldots, p_{k-1}}(x_1, \ldots, x_{(k-1)n}) \cdot x_{(k-1)n+1} \cdot \cdots \cdot x_{kn} \tag{10.3.10}$$

if p_k is distinct from all the points $p_1, \ldots, p_{k-1}$; otherwise, there is a largest integer $l \leqslant k-1$ such that $p_k = p_l$, in which case we define

$$\begin{aligned} C_{p_1, \ldots, p_k}&(x_1, \ldots, x_{kn}) \\ &= C_{p_1, \ldots, p_{k-1}}(x_1, \ldots, x_{(l-1)n}, M(x_{(l-1)n+1}, x_{(k-1)n+1}), \ldots, \\ &\qquad M(x_{ln}, x_{kn}), x_{ln+1}, \ldots, x_{(k-1)n}). \end{aligned} \tag{10.3.11}$$

For example, if $n = 2$ and $p_1 = p_2 = p_4 \neq p_3$, then we obtain

$$C_{p_1p_2p_3p_4}(x_1, \ldots, x_8) = M^2(x_1, x_3, x_7)M^2(x_2, x_4, x_8)x_5x_6. \tag{10.3.12}$$

Next, an induction shows that every $C_{p_1, \ldots, p_k}(x_1, \ldots, x_{kn})$ can be similarly expressed as a product whose factors are either of the form x_m or of the form $M^{j-1}(x_{m_1}, \ldots, x_{m_j})$. By Lemma 6.3.5 and Theorem 6.6.5 it follows that $C_{p_1, \ldots, p_k}$ is a copula. Since (10.2.4) and (10.2.5) are clearly satisfied, the proof is complete. □

Corollary 10.3.7. *An independent distribution-generated space is an E-space and hence a* PPM *space under* τ_W.

The most important independent distribution-generated spaces are the C-spaces, which we introduce via

Definition 10.3.8. A distribution-generated space over R^n is *convex* (in accordance with the terminology introduced by A. Wintner [1938]) and is called a *C-space* if the following conditions hold.

i. It is independent.
ii. The n-d.f. G_p of any nonsingular point p is absolutely continuous and has a spherically symmetric and unimodal density g_p.

It follows from condition ii that every point p in a C-space has a unique point $\mathbf{c}_p$ of R^n associated with it. If p is nonsingular, then $\mathbf{c}_p$ is the center and mode of g_p; if p is singular, then $\mathbf{c}_p$ is the point of R^n at which the probability distribution G_p is concentrated. Greatest interest attaches to C-spaces with nonsingular points. Any such C-space may be visualized as a set of clouds (another motivation for the term "C-space"), which are each spherically symmetric and unimodal.

Theorem 10.3.9. *Let p, q be distinct points of a C-space and let $\mathbf{s}_{pq} = \mathbf{c}_p - \mathbf{c}_q$. If p and q are both singular, then*

$$F_{pq} = \varepsilon_{|\mathbf{s}_{pq}|}. \tag{10.3.13}$$

If p and q are not both singular, then F_{pq} is given by

$$F_{pq}(x) = \int_{|\mathbf{u} - \mathbf{s}_{pq}| < x} g_{pq}(\mathbf{u})\, d\mathbf{u} \quad \text{for } x > 0, \tag{10.3.14}$$

where g_{pq} is the n-dimensional density given by

$$g_{pq}(\mathbf{u}) = g_{qp}(\mathbf{u}) = \begin{cases} g_p(\mathbf{c}_p + \mathbf{u}), & p \text{ nonsingular}, q \text{ singular}, \\ g_q(\mathbf{c}_q + \mathbf{u}), & p \text{ singular}, q \text{ nonsingular}, \\ \int_{R^n} g_p(\mathbf{c}_p + \mathbf{v}) g_q(\mathbf{c}_q + \mathbf{u} + \mathbf{v})\, d\mathbf{v}, & \\ & p, q \text{ both nonsingular}. \end{cases} \tag{10.3.15}$$

In each case, the density g_{pq} is spherically symmetric and unimodal with center and mode $\mathbf{0}$.

Proof. Display (10.3.13) is immediate, while (10.3.14) and (10.3.15) follow from straightforward manipulation of multiple integrals using (10.3.2). That

g_{pq} is a spherically symmetric and unimodal density is a special case of the well-known result (to the best of our knowledge, due to A. Wintner [1938] p. 30; see also [Rogozin 1965]), that the convolution of two spherically symmetric, unimodal densities is again a spherically symmetric, unimodal density.

Corollary 10.3.10. *If p and q are distinct and not both singular, then F_{pq} is continuous.*

Combining Theorem 10.3.9, Corollary 10.3.10, and Corollary 10.3.7 yields

Corollary 10.3.11. *A C-space over R^n is an E-space and hence a* PPM *space under τ_W; and if the space does not contain two distinct singular points p, q with $\mathbf{c}_p = \mathbf{c}_q$, then it is a Menger space under W.*

Since the ordinary volume of an n-sphere of radius x is given by $(\pi^{n/2}/\Gamma(n/2+1))x^n$, we also have

Corollary 10.3.12. *If p is a nonsingular point of a C-space over R^n and $q \neq p$, then for any x in R^+*

$$F_{pq}(x) \leqslant \mathrm{Min}\left(1, g_p(\mathbf{c}_p)\frac{\pi^{n/2}}{\Gamma(n/2+1)}x^n\right). \tag{10.3.16}$$

10.4. Homogeneous and Semihomogeneous C-Spaces

Suppose a C-space $(S, \mathfrak{F})$ has the following property: Every p in S is nonsingular and all the densities g_p are translates of a fixed, spherically symmetric, unimodal density g_0, with center and mode $\mathbf{0}$, so that

$$g_p(\mathbf{u}) = g_0(\mathbf{u} - \mathbf{c}_p). \tag{10.4.1}$$

Then for all pairs (p, q) of distinct points of S, (10.3.15) yields $g_{pq} = g$, where g is the density given by

$$g(\mathbf{u}) = \int_{R^n} g_0(\mathbf{v})g_0(\mathbf{u}+\mathbf{v})\,d\mathbf{v}. \tag{10.4.2}$$

This observation motivates

Definition 10.4.1. A C-space $(S, \mathfrak{F})$ over R^n is *homogeneous* if either:

i. all points in S are singular; or
ii. all points in S are nonsingular and there is a spherically symmetric,

unimodal density g with center and mode $\mathbf{0}$ such that $g_{pq} = g$ for all pairs (p, q) of distinct points of S.

In a homogeneous C-space, Corollary 10.3.11 can be strengthened as follows.

Theorem 10.4.2. *Let $(S, \mathcal{F})$ be a homogeneous C-space over R^n. Then for any p, q, r in S*

$$F_{pr} \geqslant \tau_{ZM}(F_{pq}, F_{qr}) \tag{10.4.3}$$

where ZM is the t-norm, stronger than W, given by

$$ZM(x, y) = \begin{cases} M(x, y), & x \text{ in } \left[\tfrac{1}{2}, 1\right] \text{ or } y \text{ in } \left[\tfrac{1}{2}, 1\right], \\ 0, & x \text{ in } [0, \tfrac{1}{2}) \text{ and } y \text{ in } [0, \tfrac{1}{2}). \end{cases} \tag{10.4.4}$$

PROOF. If every point is singular, then (10.4.3) is immediate. Thus let p, q, r be nonsingular. Set $\mathbf{a} = \mathbf{c}_p - \mathbf{c}_q$, $\mathbf{b} = \mathbf{c}_q - \mathbf{c}_r$, $\mathbf{c} = \mathbf{c}_p - \mathbf{c}_r$, $a = |\mathbf{a}|$, $b = |\mathbf{b}|$, $c = |\mathbf{c}|$. Let $w = u + v$, where u and v are in $(0, \infty)$. If both $F_{pq}(u) < \frac{1}{2}$ and $F_{qr}(v) < \frac{1}{2}$, then $F_{pr}(w) \geqslant 0 = ZM(F_{pq}(u), F_{qr}(v))$. Now suppose that $F_{pq}(u) \geqslant \frac{1}{2}$. If $u < a$, then the open ball $B(\mathbf{a}, u)$ is disjoint from a closed half-space H of R^n bounded by a hyperplane through $\mathbf{0}$. By virtue of the spherical symmetry of g, this implies that

$$F_{pq}(u) = \int_{B(\mathbf{a},u)} g(\mathbf{v})\, d\mathbf{v} < \int_H g(\mathbf{v})\, d\mathbf{v} = \tfrac{1}{2},$$

contrary to hypothesis. Therefore $u \geqslant a$. Since $a \geqslant |b - c|$, we have $u \geqslant b - c$ and $u \geqslant c - b$. But these inequalities are, respectively, equivalent to $c + w \geqslant b + v$ and $c - w \leqslant b - v$, which together imply that the ball $B(\mathbf{a}, v)$ can, by a rotation about $\mathbf{0}$, be made to fit inside the ball $B(\mathbf{c}, w)$. Hence the spherical symmetry of g yields $F_{pr}(w) \geqslant F_{qr}(v)$. Similarly, if $F_{qr}(v) \geqslant \frac{1}{2}$, we obtain $F_{pr}(w) \geqslant F_{pq}(u)$. Combining, we have $F_{pr}(w) \geqslant M(F_{pq}(u), F_{qr}(v))$, from which (10.4.3) readily follows. □

Note that the binary system (I, ZM) is the ordinal sum of the binary systems $([0, \frac{1}{2}], Z_1)$ and $([\frac{1}{2}, 1], M_2)$, where $Z_1(x, y) = \frac{1}{2}Z(2x, 2y)$ for all x, y in $[0, \frac{1}{2}]$ (see (5.6.6)), and M_2 is the restriction of M to $[\frac{1}{2}, 1]$.

Since $ZM(c, c) = c$ for any c in $[\frac{1}{2}, 1]$, Theorem 8.2.5 and Corollary 10.3.10 yield

Corollary 10.4.3. *Let $(S, \mathcal{F})$ be a homogeneous C-space in which all points are nonsingular. Then for any c in $[\frac{1}{2}, 1]$, the c-tiles—and in particular, the medians—of the* d.d.f.'s *form a metric space.*

In the other direction, there are homogeneous C-spaces that are not Menger spaces under Π. For example, take $n = 1$, $S = (-\infty, \infty)$, and let g_0 be defined by $g_0(u) = (2\pi)^{-1/2}\exp(-u^2/2)$. Then (10.4.2) yields $g(u) = (2\pi)^{-1/2}\exp(-u^2/4)$. Taking $p = 0$, $q = 12\sqrt{\pi}$, $r = 24\sqrt{\pi}$, straightforward estimations yield $F_{pq}(2\sqrt{\pi}) = F_{qr}(2\sqrt{\pi}) \geqslant 2\exp(-49\pi)$, and $F_{pr}(4\sqrt{\pi}) \leqslant 4\exp(-100\pi)$. Hence $F_{pr}(4\sqrt{\pi}) < \Pi(F_{pq}(2\sqrt{\pi}), F_{qr}(2\sqrt{\pi}))$, so that the space is not a Menger space under Π. By Theorem 8.1.8, this example also shows that there are homogeneous C-spaces that are not Wald spaces.

Definition 10.4.4. A C-space $(S, \mathfrak{F})$ over R^n is *semihomogeneous* if the following conditions hold.

i. There is a pseudometric $\boldsymbol{\sigma}$ on S such that $\boldsymbol{\sigma}(p,q) = 0$ if and only if $p = q$ or p and q are both singular. We write σ_{pq} instead of $\boldsymbol{\sigma}(p,q)$.
ii. There is an n-dimensional density g that is spherically symmetric and unimodal with center and mode $\mathbf{0}$ and such that if $\sigma_{pq} > 0$, then for all $\mathbf{u}$ in R^n

$$g_{pq}(\mathbf{u}) = \frac{1}{\sigma_{pq}^n}\, g\left(\frac{1}{\sigma_{pq}}\mathbf{u}\right). \tag{10.4.5}$$

A semihomogeneous C-space in which σ_{pq} is the same for every pair (p,q) of distinct points is homogeneous. Conversely, any homogeneous C-space is semihomogeneous. If a semihomogeneous C-space contains at most one singular point, then $\boldsymbol{\sigma}$ is a metric.

From (10.3.14) and (10.4.5) we readily obtain

Theorem 10.4.5. *Let $(S, \mathfrak{F})$ be a semihomogeneous C-space over R^n. Then for any p, q in S and any x in R^+, we have*

$$F_{pq}(x) = R_n(\sigma_{pq}, |\mathbf{s}_{pq}|; x). \tag{10.4.6}$$

Here $\mathbf{s}_{pq} = \mathbf{c}_p - \mathbf{c}_q$, and R_n is the three-place function defined by

i. $R_n(\sigma, s; 0) = 0,$ (10.4.7)
ii. $R_n(0, s; x) = \varepsilon_s(x),$ (10.4.8)
iii. $R_n(\sigma, s; x) = \int_{|\sigma\mathbf{u} - \mathbf{s}| < x} g(\mathbf{v})\, d\mathbf{v}$ (10.4.9)

when $\sigma > 0$ and $x > 0$, where $\mathbf{s}$ is any point in R^n with $|\mathbf{s}| = s$.

For $n = 1$, (10.4.9) reduces to

$$R_1(\sigma, s; x) = \int_{(s-x)/\sigma}^{(s+x)/\sigma} g(v)\, dv. \tag{10.4.10}$$

10.5. Moments and Metrics

Let $(S,\mathfrak{F})$ be a C-space. For β in $(0,\infty)$, let the function $d_{(\beta)}$ be defined on $S\times S$ by

$$d_{(\beta)}(p,q)=\begin{cases} m^{(\beta)}F_{pq}, & \beta \text{ in } (0,1], \\ \left(m^{(\beta)}F_{pq}\right)^{1/\beta}, & \beta \text{ in } [1,\infty), \end{cases} \tag{10.5.1}$$

where $m^{(\beta)}$ is defined in (4.4.20). Since, by Corollary 10.3.7, $(S,\mathfrak{F})$ is an E-space, it follows from Theorem 9.1.4 that the functions $d_{(\beta)}$ are pseudometrics on S. In this section we consider these pseudometrics and related topics. We omit proofs and many technical details. These may be found in [S^2 1962].

Definition 10.5.1. Let g be an n-dimensional density that is spherically symmetric about $\mathbf{0}$. Then the *profile of* g is the function h defined for $u\geqslant 0$ by

$$h(u)=A(u)\,g(\mathbf{u}) \tag{10.5.2}$$

where $\mathbf{u}$ is any point in R^n with $|\mathbf{u}|=u$ and $A(u)=(2\pi^{n/2}/\Gamma(n/2))u^{n-1}$ is the surface area of the n-sphere of radius u.

Lemma 10.5.2. *If g is a spherically symmetric n-dimensional density with center $\mathbf{0}$, then its profile h is a one-dimensional density whose* d.f. *is in* $\mathfrak{D}^+$.

Definition 10.5.3. Let n be a positive integer and β be in $(0,\infty)$. Then Q_n^β is the function defined on $[0,\infty)^2$ by

i. $Q_1^\beta(s,u)=\frac{1}{2}\left(|s-u|^\beta+(s+u)^\beta\right),$ (10.5.3)

ii. $Q_n^\beta(s,0)=s^\beta \quad \text{for } n\geqslant 2,$ (10.5.4)

iii. $Q_n^\beta(s,u)=\dfrac{\Gamma(n/2)}{2\pi^{n/2}}u^{1-n}\displaystyle\int|\mathbf{u}-\mathbf{s}|^\beta\,dS$ (10.5.5)

for $u>0$ and $n\geqslant 2$, where the integral is over the surface of the n-sphere $|\mathbf{u}|=u$, and $\mathbf{s}$ is any point in R^n with $|\mathbf{s}|=s$.

Since $|\mathbf{u}-\mathbf{s}|$ attains its maximum (resp. minimum) when $\mathbf{u}$ and $\mathbf{s}$ are antiparallel (resp., parallel), we have

$$|s-u|^\beta\leqslant Q_n^\beta(s,u)\leqslant(s+u)^\beta. \tag{10.5.6}$$

Lemma 10.5.4. *Let $(S,\mathfrak{F})$ be a C-space over R^n. If p and q in S are both singular, then*

$$m^{(\beta)}F_{pq}=|\mathbf{s}_{pq}|^\beta. \tag{10.5.7}$$

If p and q are not both singular and $m^{(\beta)}F < \infty$, then

$$m^{(\beta)}F_{pq} = \int_{R^n} |\mathbf{u} - \mathbf{s}_{pq}|^{\beta} g_{pq}(\mathbf{u})\, d\mathbf{u} = \int_0^{\infty} h_{pq}(u) Q_n^{\beta}(s_{pq}, u)\, du, \quad (10.5.8)$$

where $s_{pq} = |\mathbf{s}_{pq}|$ and h_{pq} is the profile of g_{pq}. If $(S, \mathcal{F})$ is semihomogeneous, then in view of (10.4.5), (10.5.8) *assumes the form*

$$m^{(\beta)}F_{pq} = \int_{R^n} |\sigma_{pq}\mathbf{u} - \mathbf{s}_{pq}|^{\beta} g(\mathbf{u})\, d\mathbf{u} = \int_0^{\infty} h(u) Q_n^{\beta}(s_{pq}, \sigma_{pq}u)\, du \quad (10.5.9)$$

where h is the profile of g.

Now let h be the profile of a spherically symmetric n-dimensional density g with center $\mathbf{0}$. For β in $(0, \infty)$ let M_n^{β} be the function defined on $[0, \infty)^2$ by

$$M_n^{\beta}(\sigma, s) = \int_0^{\infty} h(u) Q_n^{\beta}(s, \sigma u)\, du. \quad (10.5.10)$$

Then we have

Lemma 10.5.5. *Let $(S, \mathcal{F})$ be a semihomogeneous C-space over R^n. If p and q are not both singular, and $m^{(\beta)}F_{pq} < \infty$, then*

$$m^{(\beta)}F_{pq} = M_n^{\beta}(\sigma_{pq}, s_{pq}), \quad (10.5.11)$$

and for $\beta \geqslant 1$

$$d_{(\beta)}(p, q) = \left(M_n^{\beta}(\sigma_{pq}, s_{pq})\right)^{1/\beta}. \quad (10.5.12)$$

The principal results of this section are contained in

Theorem 10.5.6. *Let $\beta \geqslant 1$, and suppose that $\int_0^{\infty} u^{\beta} h(u)\, du < \infty$ (see* (10.5.6) *and* (10.5.10)*). Then*

i. *For a fixed s in $[0, \infty)$, M_n^{β} is a nondecreasing convex function of σ, and*

$$\inf_{\sigma} M_n^{\beta}(\sigma, s) = M_n^{\beta}(0, s) = s^{\beta}. \quad (10.5.13)$$

ii. *For a fixed σ in $[0, \infty)$, M_n^{β} is a nondecreasing convex function of s, and*

$$\inf_{s} M_n^{\beta}(\sigma, s) = M_n^{\beta}(\sigma, 0) = \sigma^{\beta} \int_0^{\infty} u^{\beta} h(u)\, du. \quad (10.5.14)$$

iii. *For any σ, s in $[0, \infty)$*

$$s \leqslant \left(M_n^{\beta}(\sigma, s)\right)^{1/\beta} \leqslant s + \sigma\left(\int_0^{\infty} u^{\beta} h(u)\, du\right)^{1/\beta}. \quad (10.5.15)$$

Hence for σ fixed,

$$\lim_{s \to \infty} \frac{1}{s}\left(M_n^{\beta}(\sigma, s)\right)^{1/\beta} = 1. \quad (10.5.16)$$

iv. *If* $\int_0^\infty u^2 h(u)\,du < \infty$, *then* (*since* $Q_n^2(s,u) = s^2 + u^2$)

$$\left(M_n^1(\sigma,s)\right)^2 \leqslant M_n^2(\sigma,s) = s^2 + \sigma^2 \int_0^\infty u^2 h(u)\,du, \tag{10.5.17}$$

whence, for $s > 0$,

$$M_n^1(\sigma,s) \leqslant s + \frac{\sigma^2}{2s} \int_0^\infty u^2 h(u)\,du. \tag{10.5.18}$$

Combining (10.5.12), (10.5.14), and (10.5.16), we see that the *Fréchet–Minkowski metrics* $d_{(\beta)}$ associated with semihomogeneous C-spaces over R^n have a remarkable structure. In the small, they are decidedly non-Euclidean—in fact, the distance between any two distinct points p, q is not smaller than a fixed positive multiple of σ_{pq}. In the large, however, they are asymptotic to the Euclidean metric of R^n itself. Looking at an individual C-space as a space of particles and clouds that may move about, we see on the one hand, that as the Euclidean distance $d(\mathbf{c}_p, \mathbf{c}_q)$ between the centers of the clouds of p and q approaches 0, any Fréchet–Minkowski distance between p and q remains greater than a positive number; and on the other hand, that when it is large, $d(\mathbf{c}_p, \mathbf{c}_q)$ is a good estimate of the distance between the particles themselves.

Now (10.5.15) and (10.5.17), when combined with (10.5.11), yield

$$0 \leqslant m^{(2)}F_{pq} - \left(m^{(1)}F_{pq}\right)^2 \leqslant \sigma_{pq}^2 \int_0^\infty u^2 h(u)\,du. \tag{10.5.19}$$

Since $m^{(2)}F_{pq} - (m^{(1)}F_{pq})^2$ is the variance of F_{pq} and its square root the standard deviation of F_{pq}, we have

Theorem 10.5.7. *Let* $(S, \mathfrak{F})$ *be a semihomogeneous* C*-space with at least one nonsingular point. Suppose that* $\int_0^\infty u^2 h(u)\,du < \infty$. *Then* $m^{(1)}F_{pq} < \infty$ *and* $m^{(2)}F_{pq} < \infty$ *for all* p, q *in* S. *If in addition there is a number* $\sigma_0 > 0$ *such that* $\sigma_{pq} < \sigma_0$ *for all* p, q *in* S, *then the variance of* F_{pq} *is bounded on* $S \times S$.

It follows at once that as s_{pq} increases, the ratio of the standard deviation of F_{pq} to the mean of F_{pq}, a quantity that measures the relative uncertainty in the probabilistic determination of the distance between the particles p and q, decreases to 0. This means that the "haziness" of the distance between p and q—which is predominant when their clouds are close together—becomes virtually insignificant when their clouds are sufficiently far apart. In this sense, the probabilistic metric of a semihomogeneous C-space, just as each of the associated Fréchet–Minkowski metrics, is asymptotically Euclidean.

10.6. Normal C-Spaces

Definition 10.6.1. A C-space over R^n is *normal* if the density g_p of a nonsingular point p is a spherically symmetric normal density, i.e., if there is a number $\sigma_p > 0$ such that

$$g_p(\mathbf{u}) = \left(2\pi\sigma_p^2\right)^{-n/2} \exp\left[-\frac{|\mathbf{u}-\mathbf{c}_p|^2}{2\sigma_p^2}\right] \tag{10.6.1}$$

for all $\mathbf{u}$ in R^n. If p is singular, then we set $\sigma_p = 0$.

A standard calculation shows that whenever p and q are distinct and not both singular, then g_{pq} is also normal and is given by

$$g_{pq}(\mathbf{u}) = \left(2\pi\sigma_{pq}^2\right)^{-n/2} \exp\left(-\frac{|\mathbf{u}|^2}{2\sigma_{pq}^2}\right) = \frac{1}{\sigma_{pq}^n}\, g\left(\frac{1}{\sigma_{pq}}\mathbf{u}\right) \tag{10.6.2}$$

where

$$\sigma_{pq}^2 = \sigma_p^2 + \sigma_q^2 \tag{10.6.3}$$

and

$$g(\mathbf{u}) = (2\pi)^{-n/2} \exp(-|\mathbf{u}|^2/2). \tag{10.6.4}$$

From (10.6.3) it follows at once that

$$\sigma_{pr}^2 \leqslant \sigma_{pq}^2 + \sigma_{qr}^2 \tag{10.6.5}$$

for any p, q, r in S; and (10.6.5) in turn implies

$$\sigma_{pr} \leqslant \sigma_{pq} + \sigma_{qr}. \tag{10.6.6}$$

Hence we have

Theorem 10.6.2. *A normal C-space is semihomogeneous.*

Note that the space appearing in the example following Corollary 10.4.3 is normal, which shows that there are normal C-spaces, even homogeneous ones, that are not Menger spaces under Π and hence not Wald spaces.

Straightforward manipulation of multiple integrals yields

Theorem 10.6.3. *With g as given by* (10.6.4), *the function R_n defined by* (10.4.9) *is given, for σ, s in* $(0, \infty)$, *x in R^+, by*

$$R_n(\sigma, s; x) = \left(\frac{s}{\sigma}\right)^{1-n/2} \exp\left(-\frac{s^2}{2\sigma^2}\right) \int_0^{x/\sigma} y^{n/2} I_{n/2-1}\left(\frac{s}{\sigma}y\right) \exp\left(-\frac{y^2}{2}\right) dy, \tag{10.6.7}$$

where $I_{n/2-1}$ is the modified Bessel function of the first kind of order $n/2-1$.

The d.f. in (10.6.7) occurs frequently in the literature, for $R_n(\sigma, s; x) = G(x^2/\sigma^2)$, where G is the d.f. of the noncentral chi-square distribution with n degrees of freedom and noncentrality parameter s^2/σ^2 (see [Johnson and Kotz 1970, Ch. 28]). Some methods for evaluating R_n are given in [Quenouille 1949] and [Amos 1978].

The random distance $d(p,q)$ between two points of a normal C-space was first studied by P. C. Mahalanobis (see [Mahalanobis 1949] and [Kendall 1961, §7]), who introduced it in 1925 as a measure of the distance between two samples from two normal populations having the same known covariance matrix. Up to a constant factor, $d^2(p,q)$ is Mahalanobis's D_1^2 (see [Mahalanobis 1936; Bose 1935]) and the "statistical field" introduced by Mahalanobis [1936] is succinctly described in our terminology as an E-space over R^n whose elements are (not necessarily spherically symmetric) Gaussian random vectors (see also [Rao 1949] and [Hotelling 1928]).

Mahalanobis's D_1^2 is closely related to Mahalanobis's D^2, which in turn is a multiple of Hotelling's T^2 (see [Wilks 1962]). All of these quantities are of importance in discriminatory analysis (see [Srivastava and Khatri 1979]); and this points to the fact that there are close connections between the theory of normal C-spaces and problems of classification and discrimination in multivariate analysis. To date these have not been exploited in either direction.

Considerations of a physical nature that lead to normal C-spaces may be found in the writings of L. de Broglie [1935], A. S. Eddington [1953], and N. Rosen [1947, 1962]. In addition, a number of related matters and further questions concerning the possible role of a stochastic geometry in the foundations of physics are dealt with in detail by D. I. Blokhintsev in his monograph [1971] and book [1973], where further references to the literature may be found.

Normal C-spaces have also appeared in psychology. For example, A. A. J. Marley [1971] has used them to develop a model applicable to the study of certain discrimination tasks.

10.7. Moments in Normal C-Spaces

The profile h of the function g of (10.6.4) is given by

$$h(u) = \frac{u^{n-1}}{2^{n/2-1}\Gamma(n/2)} \exp\left(-\frac{u^2}{2}\right). \tag{10.7.1}$$

Standard calculations then yield

Theorem 10.7.1. *For any β in $(0, \infty)$*

$$\int_0^{\infty} u^{\beta} h(u)\,du = 2^{\beta/2} \frac{\Gamma((n+\beta)/2)}{\Gamma(n/2)} \tag{10.7.2}$$

and

$$M_n^{\beta}(\sigma, s) = 2^{\beta/2} \frac{\Gamma((n+\beta)/2)}{\Gamma(n/2)} \sigma^{k} {}_1F_1\left(-\frac{\beta}{2}; \frac{n}{2}; -\frac{s^2}{2\sigma^2}\right), \tag{10.7.3}$$

where ${}_1F_1$ is the confluent hypergeometric function. In particular,

$$M_n^2(\sigma, s) = n\sigma^2 + s^2 \tag{10.7.4}$$

and

$$M_n^1(\sigma, 0) = \sqrt{2}\,\sigma \frac{\Gamma((n+1)/2)}{\Gamma(n/2)}. \tag{10.7.5}$$

If s/σ is large, then the following asymptotic approximations hold:

$$M_n^1(\sigma, s) \sim s + \frac{n-1}{2} \frac{\sigma^2}{s} - \frac{(n-1)(n-3)}{8} \frac{\sigma^4}{s^3}, \tag{10.7.6}$$

$$\left(M_n^2(\sigma, s) - \left(M_n^1(\sigma, s)\right)^2\right)^{1/2} \sim \sigma - \frac{n-1}{4} \frac{\sigma^3}{s^2}, \tag{10.7.7}$$

$$M_1^1(\sigma, s) \sim s + \sqrt{\frac{2}{\pi}} \frac{\sigma^3}{s^2} \exp\left(-\frac{s^2}{2\sigma^2}\right), \tag{10.7.8}$$

$$\left(M_1^2(\sigma, s) - \left(M_1^1(\sigma, s)\right)^2\right)^{1/2} \sim \sigma - \sqrt{\frac{2}{\pi}} \frac{\sigma^2}{s} \exp\left(-\frac{s^2}{2\sigma^2}\right). \tag{10.7.9}$$

Finally, upon combining (10.7.3) with (10.5.11), (10.5.1), and the fact that for any C-space $(S, \mathfrak{F})$, $d_{(\beta)}$ is a pseudometric on S, we obtain

Theorem 10.7.2. *Let s_1, s_2, s_3 be nonnegative numbers and σ_1, σ_2, σ_3 positive numbers such that there exist triangles with sides of lengths s_1, s_2, s_3 and σ_1^2, σ_2^2, σ_3^2, respectively. Let n be a positive integer. If $\beta \geqslant 1$, then*

$$\sigma_3\left[{}_1F_1\left(-\frac{\beta}{2}; \frac{n}{2}; -\frac{s_3^2}{2\sigma_3^2}\right)\right]^{1/\beta} \leqslant \sigma_1\left[{}_1F_1\left(-\frac{\beta}{2}; \frac{n}{2}; -\frac{s_1^2}{2\sigma_1^2}\right)\right]^{1/\beta} + \sigma_2\left[{}_1F_1\left(-\frac{\beta}{2}; \frac{n}{2}; -\frac{s_2^2}{2\sigma_2^2}\right)\right]^{1/\beta},$$

while if β *is in* $(0,1]$, *then*

$$\sigma_3^\beta\,{}_1F_1\left(-\frac{\beta}{2};\frac{n}{2};-\frac{s_3^2}{2\sigma_3^2}\right)\leqslant\sigma_1^\beta\,{}_1F_1\left(-\frac{\beta}{2};\frac{n}{2};-\frac{s_1^2}{2\sigma_1^2}\right)$$
$$+\sigma_2^\beta\,{}_1F_1\left(-\frac{\beta}{2};\frac{n}{2};-\frac{s_2^2}{2\sigma_2^2}\right).$$

These inequalities were established in [S² 1963a]. Since they are of a purely function-theoretic nature, it is of some interest to find direct analytic proofs. We have found such proofs in the following cases: (i) $\beta=n=1$; (ii) β in $[1,2]$, and n a real number satisfying $n\geqslant 4-\beta$.

10.8. Open Problems

In a sense, distribution-generated spaces are to general PM spaces as Euclidean spaces are to metric spaces. Thus, following Menger [1928] (see also [Blumenthal 1953]), it is natural to ask:

Problem 10.8.1. Under what conditions is a PM space isometric to a distribution-generated space? To a C-space? To a distribution-generated space of given dimension? To a C-space of given dimension?

Theorem 10.4.2 represents a strengthening of Corollary 10.3.11 for a particular class of C-spaces. One can look for a similar strengthening for other classes of C-spaces, and for a strengthening of Theorem 10.4.2 itself. Thus we have

Problem 10.8.2. Is there a positive t-norm T such that every semihomogeneous C-space (not containing distinct singular points p, q with $\mathbf{c}_p=\mathbf{c}_q$) is a Menger space under T? Every homogeneous C-space? Every normal C-space? Every homogeneous normal C-space? Is there a best-possible t-norm for the classes of semihomogeneous, homogeneous, normal, and homogeneous normal C-spaces?

Problem 10.8.3. (i) Find purely analytic proofs of the hypergeometric function inequalities in Section 10.7. (ii) Find corresponding inequalities for other special functions by considering C-spaces that are not normal.

11

Transformation-Generated Spaces

11.1. Transformation-Generated Spaces

In this chapter we consider two classes of PM spaces that are limits of sequences of E-spaces. To construct the first class, let (S, d) be a metric space and ψ a transformation on S, i.e., a function from S into S. For brevity, we denote $\psi^m(p)$, the value of the mth iterate of ψ at p, by $\psi^m p$. For any p, q in S and any integer $n \geqslant 1$, let

$$F_{pq}^{(n)} = \frac{1}{n} \sum_{m=0}^{n-1} \varepsilon_{d(\psi^m p, \psi^m q)} . \tag{11.1.1}$$

Thus $F_{pq}^{(1)} = \varepsilon_{d(p,q)}$, $F_{pq}^{(2)} = \frac{1}{2}(\varepsilon_{d(p,q)} + \varepsilon_{d(\psi p, \psi q)})$; and, in general, for any $x > 0$, $F_{pq}^{(n)}(x)$ is the average number of times in the first $(n-1)$ iterations of ψ that the distance $d(\psi^m p, \psi^m q)$ is less than x. Clearly, $F_{pq}^{(n)}$ is in Δ^+ and if the metric d never assumes the value ∞, then $F_{pq}^{(n)}$ is in $\mathcal{D}^+$. Moreover, for any β in $(0, \infty)$ we have

$$m^{(\beta)} F_{pq}^{(n)} = \frac{1}{n} \sum_{m=0}^{n-1} (d(\psi^m p, \psi^m q))^{\beta}. \tag{11.1.2}$$

For any n the function $\mathcal{F}^{(n)}$ defined on $S \times S$ by

$$\mathcal{F}^{(n)}(p, q) = F_{pq}^{(n)} \tag{11.1.3}$$

satisfies (8.1.1) and (8.1.3). Thus each of the spaces $(S, \mathcal{F}^{(n)})$ is a pre-PM

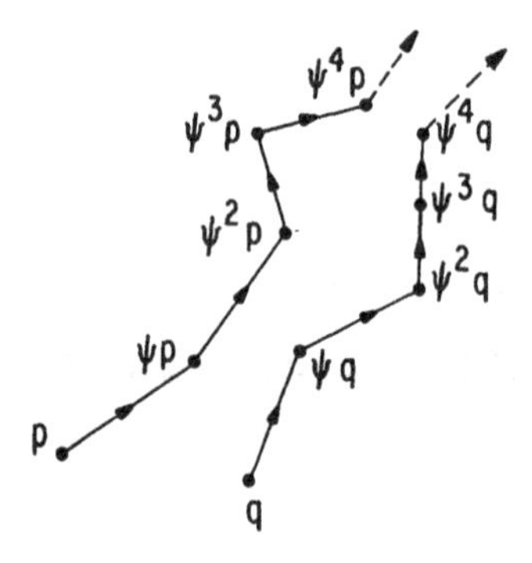

Figure 11.1.1. Trajectories of p and q under the transformation ψ.

space. But more is true:

Theorem 11.1.1. *For every $n \geqslant 1$ the space $(S, \mathscr{F}^{(n)})$ is isometric to an E-space and hence is a* PPM*-space under τ_W.*

PROOF. Given $n \geqslant 1$, let $(\Omega_n, \mathscr{C}_n, P_n)$ be the probability space for which $\Omega_n = \{0, 1, \ldots, n-1\}$, $\mathscr{C}_n = \mathscr{P}(\Omega_n)$, and P_n is the uniform probability measure determined by assigning the weight $1/n$ to each point in Ω_n. For each p in S let p_n be the function from Ω_n into S defined by $p_n(m) = \psi^m p$; let $S_n = \{p_n \mid p \text{ in } S\}$; and let $\mathscr{F}_n$ be the function from $S_n \times S_n$ into Δ^+ defined by

$$\mathscr{F}_n(p_n, q_n) = \mathscr{F}^{(n)}(p, q). \tag{11.1.4}$$

By construction, $(S, \mathscr{F}^{(n)})$ and $(S_n, \mathscr{F}_n)$ are isometric. But it is immediate from Definition 9.1.1 that $(S_n, \mathscr{F}_n)$ is an E-space (in fact, a random variable generated E-space), whence by Theorem 9.1.2, $(S_n, \mathscr{F}_n)$, and therefore $(S, \mathscr{F}^{(n)})$ as well, is a PPM space under τ_W. □

Note that $(S, \mathscr{F}^{(n)})$ is pseudometrically generated by the set of pseudometrics $\{d_m \mid m \text{ in } \Omega_n\}$ where $d_m(p,q) = d(\psi^m p, \psi^m q)$ for all p, q in S.

Theorem 11.1.1 gives us a sequence $\{(S, \mathscr{F}^{(n)})\}$ of PPM spaces, each of which is, up to isometry, an E-space. Our principal concern, however, is not with the sequence itself but with its limit, which yields information about the asymptotic average behavior of the sequence of distances $\{d(\psi^m p, \psi^m q)\}$ (see Figure 11.1.1). The next theorem, which is due to H. Sherwood [1976], shows that, in a weak sense, this limit always exists.

Theorem 11.1.2. *Let (S, d) be a metric space and ψ a transformation on S. For all p, q in S and all $n \geqslant 1$, let $F_{pq}^{(n)}$ be given by (11.1.1). Let φ_{pq} be* *defined on* R^+ *by*

$$\varphi_{pq}(x) = \liminf_{n\to\infty} F_{pq}^{(n)}(x) = \lim_{n\to\infty}\left(\inf\left\{F_{pq}^{(m)}(x) \mid m \geqslant n\right\}\right); \tag{11.1.5}$$

and define $\mathfrak{F}$ *on* $S \times S$ *via*

$$F_{pq} = l^- \varphi_{pq}. \tag{11.1.6}$$

Then F_{pq} *is in* Δ^+ *and* $(S, \mathfrak{F})$ *is a* PPM *space under* τ_W.

PROOF. It is immediate that $(S, \mathfrak{F})$ is a pre-PM space. To establish the triangle inequality, let p, q, r be in S. Then, since each $(S, \mathfrak{F}^{(n)})$ is a PPM space under τ_W, for all u, v in R^+, for a fixed n, and for all $k \geqslant n$ we have

$$\begin{aligned} F_{pr}^{(k)}(u+v) &\geqslant W\left(F_{pq}^{(k)}(u), F_{qr}^{(k)}(v)\right) \\ &\geqslant W\left(\inf\left\{F_{pq}^{(m)}(u) \mid m \geqslant n\right\}, \inf\left\{F_{qr}^{(m)}(v) \mid m \geqslant n\right\}\right). \end{aligned}$$

Therefore

$$\begin{aligned} &\inf\left\{F_{pr}^{(m)}(u+v) \mid m \geqslant n\right\} \\ &\qquad \geqslant W\left(\inf\left\{F_{pq}^{(m)}(u) \mid m \geqslant n\right\}, \inf\left\{F_{qr}^{(m)}(v) \mid m \geqslant n\right\}\right). \end{aligned}$$

Applying (11.1.5) and using the continuity of W, we obtain

$$\varphi_{pr}(u+v) \geqslant W(\varphi_{pq}(u), \varphi_{qr}(v))$$

for all u, v in R^+, whence a standard argument (see [Sherwood 1976]) yields

$$F_{pr}(u+v) \geqslant W(F_{pq}(u), F_{qr}(v))$$

for all u, v in R^+, which is equivalent to

$$F_{pr} \geqslant \tau_W(F_{pq}, F_{qr}).$$

Hence $(S, \mathfrak{F})$ is a PPM space under τ_W and the theorem is proved. □

The space $(S, \mathfrak{F})$ constructed in Theorem 11.1.2 is called the *transformation-generated space* determined by the metric space (S, d) and the transformation ψ, and is denoted by $[S, d, \psi]$. For such spaces we have the following result, which is also due to Sherwood [1976]:

Theorem 11.1.3. *If* $(S, \mathfrak{F})$ *is the transformation-generated space* $[S, d, \psi]$, *then*

$$\mathfrak{F}(\psi p, \psi q) = \mathfrak{F}(p, q) \tag{11.1.7}$$

for all p, q *in* S. *Hence* ψ *preserves probabilistic distances and, if one to one, is an isometry of* $(S, \mathfrak{F})$.

PROOF. Using (11.1.1) and the fact that $\psi^m\psi = \psi^{m+1}$, we find after a short calculation that for any p, q in S and any $n \geqslant 1$

$$F^{(n)}_{\psi p \psi q} = F^{(n+1)}_{pq} + \frac{1}{n}\left[F^{(n+1)}_{pq} - F^{(1)}_{pq}\right],$$

whence $\varphi_{\psi p \psi q} = \varphi_{pq}$, which in turn implies (11.1.7). □

Corollary 11.1.4. *If p and q are distinct points in a transformation-generated space $[S, d, \psi]$ and there is a positive integer m such that $\psi^m p = \psi^m q$, then $F_{pq} = \varepsilon_0$.*

A simple example of a transformation-generated space is the space $[S, d, \psi]$, where (S, d) is the Euclidean line and ψ is defined by

$$\psi p = \begin{cases} 1/p, & p \neq 0, \\ 0, & p = 0. \end{cases} \tag{11.1.8}$$

In this case,

$$F^{(n)}_{pq} = \begin{cases} \dfrac{n+1}{2n}\varepsilon_{d(p,q)} + \dfrac{n-1}{2n}\varepsilon_{d(\psi p, \psi q)} & \text{for } n \text{ odd}, \\ \frac{1}{2}\varepsilon_{d(p,q)} + \frac{1}{2}\varepsilon_{d(\psi p, \psi q)} & \text{for } n \text{ even}, \end{cases}$$

whence

$$F_{pq} = \tfrac{1}{2}\varepsilon_{d(p,q)} + \tfrac{1}{2}\varepsilon_{d(\psi p, \psi q)}$$

for all p, q in S. Moreover, ψ is an isometry of $[S, d, \psi]$.

11.2. Measure-Preserving Transformations

For transformation-generated spaces it is of course important to know when the limit inferior in (11.1.5) can be replaced by a bona fide limit. This can be done trivially when ψ is an isometry of (S, d)—in which case $F_{pq} = \varepsilon_{d(p,q)}$—or when ψ is a contraction—in which case $F_{pq} = \varepsilon_0$. It can also be done in a more interesting setting (see [S² 1973]).

Definition 11.2.1. Let $(S, \mathcal{Q}, P)$ be a probability space and ψ a transformation on S. Then ψ is *measure preserving with respect to P* if for all A in $\mathcal{Q}$, $\psi^{-1}A$ is in $\mathcal{Q}$ and $P(\psi^{-1}A) = P(A)$.

For example, the transformation ψ defined in (11.1.8) is measure preserving with respect to the Lebesgue-Stieltjes F-measure (see Section 2.3) on R given by $F(x) = \frac{1}{2} + (1/\pi)\arctan x$.

The next lemma is readily established.

Lemma 11.2.2. *Let $(S, \mathcal{Q}, P)$ be a probability space. For any $n \geqslant 1$, let $(S^n, \mathcal{Q}^{(n)}, P^{(n)})$ be the probability space in which $\mathcal{Q}^{(n)}$ is the product sigma field $\mathcal{Q} \times \mathcal{Q} \times \cdots \times \mathcal{Q}$ and $P^{(n)}$ the corresponding product measure* (*see* [*Kingman and Taylor* 1966, §6.2]). *For any transformation ψ on S, let $\psi^{(n)}$ be the transformation on S^n defined by*

$$\psi^{(n)}(p_1, \ldots, p_n) = (\psi p_1, \ldots, \psi p_n). \tag{11.2.1}$$

If ψ is measure preserving with respect to P, then $\psi^{(n)}$ is measure preserving with respect to $P^{(n)}$.

The principal property of measure-preserving transformations that we require is contained in the following theorem, which is part of the Birkhoff ergodic theorem (see [Kingman and Taylor 1966, Th. 7.9]).

Theorem 11.2.3. *Let $(\Omega, \mathcal{B}, P^*)$ be a probability space and ρ a transformation on Ω that is measure preserving with respect to P^*. Let f be an integrable function from Ω into R. Then there exists an integrable function Φ from Ω into R such that*

i. $$\lim_{n \to \infty} \frac{1}{n} \sum_{m=0}^{n-1} f(\rho^m p) = \Phi(p) \quad \text{a.s.}; \tag{11.2.2}$$

ii. $$\int_\Omega \Phi \, dP^* = \int_\Omega f \, dP^*. \tag{11.2.3}$$

To apply Theorem 11.2.3 to transformation-generated spaces, we begin with

Lemma 11.2.4. *Suppose that the metric space (S, d) is separable and that there is a probability space $(S, \mathcal{Q}, P)$ such that every open ball $B(r, x)$ belongs to $\mathcal{Q}$. For any $x > 0$ let $D(x)$ be the subset of $S \times S$ defined by*

$$D(x) = \{(p, q) \mid d(p, q) < x\}. \tag{11.2.4}$$

Then $D(x)$ is in $\mathcal{Q}^{(2)}$.

PROOF. Let S_0 be a dense denumerable subset of S. It is easily seen that $D(x)$ is the union of the sets $(B(r, x))^2$ for all r in S_0. Since S_0 is denumerable and each set $(B(r, x))^2$ is in $\mathcal{Q}^{(2)}$, the conclusion follows. □

If (S, d) is not separable, then Lemma 11.2.4 may not be valid (see [Billingsley 1968, pp. 224–225]).

Now for any set A let χ_A denote the *indicator function* of A, so that $\chi_A(\xi) = 1$ if ξ is in A and $\chi_A(\xi) = 0$ otherwise. Then starting with (11.1.1) and using (11.2.1), we obtain

$$F_{pq}^{(n)}(x) = \frac{1}{n}\sum_{m=0}^{n-1} \chi_{D(x)}(\psi^m p, \psi^m q)$$
$$= \frac{1}{n}\sum_{m=0}^{n-1} \chi_{D(x)}\left(\psi^{(2)}\right)^m (p,q). \tag{11.2.5}$$

Hence, if in Theorem 11.2.3 we take $\Omega = S^2$, $\mathfrak{B} = \mathcal{Q}^{(2)}$, $P^* = P^{(2)}$, $\rho = \psi^{(2)}$, and $f = \chi_{D(x)}$, then using (11.2.5), we immediately have

Lemma 11.2.5. *Suppose that* (S,d) *and* $(S,\mathcal{Q},P)$ *satisfy the hypotheses of Lemma* 11.2.4. *Let* ψ *be a transformation on* S *that is measure preserving with respect to* P. *For any* p, q *in* S *and any positive integer* n, *let* $F_{pq}^{(n)}$ *be defined by* (11.1.1) *or, equivalently, by* (11.2.5); *and let* φ_{pq} *be defined by* (11.1.5). *Then for every* x *in* R^+, *there is a subset* $A(x)$ *of* $S \times S$ *with* $P^{(2)}(A(x)) = 1$ *such that for all* (p,q) *in* $A(x)$

$$\lim_{n\to\infty} F_{pq}^{(n)}(x) = \varphi_{pq}(x). \tag{11.2.6}$$

The function F_{pq} defined in (11.1.6) coincides with φ_{pq} at every point of continuity of either function, and is completely determined by its values on any denumerable dense subset of R^+. Thus, as an easy consequence of Lemma 11.2.5 we obtain the principal result of this section (for details, see [S² 1973]):

Theorem 11.2.6. *Let* $[S,d,\psi]$ *be a transformation-generated space. Suppose that the following conditions hold:*

i. *The metric space* (S,d) *is separable.*
ii. *There is a probability measure* P, *defined on a sigma field* $\mathcal{Q}$ *of subsets of* S.
iii. *Every open ball in* S *belongs to* $\mathcal{Q}$.
iv. ψ *is measure preserving with respect to* P.

Then there is a subset A_0 *of* $S \times S$ *with* $P^{(2)}(A_0) = 1$ *such that if* $\{F_{pq}^{(n)}\}$ *is the sequence defined in* (11.1.1) *and if* F_{pq} *is the function defined in* (11.1.6), *then for all* (p,q) *in* A_0,

$$F_{pq}^{(n)} \xrightarrow{w} F_{pq}. \tag{11.2.7}$$

Thus, under the hypotheses of Theorem 11.2.6, "lim inf" in (11.1.5) may be replaced almost everywhere by "lim," as in (11.2.6). Hence the resultant transformation-generated space $[S,d,\psi]$ is, up to a set of measure zero, a limit of E-spaces.

The $P^{(2)}$-measurability of the sets $D(x)$ and the continuity of the metric d (Theorem 3.1.2) together imply that for any β in $(0, \infty)$, the function d^{β} is $P^{(2)}$-measurable. Thus Theorem 11.2.3 can be applied to the sums in (11.1.2) to yield the following:

Corollary 11.2.7. *Under the hypotheses of Theorem* 11.2.6, *for any* $\beta > 0$ *there is a subset* A_{β} *of* $S \times S$ *with* $P^{(2)}(A_{\beta}) = 1$ *such that the sequence* $\{m^{(\beta)}F_{pq}^{(n)}\}$ *converges (possibly to* ∞*) for all pairs* (p,q) *in* A_{β}. *If in addition there is a* β_0 *in* $(0, \infty)$ *such that* $|x|^{\beta_0}$ *is uniformly integrable in* $F_{pq}^{(n)}$ *(see [Loève* 1977, §11.4]), *then for all* β *in* $(0, \beta_0]$

$$\lim_{n\to\infty} m^{(\beta)}F_{pq}^{(n)} = m^{(\beta)}F_{pq} < \infty \tag{11.2.8}$$

for all pairs (p,q) *in* $A_{\beta} \cap A_0$.

Finally, for any fixed $x \geqslant 0$ the function F_x from S^2 into R^+ defined by $F_x(p,q) = F_{pq}(x)$ is $P^{(2)}$-measurable, whence (11.2.3) yields

$$\int_{S^2} F_x \, dP^{(2)} = \int_{S^2} \chi_{D(x)} \, dP^{(2)} = P^{(2)}(D(x)). \tag{11.2.9}$$

11.3. Mixing Transformations

The more one assumes about the transformation ψ, the more one can say about the structure of the transformation-generated space $[S, d, \psi]$. The case when ψ is mixing is of particular interest (see [Erber, Schweizer, and Sklar 1973]).

Definition 11.3.1. Let $(S, \mathcal{Q}, P)$ be a probability space. A transformation ψ on S is *mixing with respect to* P if ψ is measure preserving and if, for all A, B in $\mathcal{Q}$,

$$\lim_{n\to\infty} P(\psi^{-n}A \cap B) = P(A)P(B). \tag{11.3.1}$$

Lemma 11.3.2. *Let* $(S, \mathcal{Q}, P)$ *be a probability space, and* ψ *a transformation on S. For any* $n \geqslant 1$, *let* $\mathcal{Q}^{(n)}$, $P^{(n)}$, *and* $\psi^{(n)}$ *be as in Lemma* 11.2.2. *If* ψ *is mixing with respect to P, then* $\psi^{(n)}$ *is mixing with respect to* $P^{(n)}$.

For mixing transformations, the Birkhoff ergodic theorem admits the following extension:

Theorem 11.3.3. *Let* $(\Omega, \mathfrak{B}, P^*)$ *be a probability space, and* ρ *a transformation on* Ω *that is mixing with respect to* P^*. *Let f be an integrable function from* Ω *into R. Then the limit function* Φ *of Theorem* 11.2.3 *is a.s. constant;*

i.e., there is a number Φ_0 *such that* $\Phi(p) = \Phi_0$ *a.s. Hence* (11.2.2) *and* (11.2.3) *combine to yield*

$$\lim_{n\to\infty} \frac{1}{n} \sum_{m=0}^{n-1} f(\rho^m p) = \Phi_0 = \int_\Omega f\,dP^* \quad a.s. \tag{11.3.2}$$

If one considers iteration as a process taking place in time, then the left-hand side of (11.3.2) is the average over future time of the function f, given the initial point p. The right-hand side of (11.3.2), on the other hand, is the average over the space Ω of the function f. Thus (11.3.2) states that for almost all initial positions, time averages can be replaced by physically more accessible space averages. The desirability of this interchange was and remains the stimulus for the development of ergodic theory.

When applied to transformation-generated spaces, Theorem 11.3.3 yields the following extension of Theorem 11.2.6:

Theorem 11.3.4. *Suppose that the hypotheses of Theorem* 11.2.6 *are satisfied and that* ψ *is mixing with respect to* P. *Then there is a function* F *in* Δ^+ *such that*

i. $F(x) = P(D(x))$ *for all* x *in* R^+; (11.3.3)
ii. *there is a subset* A_1 *of* $S \times S$ *with* $P^{(2)}(A_1) = 1$ *such that for all* (p,q) *in* A_1 *the sequence* $\{F_{pq}^{(n)}\}$ *defined in* (11.1.1) *converges weakly to* F, *i.e., the function* F_{pq} *in* (11.2.7) *is constant on* A_1.

It follows that under the hypotheses of Theorem 11.3.4, the associated transformation-generated space $[S,d,\psi]$ is almost an equilateral PPM space; and if there is an $x > 0$ such that $P^{(2)}(D(x)) < 1$, then $F \neq \varepsilon_0$ and $[S,d,\psi]$ is almost an equilateral PM space. Note that in this case A_1 cannot contain any pair of the form (p,p), whence

$$P^{(2)}\{(p,p)\,|\,p \text{ in } S\} = 0. \tag{11.3.4}$$

By working with $\psi^{(2k)}$ rather than $\psi^{(2)}$ (see (11.2.5)) we can extend Theorem 11.3.4 to k-tuples of pairs of points as follows:

Theorem 11.3.5. *Under the hypotheses of Theorem* 11.3.4, *for any* $k \geqslant 2$ *there is a subset* A_k *of* S^{2k} *with* $P^{(2k)}(A_k) = 1$ *such that, for any* $2k$*-tuple* $(p_1, q_1, \ldots, p_k, q_k)$ *in* A_k

$$\begin{aligned}\lim_{n\to\infty} \frac{1}{n} \sum_{m=0}^{n-1} \chi_{D(x_1)\times\cdots\times D(x_k)}\left[\left(\psi^{(2)}\right)^m (p_1,q_1), \ldots, \left(\psi^{(2)}\right)^m (p_k,q_k)\right] \\ = F(x_1)\cdot\,\cdots\,\cdot F(x_k).\end{aligned} \tag{11.3.5}$$

Thus, loosely speaking, under iteration of ψ, k-tuples of distinct distances $d(p,q)$ behave asymptotically as independent, identically distributed random variables.

These results are illustrated by the following example, which is discussed in detail in [Johnson and Sklar 1976]. Let (S,d) be the interval $[-2,2]$ endowed with the usual Euclidean distance, and let P_c be the Lebesgue-Stieltjes F_c-measure on S determined by

$$F_c(x) = \frac{1}{2} + \frac{1}{\pi}\arcsin\frac{x}{2}. \qquad (11.3.6)$$

For any $n \geqslant 0$ let C_n be the function defined on S by

$$C_n(x) = 2\cos(n\arccos(x/2)). \qquad (11.3.7)$$

It follows from (11.3.7) that $C_m \circ C_n = C_n \circ C_m = C_{mn}$ for all m, $n \geqslant 0$, whence $C_n^m = C_{n^m}$. Each C_n is the restriction to S of an nth-degree polynomial that is monic for $n \geqslant 1$ and is related to the standard nth-degree Čebyšev polynomial T_n by $C_n(x) = 2T_n(x/2)$ for all x in S. In particular, $C_2(x) = x^2 - 2$.

It is a known fact that for each $n \geqslant 2$, C_n is mixing with respect to P_c (see [Adler and Rivlin 1964] or [Johnson and Sklar 1976]). Using (11.3.3), we find, after some manipulation of double integrals, that

$$F(x) = \begin{cases} \dfrac{8}{\pi^2}\displaystyle\int_0^x \frac{1}{4+y} K\left(\frac{4-y}{4+y}\right) dy, & x \text{ in } [0,4], \\ 1, & x \geqslant 4, \end{cases} \qquad (11.3.8)$$

where K is the complete elliptic integral of the first kind. Similar calculations yield

$$m^{(1)}F = 16/\pi^2 = 1.621\ldots, \qquad m^{(2)}F = 4,$$

whence the standard deviation of F is $(4 - (16/\pi^2)^2)^{1/2} = 1.171\ldots$. Computer runs on selected pairs of points, using (11.1.2) with C_2 as ψ, indicate that the convergence of $m^{(1)}F_{pq}^{(n)}$ to $m^{(1)}F$ and $m^{(2)}F_{pq}^{(n)}$ to $m^{(2)}F$ is fairly rapid (see [Johnson and Sklar 1976, §6]). Closely related results may be found in [Erber, Everett, and Johnson 1979], where, inter alia, the mixing character of the polynomials C_n and (11.3.5) are used to develop an efficient method for generating sequences of pseudo-random numbers.

11.4. Recurrence

Mixing transformations exhibit recurrence and dispersive effects. Thus, for example, the dispersive character of the functions C_n defined by (11.3.7) is brought out by the fact that if A is a subinterval of $[-2,2]$ with $P_c(A) > 0$, if $n \geqslant 2$ and m is any integer such that $m \geqslant (\log(2/P_c(A))/\log n) + 2$, then $C_n^m(A) = [-2,2]$; i.e., the mth image of A is the entire space (see [Johnson and Sklar 1976, Th. 7]).

A mixing transformation—indeed any measure-preserving transformation—ψ that is one to one cannot exhibit such extreme behavior, for in this case ψ^{-1} is also measure preserving. Hence for any measurable set A, $P(\psi A) = P(A)$; this in turn implies that $P(\psi^m A) = P(A)$ for all m. Nevertheless, as was shown in [Erber, Schweizer, and Sklar 1973] and [Rice 1978], under iteration, all mixing transformations tend to spread sets out. To discuss this behavior, we begin with

Definition 11.4.1. Let (S,d) be a metric space. For any $A \subseteq S$, the *diameter* of A is the number

$$\operatorname{diam} A = \sup\{d(p,q) \mid p,q \text{ in } A\}. \tag{11.4.1}$$

If, in addition, $(S,\mathcal{Q},P)$ is a probability space and A is in $\mathcal{Q}$, then the *essential diameter* of A is the number

$$\operatorname{ess\,diam} A = \inf\{\operatorname{diam} B \mid B \subseteq A, P(B) = P(A)\}. \tag{11.4.2}$$

We then have the following result, which is Theorem 2 in [Rice 1978] and is a substantial improvement of Theorems 1 and 2 in [Erber, Schweizer, and Sklar 1973]:

Theorem 11.4.2. *Let (S,d) be a metric space and $(S,\mathcal{Q},P)$ a probability space such that every open ball of (S,d) has positive measure. Let ψ be a transformation on S that is mixing with respect to P. If A is in $\mathcal{Q}$ and $P(A) > 0$, then*

$$\lim_{n\to\infty} \operatorname{diam} \psi^n A = \operatorname{diam} S. \tag{11.4.3}$$

If in addition ψB is in $\mathcal{Q}$ whenever B is in $\mathcal{Q}$, then

$$\lim_{n\to\infty} \operatorname{ess\,diam} \psi^n A = \operatorname{diam} S. \tag{11.4.4}$$

Note that the hypotheses of Theorem 11.4.2 are such that $\operatorname{diam} S = \operatorname{ess\,diam} S$.

The mixing condition (11.3.1) shows that any set A of positive measure eventually penetrates every other set of positive measure; and Theorem 11.4.2 shows that A necessarily spreads out in diameter. Thus, even though A may not spread out in "volume" (measure), there is a very definite sense in which A does not remain small.

Theorem 11.4.2 is closely related to a number of results involving various notions of recurrence.

Definition 11.4.3. Let $(S, \mathcal{Q}, P)$ be a probability space and ψ a transformation on S. Then a point p in S is *Poincaré recurrent under* ψ if, for every A in $\mathcal{Q}$ such that p is in A and $P(A) > 0$, there are infinitely many distinct positive integers m such that $\psi^m p$ is in A. Suppose further that for any B in $\mathcal{Q}$, ψB is in $\mathcal{Q}$ and $P(\psi B) \geqslant P(B)$. Then a set A in $\mathcal{Q}$ for which $0 < P(A) < 1$ is *coherently recurrent under* ψ if there is a positive number a such that

$$P(A \cap \psi^m A) \geqslant (P(A) + a)P(\psi^m A) \tag{11.4.5}$$

for infinitely many distinct positive integers m.

A basic result going back to Poincaré (see [Oxtoby 1971, Ch. 17]) is

Theorem 11.4.4. *If $(S, \mathcal{Q}, P)$ is a probability space and ψ is a measure-preserving transformation on S, then almost every point p in S is Poincaré recurrent under ψ.*

If ψ is mixing and one to one, then ψ^{-1} is also mixing on S and (11.3.1) yields

$$\lim_{n \to \infty} P(\psi^n A \cap B) = P(A)P(B) \tag{11.4.6}$$

for all A, B in $\mathcal{Q}$. Since (11.4.6) contradicts (11.4.5), it follows at once that coherent recurrence cannot occur when ψ is mixing and one to one. Indeed, coherent recurrence cannot occur under any mixing transformation; for as Rice [1978] has shown, we have

Theorem 11.4.5. *Let $(S, \mathcal{Q}, P)$ be a probability space and ψ a transformation on S such that for any B in $\mathcal{Q}$, ψB is in $\mathcal{Q}$, and $P(\psi B) \geqslant P(B)$. If ψ is mixing with respect to P, then no set A in $\mathcal{Q}$ with $0 < P(A) < 1$ is coherently recurrent under P.*

These results have physical significance. To paraphrase from [Erber, Schweizer, and Sklar 1973]: Perhaps the most significant implications of Theorems 11.4.2 and 11.4.5 arise in connection with a question that goes back to the very origins of ergodic theory, the famous controversy between E. Zermelo [1896ab] and L. Boltzmann [1896] about the relevance of the conclusion of Theorem 11.4.4. (For further discussion, see [Erber and Sklar 1974]; and for a review of recent developments in the foundation of statistical mechanics, see [Penrose 1979].) Since states of physical systems are often represented as points in appropriate phase spaces, and the transformations these spaces undergo are measure preserving (Liouville's theorem: see [Oxtoby 1971, Ch. 17]), Theorem 11.4.4 holds, with an effect that seems to contradict the observed manifestations of fundamental thermodynamic principles; but to represent a state of a physical system as a point, the state must be known exactly. Since this is effectively never the case, an obviously more realistic model—which reflects the imprecision of measurements—is one in which the initial configuration of a system does not comprise a single state but rather an assembly of experimentally indistinguishable states that form a set of (possibly very small but still) positive diameter and (possibly very small but still) positive measure. If now the physical transformation involved can be represented by a mixing transformation of the phase space (a far-reaching, but in view of (11.3.2) not altogether unnatural assumption), then Theorems 11.4.2 and 11.4.5 apply and yield the fact that the initial set will spread out in diameter and that there will be *no* recurrence. In this sense, Theorems 11.4.2 and 11.4.5 constitute a partial, if belated, vindication of the physical intuition of Ludwig Boltzmann.

11.5. *E*-Processes: The Case of Markov Chains

In transformation-generated spaces we studied the distribution of the distances between points or "particles" as they "move about" in a metric space, subject to a "law of motion" given by a transformation acting on the space. R. Moynihan [1976] considered the same question when the motion of each point is governed, instead, by a Markov chain. It is convenient to present his results by beginning with a more general setting and then specializing.

Definition 11.5.1. Let $(\Omega, \mathcal{A}, P)$ be a probability space, (M, d) a metric space, and Z either the half-line $[0, \infty)$ or the set of nonnegative integers. Let S be a set of functions from $Z \times \Omega$ into M such that the following conditions hold.

i. For every p in S and t in Z the function p_t from Ω into M defined by $p_t(\omega) = p(t, \omega)$ is an M-valued random variable, i.e., is P-measurable with respect to a suitable sigma-field in (M, d).
ii. For all p, q in S, t in Z, and x in R^+ the set

$$D(p,q,t,x) = \{\omega \text{ in } \Omega \mid d(p_t(\omega), q_t(\omega)) < x\} \tag{11.5.1}$$

is in $\mathcal{A}$.
iii. For p, q in S and t in Z, $F_{pq}^{(t)}$ is the function in Δ^+ given by

$$F_{pq}^{(t)}(x) = P(D(p,q,t,x)). \tag{11.5.2}$$

iv. For each t in Z, $\mathcal{F}^{(t)}$ is the function from $S \times S$ into Δ^+ given by

$$\mathcal{F}^{(t)}(p,q) = F_{pq}^{(t)}. \tag{11.5.3}$$

v. $\mathcal{F}^*$ is the function defined on Z by

$$\mathcal{F}^*(t) = \mathcal{F}^{(t)}. \tag{11.5.4}$$

Then $(S, \mathcal{F}^*)$ is an *E-space process* (briefly, an *E-process*) *with base* $(\Omega, \mathcal{Q}, P)$ *and target* (M, d).

Note that for every t in Z, the space $(S, \mathcal{F}^{(t)})$ is a random variable generated E-space.

In the case studied by Moynihan, the target is denumerable and the movement of the points p from one E-space to the next is governed by a set of Markov chain transition probabilities. Moynihan defined his processes directly as Markov chains. Using the regular Markov existence theorem (see [Loève 1977, §38.2]) we achieve the same result by introducing appropriate transition probabilities and initial distributions, as follows.

Definition 11.5.2. Let M be a denumerable set and Z either $[0, \infty)$ or $\{0, 1, 2, \dots\}$. For every i, j in M, let P_{ij} be a function from Z into I such that, for any s, t in Z,

$$\sum_{j \text{ in } M} P_{ij}(t) = P_{ii}(0) = 1, \tag{11.5.5}$$

$$P_{ik}(s+t) = \sum_{j \text{ in } M} P_{ij}(s) P_{jk}(t). \tag{11.5.6}$$

Then $\{P_{ij}\}$ is a family of *Markov transition probabilities on* M.

Definition 11.5.3. An *ME-chain* is an E-process $(S, \mathcal{F}^*)$ with base $(\Omega, \mathcal{Q}, P)$ and target (M, d) such that the following conditions are satisfied:

i. M is denumerable.
ii. For each p in S there is an i in M such that

$$p_0 = i \quad \text{a.s.} \tag{11.5.7}$$

iii. There is a family $\{P_{ij}\}$ of Markov transition probabilities on M such that for every p in S, j in M, and t in Z

$$P\{\omega \mid p_t(\omega) = j\} = P_{ij}(t) \tag{11.5.8}$$

where i is given by (11.5.7).

Each point p in an ME-chain is a Markov chain (see [Chung 1967]). Note that in any ME-chain, once the family $\{P_{ij}\}$ is specified, it becomes unnecessary to refer to $(\Omega, \mathcal{Q}, P)$. We shall therefore speak of an ME-chain with *state space* (M, d) and *Markov matrix* $\{P_{ij}\}$. Furthermore, we have

Theorem 11.5.4. *Let $(S,\mathcal{F}^*)$ be an ME-chain with state space (M,d) and Markov matrix $\{P_{ij}\}$. For any i,j in M and t in Z let the function $G_{ij}^{(t)}$ be defined on R^+ by*

$$G_{ij}^{(t)}(x) = \sum_{k \text{ in } M} P_{ik}(t) \sum_{\substack{l \text{ in } M \\ d(k,l)<x}} P_{jl}(t). \tag{11.5.9}$$

Then each $G_{ij}^{(t)}$ is in $\mathcal{D}^+$, and for distinct p, q in S

$$F_{pq}^{(t)} = G_{ij}^{(t)} \tag{11.5.10}$$

for all t in Z, where $i = p_0$ and $j = q_0$.

Combining (11.5.9) and (11.5.6) yields

$$G_{ij}^{(s+t)}(x) = \sum_{k,l \text{ in } M} P_{ik}(s)P_{jl}(s)G_{kl}^{(t)}(x) \tag{11.5.11}$$

for all i,j in M, s, t in Z, and x in R^+.

Theorem 11.5.5. *Let $(S,\mathcal{F}^*)$ be an ME-chain with state space (M,d) and Markov matrix $\{P_{ij}\}$. Suppose that for every i,j in M there is a function G_{ij} in Δ^+ such that one of the following two conditions holds:*

i. *For all i,j in M*

$$G_{ij}^{(t)} \xrightarrow{w} G_{ij}. \tag{11.5.12}$$

ii. *The set Z is the set of all nonnegative integers (the discrete time parameter case), and for all i,j in M*

$$\frac{1}{n}\sum_{m=1}^{n} G_{ij}^{(m)} \xrightarrow{w} G_{ij}. \tag{11.5.13}$$

Let $\mathcal{F}$ be defined on $S \times S$ by $\mathcal{F}(p,q) = \varepsilon_0$ for $p = q$, while for $p \neq q$

$$\mathcal{F}(p,q) = G_{ij} \tag{11.5.14}$$

where $i = p_0$ and $j = q_0$. Then $(S,\mathcal{F})$ is a PPM *space under τ_W.*

Note that in the discrete time parameter case (11.5.12) implies (11.5.13). It can be shown that under the hypotheses of Theorem 11.5.5, the stationarity equation

$$G_{ij}(x) = \sum_{k,l \text{ in } M} P_{ik}(s)P_{jl}(s)G_{kl}(x) \tag{11.5.15}$$

is satisfied for all i,j in M, s in Z, and x in R^+.

The hypotheses of Theorem 11.5.5 are satisfied in a wide variety of circumstances. In particular, we have

Theorem 11.5.6. *Let* $(S, \mathfrak{F}^*)$ *be a discrete time parameter ME-chain with state space* (M, d) *and Markov matrix* $\{P_{ij}\}$. *Let the Markov matrix represent an irreducible and positive recurrent Markov chain. Then*:

i. *If the period of the chain is* 1, *then condition* i *of Theorem* 11.5.5 *holds. In fact, there is a* G *in* $\mathfrak{D}^+$ *such that* $G_{ij} = G$ *for all* i, j *in* M, *i.e., the limit space* $(S, \mathfrak{F})$ *is equilateral.*
ii. *If the period of the chain is* $b \geqslant 1$, *then condition* ii *of Theorem* 11.5.5 *holds with all* G_{ij} *in* $\mathfrak{D}^+$, *and in the limit space* $(S, \mathfrak{F})$ *the number of distinct distribution functions in* Ran $\mathfrak{F}$ *(including* ε_0*) does not exceed* $(b/2) + 2$; *the bound is sharp.*

It is noteworthy that these results can be extended to many *ME*-chains in which the underlying Markov states are transient. For any *ME*-chain in this class, we have $\lim_{n \to \infty} P_{ii}(n) = 0$ for every i in M, whereas the sequences $\{F_{pq}^{(n)}\}$ have limits in $\mathfrak{D}^+$ for all p, q in S. Thus, although the limits of the transition probabilities are trivial in such a case, the limiting PM space is not.

These and other results can be extended to the case where distinct points in S are governed by differing sets of Markov chain transition probabilities.

11.6. Open Problems

While a PM space generated by a mixing transformation is essentially equilateral (Theorem 11.3.4), spaces generated by arbitrary measure-preserving transformations are more varied. But just how varied? Specifically, we have

Problem 11.6.1. What PM spaces are isometric to transformation-generated spaces?

Corresponding to this problem, there is

Problem 11.6.2. What PM spaces are limits of *ME*-chains or, more generally, of *E*-processes?

In the other direction, we have

Problem 11.6.3. Find necessary and sufficient conditions on the state space and Markov matrix of an *ME*-chain for the chain to have a nontrivial PPM space limit.

A related problem is

Problem 11.6.4. Under what conditions on the state space and Markov matrix of an *ME*-chain $(S, \mathfrak{F}^*)$ does the stationarity equation (11.5.15) admit solutions G_{ij} distinct from ε_∞? If there are such nontrivial solutions G_{ij}, and $\mathfrak{F}$ is defined via (11.5.14), is the pair $(S, \mathfrak{F})$ necessarily the limit PPM space of the *ME*-chain?

Even though a large set cannot have a small diameter and a set with a large diameter cannot be truly small, it is still true that the diameter of a set need not be a good measure of its size and shape. A better measure is furnished by the *transfinite diameter*. For a compact subset E of the complex plane this is the number $\rho(E) = \lim_{n\to\infty} \rho_n(E)$, where

$$\rho_n(E) = \sup \prod_{i<j} |z_i - z_j|^{2/n(n-1)} \tag{11.6.1}$$

and the supremum is over all n-tuples $(z_1, \ldots, z_n)$ of points in E (see [Hille 1962, §16.2]); and for a compact subset of a metric space there is a similar definition (see [Hille 1965]). Comparison of (11.6.1) with (11.3.5) and the proof of Theorem 11.4.2 quickly leads to the following conjecture:

Problem 11.6.5. Does Theorem 11.4.2 remain valid when "diameter" is replaced by "transfinite diameter"?

The transfinite diameter of a compact subset A of a metric space is closely related, and often equal, to the *capacity* of A (see [Hille 1965; Landkof 1972]). Thus we may also pose

Problem 11.6.6. Does Theorem 11.4.2 remain valid when "diameter" is replaced by "capacity"?

Theorem 11.4.2 also motivates the following

Problem 11.6.7. Prove or disprove: If ψ is continuous and mixing with respect to P on S, and if A is an arc in S, then $\lim_{n\to\infty} \psi^n(A)$ is a space-filling curve in S (cf. [Holbrook 1981]).

12

The Strong Topology

12.1. The Strong Topology and Strong Uniformity

Many different topological structures may be defined on a PM space. The one that has received the most attention to date is the strong topology. It is the principal topic of this chapter.

Definition 12.1.1. Let $(S, \mathcal{F})$ be a PSM space. For p in S and $t > 0$, the *strong t-neighborhood of p* is the set

$$N_p(t) = \{q \mid F_{pq}(t) > 1 - t\}. \tag{12.1.1}$$

The *strong neighborhood system at p* is the collection $\mathfrak{N}_p = \{N_p(t) \mid t > 0\}$, and the *strong neighborhood system for S* is the union $\mathfrak{N} = \bigcup_{p \text{ in } S} \mathfrak{N}_p$.

Note that if $t > 1$, then $N_p(t) = S$.

It is immediate that $\mathfrak{N}$ is in fact a neighborhood system for S, i.e., satisfies N1 of Section 3.5; and since $\mathfrak{N}_p$ and the collection of neighborhoods $\{N_p(1/n) \mid n = 1, 2, \dots\}$ are equivalent at every point p in S, $\mathfrak{N}$ is first-countable. Furthermore, since $N_p(t_1) \subseteq N_p(t_2)$ whenever $t_1 \leqslant t_2$, $\mathfrak{N}$ satisfies N2 as well. Thus $(S, \mathfrak{N})$ is a V_D-space. Comparing (12.1.1) with (4.3.4), we see that q is in $N_p(t)$ if and only if $d_L(F_{pq}, \varepsilon_0) < t$, whence

$$N_p(t) = \{q \mid d_L(F_{pq}, \varepsilon_0) < t\}. \tag{12.1.2}$$

Thus the function δ defined on $S \times S$ by

$$\delta(p, q) = d_L(F_{pq}, \varepsilon_0) \tag{12.1.3}$$

is a semimetric on S, and a strong neighborhood in $(S, \mathfrak{F})$ is an open ball in (S, δ). In general, δ is not a metric on S. However, if $(S, \mathfrak{F})$ is a PM space under τ, then in view of Lemma 4.3.4 and the triangle inequality (8.1.4), we do have

$$d_L(F_{pr}, \varepsilon_0) \leqslant d_L(\tau(F_{pq}, F_{qr}), \varepsilon_0) \tag{12.1.4}$$

for any p, q, r in S. The inequality (12.1.4) can be regarded as a weak form of triangle inequality for δ. It plays a decisive role in the proof of

Theorem 12.1.2. *Let $(S, \mathfrak{F}, \tau)$ be a* PM *space. If τ is continuous, then the strong neighborhood system $\mathfrak{N}$ satisfies* N3 *and* N4, *and thus determines a Hausdorff topology for S.*

PROOF. Let q be in $N_p(t)$ and let $d_L(F_{pq}, \varepsilon_0) = \alpha$. Then $t - \alpha > 0$ and the uniform continuity of τ implies that there exists a $t' > 0$ such that $d_L(\tau(F_{pq}, G), F_{pq}) < t - \alpha$ whenever $d_L(G, \varepsilon_0) < t'$. Now let r be in $N_q(t')$. Then $d_L(F_{qr}, \varepsilon_0) < t'$, and using (12.1.4) we have

$$\begin{aligned} d_L(F_{pr}, \varepsilon_0) &\leqslant d_L(\tau(F_{pq}, F_{qr}), \varepsilon_0) \\ &\leqslant d_L(\tau(F_{pq}, F_{qr}), F_{pq}) + d_L(F_{pq}, \varepsilon_0) \\ &< t - \alpha + \alpha = t. \end{aligned}$$

Hence r is in $N_p(t)$. Thus $N_q(t') \subseteq N_p(t)$ and N3 holds.

To prove N4, note that since $p \neq r$, $F_{pr} \neq \varepsilon_0$, whence $\eta = d_L(F_{pr}, \varepsilon_0) > 0$. By the uniform continuity of τ, there exists a $t > 0$ such that

$$d_L(\tau(G_1, G_2), \varepsilon_0) < \eta \tag{12.1.5}$$

whenever $d_L(G_1, \varepsilon_0) < t$ and $d_L(G_2, \varepsilon_0) < t$. Suppose q is in $N_p(t) \cap N_r(t)$. Then $d_L(F_{pq}, \varepsilon_0) < t$ and $d_L(F_{qr}, \varepsilon_0) < t$, whence (12.1.4) and (12.1.5) yield

$$d_L(F_{pr}, \varepsilon_0) \leqslant d_L(\tau(F_{pq}, F_{qr}), \varepsilon_0) < \eta = d_L(F_{pr}, \varepsilon_0),$$

an impossibility. Thus $N_p(t) \cap N_r(t)$ is empty and the proof is complete. □

Note that to prove N3 we only need *continuity of τ on the boundary*, i.e., that $\tau(F, G) \xrightarrow{w} F$ as $G \xrightarrow{w} \varepsilon_0$; and to prove N4 we only need continuity of τ at $(\varepsilon_0, \varepsilon_0)$, i.e., that $\tau(G, G) \xrightarrow{w} \varepsilon_0$ as $G \xrightarrow{w} \varepsilon_0$. If τ is not continuous at $(\varepsilon_0, \varepsilon_0)$ then, as Muštari and Šerstnev [1966] have shown, the system of strong neighborhoods may fail to satisfy N3.

Definition 12.1.3. The Hausdorff topology determined by the strong neighborhood system $\mathfrak{N}$ is the *strong topology* for S.

Here and in the sequel, when we speak about "the strong topology" for a PM *space* $(S, \mathfrak{F}, \tau)$, *we always assume that* τ *is continuous.*

It follows from Lemma 3.5.14 that under the hypotheses of Theorem 12.1.2, the strong neighborhood system $\mathfrak{N}$ determines a Kuratowski closure operation. We call this operation the *strong closure*; and, for any subset A of S, denote the strong closure of A by $k(A)$. It follows from Definitions 3.5.4 and 3.5.5, Lemma 4.3.3, and (12.1.2) that for any nonempty subset A of S, $k(A)$ is the set of all p in S such that, for any $t > 0$, there is a q in A such that

$$F_{pq}(t) > 1 - t. \tag{12.1.6}$$

Now for any p in S and any nonempty subset A of S, let F_{pA} be the function defined on R by

$$F_{pA}(x) = \sup\{F_{pq}(x) \mid q \text{ in } A\}. \tag{12.1.7}$$

It is readily verified that F_{pA} is a d.d.f. We call F_{pA} the *probabilistic distance from the point* p *to the set* A. With this notation, (12.1.6) is equivalent to

$$k(A) = \{p \text{ in } S \mid F_{pA} = \varepsilon_0\}. \tag{12.1.8}$$

Our next aim is to show that the strong topology for a PM space is metrizable. We begin with

Definition 12.1.4. Let $(S, \mathfrak{F})$ be a PSM space. Then for any $t > 0$ the *strong t-vicinity* is the subset $\mathfrak{U}(t)$ of $S \times S$ given by

$$\mathfrak{U}(t) = \{(p, q) \mid F_{pq}(t) > 1 - t\}, \tag{12.1.9}$$

and the *strong vicinity system for S* is the union

$$\mathfrak{U} = \bigcup_{t>0} \mathfrak{U}(t). \tag{12.1.10}$$

Theorem 12.1.5. *Let* $(S, \mathfrak{F})$ *be a* PSM *space, and* $\mathfrak{U}$ *the strong vicinity system for* S. *Then* $\mathfrak{U}$ *satisfies the following conditions*:

U1. *For any* $t > 0$, *the diagonal set* $\{(p, p) \mid p \text{ in } S\}$ *is a subset of* $\mathfrak{U}(t)$.

U2. *For any* $t > 0$, $\mathfrak{U}(t) = \mathfrak{U}^{-1}(t)$, *where*

$$\mathfrak{U}^{-1}(t) = \{(q, p) \mid (p, q) \text{ in } \mathfrak{U}(t)\}.$$

U3. *If* $t_1 \leqslant t_2$, *then* $\mathfrak{U}(t_1) \subseteq \mathfrak{U}(t_2)$, *whence the intersection of two strong vicinities contains (indeed is) a strong vicinity.*

If in addition $(S, \mathfrak{F})$ *is a* PM *space under a continuous triangle function* τ, *then* $\mathfrak{U}$ *also satisfies the following conditions*:

U4. *For any $t > 0$, there is an $\eta > 0$ such that*

$$\mathfrak{U}(\eta) \circ \mathfrak{U}(\eta) \subseteq \mathfrak{U}(t)$$

where $\mathfrak{U}(\eta) \circ \mathfrak{U}(\eta)$ is the set

$$\{(p,r) \mid \textit{for some } q, (p,q) \textit{ and } (q,r) \textit{ are in } \mathfrak{U}(\eta)\}.$$

U5. $\bigcap_{t>0} \mathfrak{U}(t) = \{(p,p) \mid p \textit{ in } S\}$.

PROOF. Conditions U1, U2, and U3 follow directly from Definition 12.1.4. As for U4, given that $t > 0$, choose η so that (12.1.5) holds. Then for any p, q such that (p,q) and (q,r) are in $\mathfrak{U}(\eta)$, it follows that (p,r), which is in $\mathfrak{U}(\eta) \circ \mathfrak{U}(\eta)$, is also in $\mathfrak{U}(t)$. This proves U4. As for U5, suppose that $p \neq q$. Then $d_L(F_{pq}, \varepsilon_0) > 0$, whence (p,q) is not in $\mathfrak{U}(d_L(F_{pq}, \varepsilon_0))$ and so not in $\cap \mathfrak{U}(t)$. The rest follows from U1. □

In the usual topological terminology, Theorem 12.1.5 states that in a PM space $(S, \mathfrak{F}, \tau)$ where τ is continuous, the family $\mathfrak{U}$ is a *base* for a *Hausdorff uniformity for S* (see [Kelley 1955, pp. 176–180]). This uniformity is called the *strong uniformity*. Since the collection of vicinities $\{\mathfrak{U}(1/n \mid n = 1, 2, \ldots\}$ is a countable base for $\mathfrak{U}$, a standard theorem (see [Kelley 1955, p. 186]) yields the following result, which was originally established for Menger spaces in [Schweizer, Sklar, and Thorp 1960]:

Theorem 12.1.6. *If $(S, \mathfrak{F}, \tau)$ is a PM space and τ is continuous, then the strong uniformity, or equivalently, the strong topology, is metrizable.*

We have already seen that by using means, medians, c-tiles, conjugate transforms, etc., metrics can often be extracted from probabilistic metrics. Theorem 12.1.6 shows that this situation is the rule rather than the exception.

A sequence $\{p_n\}$ in S *converges strongly to* a point p in S, and we write $p_n \to p$ or $\lim p_n = p$, if for any $t > 0$ there is an integer m such that p_n is in $N_p(t)$ whenever $n \geqslant m$. Similarly, a sequence $\{p_n\}$ in S is a *strong Cauchy sequence* if for any $t > 0$ there is an integer i such that (p_m, p_n) is in $\mathfrak{U}(t)$ whenever m, $n \geqslant i$. Since the strong topology is first-countable and Hausdorff, it can be completely specified in terms of the strong convergence of sequences. An immediate consequence of (12.1.2) and (12.1.3) is

Theorem 12.1.7. *Let $\{p_n\}$ be a sequence in S. Then*

$$p_n \to p \quad \text{iff} \quad d_L(F_{p_n p}, \varepsilon_0) \to 0 \quad \text{iff} \quad \delta(p_n, p) \to 0. \tag{12.1.11}$$

Similarly, $\{p_n\}$ *is a strong Cauchy sequence if and only if*

$$\lim_{m,n\to\infty} d_L(F_{p_m p_n},\varepsilon_0)=0, \tag{12.1.12}$$

which in turn is equivalent to saying that $\{p_n\}$ *is a Cauchy sequence in the semimetric space* (S,δ).

In many cases, strong convergence is equivalent to convergence in various naturally defined metrics. Thus, for example, Theorems 7.8.12 and 8.2.8 yield

Theorem 12.1.8. *Let* T *be a continuous Archimedean t-norm. If* $(S,\mathcal{F},\tau_T)$ *is a Menger space,* $\{p_n\}$ *a sequence in* S, *and* $x>0$, *then* $p_n\to p$ *if and only if* $d_{T,x}(p_n,p)\to 0$.

In [S² 1960] we introduced a two-parameter family of neighborhoods of a point p in S by defining

$$N_p(\varepsilon,\lambda)=\{q \text{ in } S \mid F_{pq}(\varepsilon)>1-\lambda\} \tag{12.1.13}$$

for $\varepsilon>0$ and $\lambda>0$; and in subsequent papers we and others worked with these ε,λ-*neighborhoods*. Since $N_p(t,t)=N_p(t)$ for $t>0$, and $N_p(\operatorname{Min}(\varepsilon,\lambda))\subseteq N_p(\varepsilon,\lambda)$ for any $\varepsilon,\lambda>0$, the strong neighborhood system is equivalent to the system of ε,λ-neighborhoods. Hence the strong topology coincides with the ε,λ-*topology* generated by the system of ε,λ-neighborhoods.

Similarly, the strong vicinity system is equivalent to the ε,λ-*vicinity* system defined by

$$\mathcal{U}(\varepsilon,\lambda)=\{(p,q)\mid F_{pq}(\varepsilon)>1-\lambda\} \tag{12.1.14}$$

for $\varepsilon,\lambda>0$.

12.2. Uniform Continuity of the Distance Function

It is a basic fact that in a metric space (S,d) the distance function d is a uniformly continuous mapping from $S\times S$ into R^+. This result is an immediate consequence of the triangle inequality (see Theorem 3.1.2 and the discussion that follows it). Our next objective is to establish the corresponding result for PM spaces. Since (Δ^+,τ) is a semigroup rather than a group, the subtractions used in going from (3.1.10) and (3.1.11) to (3.1.12) have to be circumvented. We achieve this with the aid of the following lemma, which is the analog of the fact that if $a\leqslant b+c$ and $b\leqslant a+c$, then $|a-b|$ is small whenever c is small.

Lemma 12.2.1. *Let τ be a continuous triangle function and $\mathcal{S}$ the set of all triples (F,G,H) in $(\Delta^+)^3$ such that*

$$F \geqslant \tau(H,G) \quad \text{and} \quad G \geqslant \tau(H,F). \tag{12.2.1}$$

Then for every $h>0$ there is an $\eta>0$ such that if (F,G,H) is in $\mathcal{S}$ and $d_L(H,\varepsilon_0)<\eta$, then $d_L(F,G)<h$.

PROOF. Let $h>0$ be given. Since τ is uniformly continuous, there is an $\eta>0$ such that $d_L(\tau(H,G),G)<h$ and $d_L(\tau(H,F),F)<h$ whenever $d_L(H,\varepsilon_0)<\eta$. Hence, using (4.3.1) and (4.3.2), for any x in $(0,1/h)$ we have

$$F(x+h)+h \geqslant \tau(H,G)(x+h)+h \geqslant G(x),$$

$$G(x+h)+h \geqslant \tau(H,F)(x+h)+h \geqslant F(x).$$

Thus $d_L(F,G)<h$ and the lemma is proved. □

Theorem 12.2.2. *Suppose the hypotheses of Theorem 12.1.2 are satisfied. Let S be endowed with the strong topology and $S\times S$ with the corresponding product topology. Then $\mathcal{F}$ is a uniformly continuous mapping from $S\times S$ into Δ^+.*

PROOF. For any two pairs (p,q) and (p',q') in $S\times S$ we have

$$F_{p'q'} \geqslant \tau(F_{p'q},F_{qq'}) \geqslant \tau(\tau(F_{p'p},F_{pq}),F_{qq'})$$
$$= \tau(\tau(F_{pp'},F_{qq'}),F_{pq}),$$

and similarly,

$$F_{pq} \geqslant \tau(\tau(F_{pp'},F_{qq'}),F_{p'q'}).$$

Thus, from (12.1.5) and Lemma 12.2.1, it follows that for any $h>0$ there is an $\eta>0$ such that $d_L(F_{pq},F_{p'q'})<h$ whenever p' is in $N_p(\eta)$ and q' is in $N_q(\eta)$. □

Theorem 12.2.2 extends the principal result of [Schweizer 1966] from Menger spaces to general PM spaces.

As direct consequences of Theorem 12.2.2 we have

Theorem 12.2.3. *If $\{p_n\}$ and $\{q_n\}$ are sequences such that $p_n\to p$ and $q_n\to q$, then $d_L(F_{p_nq_n},F_{pq})\to 0$, i.e., $F_{p_nq_n}\xrightarrow{w}F_{pq}$.*

Corollary 12.2.4. *Every strongly convergent sequence in S is a strong Cauchy sequence.*

Corollary 12.2.5. *If $\{p_n\}$ and $\{q_n\}$ are strong Cauchy sequences in S, then $\{F_{p_n q_n}\}$ is a Cauchy sequence in (Δ^+, d_L).*

To compare the strong topology with other topologies, we begin with

Lemma 12.2.6. *Let $(S, \mathcal{F})$ be a* PSM *space and $\mathfrak{N}$ the strong neighborhood system for S. Then each set in $\mathfrak{N}$ is open relative to any topology for S with respect to which the functions $\mathcal{F}_p$ from S into Δ^+ given by $\mathcal{F}_p(q) = \mathcal{F}(p,q)$ are continuous.*

PROOF. Let $B(\varepsilon_0, t) = \{F \text{ in } \Delta^+ \mid d_L(F, \varepsilon_0) < t\}$ be the open ball in (Δ^+, d_L) with center ε_0 and radius $t > 0$. Then for any p in S, we have

$$\begin{aligned} N_p(t) &= \{q \text{ in } S \mid d_L(F_{pq}, \varepsilon_0) < t\} \\ &= \{q \text{ in } S \mid \mathcal{F}_p(q) \text{ is in } B(\varepsilon_0, t)\} \\ &= \mathcal{F}_p^{-1}(B(\varepsilon_0, t)). \end{aligned}$$

Since $B(\varepsilon_0, t)$ is open relative to the metric topology for (Δ^+, d_L), the conclusion of the lemma follows from the definition of a continuous function (see [Kelley 1955, p. 85]). □

It follows that whenever $\mathfrak{N}$ determines a topology $\mathfrak{T}$ for S, and $\mathfrak{T}'$ is any topology with respect to which all functions $\mathcal{F}_p$ are continuous, then $\mathfrak{T}$ is *coarser* than $\mathfrak{T}'$, i.e., $\mathfrak{T} \subseteq \mathfrak{T}'$. The inclusion can be proper: Muštari and Šerstnev [1966] have constructed an example of a PM space $(S, \mathcal{F}, \tau)$ in which τ is not continuous, the strong neighborhood system $\mathfrak{N}$ for S satisfies N3, and the topology determined by $\mathfrak{N}$ is strictly coarser than any topology with respect to which all the functions $\mathcal{F}_p$ are continuous. On the other hand, when τ is continuous, an immediate consequence of Theorem 12.2.2 and Lemma 12.2.3 is

Theorem 12.2.7. *If $(S, \mathcal{F})$ is a* PM *space under a continuous τ, then the strong topology is the coarsest topology for S with respect to which all the functions $\mathcal{F}_p$ are continuous.*

We conclude this section by noting that, given any PSM space $(S, \mathcal{F})$, I. Kurosaki [1964], using quasi-inverses, defined and studied a neighborhood system that is effectively equivalent to the system of neighborhoods determined by the sets

$$K_p(t) = \{q \text{ in } S \mid d_L(F_{pr}, F_{qr}) < t \text{ for all } r \text{ in } S\}. \qquad (12.2.2)$$

This neighborhood system determines a Hausdorff topology on S and the mapping $\mathcal{F}$ is uniformly continuous with respect to this topology. Letting $r = q$ in (12.2.2) yields $d_L(F_{pq}, \varepsilon_0) < t$, whence $K_p(t) \subseteq N_p(t)$, so that $\mathfrak{N}$ is coarser than Kurosaki's neighborhood system. In the other direction, if $(S, \mathcal{F})$ is a PM space under a continuous τ, then it follows from Lemma 12.2.1 that every neighborhood $K_p(t)$ contains a strong neighborhood, whence in this case, the topology determined by Kurosaki's neighborhoods and the strong topology are equivalent.

Similar techniques were also employed by E. Nishiura [1970] to introduce and study topological and uniform structures on PSM and PM spaces.

12.3. Examples

If p is a point of a PM space—or just a PSM space—and if there is a number $\eta > 0$ such that $F_{pq}(\eta) \leqslant 1 - \eta$ for all $q \neq p$, then $N_p(\eta) = \{p\}$ and p is an *isolated point*. Hence if for every p there is a Euclidean neighborhood of the point (0, 1) (in the plane) whose intersection with the graph of F_{pq} is empty for all q distinct from p, then the strong neighborhood system determines the discrete topology. This criterion is useful. It immediately yields the fact that the strong topology of an equilateral PM space is discrete. It also yields the following result, whose proof, given in [S² 1961b], is an amusing application of Čebyšev's inequality.

Theorem 12.3.1. *Let $(S, \mathcal{F}, \tau)$ be a PM space. Let p in S be such that for all $q \neq p$ both the mean m_{pq} and the variance σ_{pq}^2 of F_{pq} are finite. Suppose further that there exist numbers $a > 0$ and $b > 1$ such that for all $q \neq p$*

$$m_{pq} - b\sigma_{pq} > a. \tag{12.3.1}$$

Then p is an isolated point of S in the strong topology. If (12.3.1) *holds for all distinct pairs p, q in S, then the strong topology is the discrete topology.*

Turning to simple spaces, we have

Theorem 12.3.2. *In the simple space (S, d, G), the strong topology is equivalent to the d-metric topology.*

PROOF. We shall show that the strong neighborhood system is equivalent to the system of spherical neighborhoods $\{B(p,t)\}$ in (S, d). First, if $N_p(t)$ is given, choose $\eta > 0$ so that $G(t/\eta) > 1 - t$, and let q be in $B(p, \eta)$. Then $0 \leqslant d(p,q) < \eta$, whence

$$F_{pq}(t) = G(t/d(p,q)) \geqslant G(t/\eta) > 1 - t$$

and q is in $N_p(t)$. Hence $B(p, \eta) \subseteq N_p(t)$.

In the other direction, let $B(p, \eta)$ be given. Since $G \neq \varepsilon_0$, there is an x

in $[0, 1)$ such that $G^\wedge(x) > 0$, and consequently a $t > 0$ such that $t/G^\wedge(1-t) < \eta$. Let q be in $N_p(t)$. Then $F_{pq}(t) > 1 - t$ and, using (4.4.12), we have

$$F_{pq}^\wedge(1-t) = d(p,q)G^\wedge(1-t) < t,$$

whence $d(p,q) < \eta$ and q is in $B(p,\eta)$. Hence $N_p(t) \subseteq B(p,\eta)$. □

In Section 8.4 we noted that every simple space is a Menger space under M, but not conversely. If $(S, \mathfrak{F})$ is a Menger space under M and if, for any c in $(0, 1)$, d_c is the pseudometric given by (8.2.5), then the strong topology is finer than the topology determined by d_c and equivalent to the topology determined by all the pseudometrics d_c taken together. The following example, due to C. Alsina, shows that the strong topology need not be equivalent to the topology determined by any finite collection of the pseudometrics d_c.

Let $(S, \mathfrak{F}, \tau_M)$ be the space for which $S = I$ and $\mathfrak{F}$ is defined via (8.4.5). Then for any t in $(0, 1)$ the strong t-neighborhood of 1 is given by

$$\begin{aligned} N_1(t) &= \{q \mid F_{1q}(t) > 1 - t\} = \{q \mid F_{1q}^\wedge(1-t) < t\} \\ &= [1-t, 1], \end{aligned}$$

and the d_c-pseudometric t-neighborhood of 1 is given by

$$N_1^c(t) = \{q \mid F_{1q}^\wedge(c) < t\} = [c, 1].$$

Now suppose that the strong topology is equivalent to the topology determined by the metrics $d_{c_1}, \ldots, d_{c_n}$. Choose $0 < \eta < \mathrm{Min}(1 - c_1, \ldots, 1 - c_n)$ and consider the neighborhood $N_1(\eta)$. By hypothesis, there exists a $t > 0$ such that for some k, $N_1^{c_k}(t) \subseteq N_1(\eta)$. Consequently, $[c_k, 1] \subseteq [1 - \eta, 1]$, i.e., $\eta \geqslant 1 - c_k$, which is a contradiction.

In any α-simple space $(S, d, G; \alpha)$, the strong neighborhood system and the d-metric topology are also equivalent. This follows from Lemma 4.4.5, the fact that $F_{pq}^\wedge(c) = (d(p,q))^\alpha G^\wedge(c)$ for all c in I, and the existence of a c in $(0, 1)$ such that $G^\wedge(c) > 0$. Furthermore, we have

Theorem 12.3.3. *Let $(S, \mathfrak{F}, \tau_T)$ be a Menger space under a strict t-norm T. Then for any $\alpha > 1$ there exists an α-simple space $(S, d, G; \alpha)$ such that the following conditions hold.*

i. *$(S, d, G; \alpha)$ is a Menger space under T.*
ii. *The strong topologies of the spaces $(S, \mathfrak{F}, \tau_T)$ and $(S, d, G; \alpha)$ are equivalent.*

PROOF. Since T is continuous, τ_T is also continuous. Thus by Theorem 12.1.5, the strong topology of $(S, \mathfrak{F}, \tau_T)$ is metrizable. Let d denote one such metric. Let g be an additive generator of T. Then for any $\alpha > 1$, (S, d) and the function G defined by (8.6.5) generate an α-simple space $(S, d, G; \alpha)$. By Theorem 8.6.5, this space is a Menger space under T. Furthermore, the strong topology of $(S, d, G; \alpha)$ is equivalent to the metric topology of (S, d), which in turn is equivalent to the strong topology of $(S, \mathcal{F}, \tau_T)$. □

Note that the relation between $(S, \mathcal{F}, \tau_T)$ and $(S, d, G; \alpha)$ is more than a topological equivalence: The spaces share the same underlying set S and are Menger spaces under the same t-norm T.

If $(S, \mathcal{F})$ is an E-space with base $(\Omega, \mathcal{Q}, P)$ and target (M, d), then q is in $N_p(t)$ if and only if

$$F_{pq}(t) = P\{\omega \mid d(p(\omega), q(\omega)) < t\} > 1 - t.$$

Thus if $(S, \mathcal{F})$ is a random variable generated E-space—so that the functions $p, q, \ldots,$ in S are all P-measurable—then the strong topology is equivalent to the topology of convergence in probability on $(\Omega, \mathcal{Q}, P)$. Similar remarks apply to pseudometrically generated PM spaces and random metric spaces.

If p is a nonsingular point of a C-space over R^n, then the inequality (10.3.16), coupled with the discreteness criterion discussed at the beginning of this section, implies that p is an isolated point. Thus if all the singular points of the C-space are isolated points of R^n, then the strong neighborhood system determines the discrete topology. This result is in harmony with the fact that the Fréchet-Minkowski metrics associated with a semi-homogeneous C-space over R^n also determine the discrete topology.

Finally, let $(S, \mathcal{F}, \tau)$ be a PM space, and $\mathfrak{M}$ the set of all functions from a given set A into S. For any f, g in $\mathfrak{M}$, let

$$\delta(f, g) = l^-(\inf\{\mathcal{F}(f(x), g(x)) \mid x \text{ in } A\}).$$

Then $(\mathfrak{M}, \delta, \tau)$ is a PM space.

12.4. The Probabilistic Diameter

Throughout the rest of this chapter, all triangle functions are assumed to be continuous.

The diameter of a nonempty set in a metric space was defined in (11.4.1). R. Egbert [1968] introduced a probabilistic generalization of this notion as follows.

Definition 12.4.1. Let $(S, \mathcal{F}, \tau)$ be a PM space and A a nonempty subset of S. The *probabilistic diameter* of A is the function D_A defined on R^+ by $D_A(\infty) = 1$ and

$$D_A = l^- \varphi_A \quad \text{on } [0, \infty) \tag{12.4.1}$$

where

$$\varphi_A(x) = \inf\{F_{pq}(x) \mid p, q \text{ in } A\}. \tag{12.4.2}$$

It is immediate that D_A is in Δ^+ for any $A \subseteq S$.

Theorem 12.4.2. *The probabilistic diameter D_A has the following properties:*

i. $D_A = \varepsilon_0$ *iff A is a singleton set.*
ii. *If $A \subseteq B$, then $D_A \geqslant D_B$.*
iii. *For any p, q in A, $F_{pq} \geqslant D_A$.*
iv. *If $A = \{p, q\}$, then $D_A = F_{pq}$.*
v. *If $A \cap B$ is nonempty, then $D_{A \cup B} \geqslant \tau(D_A, D_B)$.*
vi. $D_A = D_{k(A)}$, *where $k(A)$ is the strong closure of A.*

PROOF. Properties i–iv are immediate; and property v is straightforward. To prove property vi, note that since $A \subseteq k(A)$, $D_A \geqslant D_{k(A)}$. Next, let $h > 0$ be given and let p', q' be in $k(A)$. Since $\mathcal{F}$ is uniformly continuous, there exist p, q in A such that $d_L(F_{pq}, F_{p'q'}) < h$. Thus, using (4.3.2) and property iii, for any x in $(0, 1/h)$ we have

$$F_{p'q'}(x+h) + h \geqslant F_{pq}(x) \geqslant D_A(x).$$

Hence, using (12.4.2) and (12.4.1), we obtain $D_{k(A)}(x+h) + h \geqslant D_A(x)$. Similarly, $D_A(x+h) + h \geqslant D_{k(A)}(x)$. Thus $d_L(D_A, D_{k(A)}) \leqslant h$ for any $h > 0$, whence $D_A = D_{k(A)}$. □

The diameter of a nonempty set A in a metric space is either finite or infinite; accordingly, A is either bounded or unbounded. In a PM space, on the other hand, there are three distinct possibilities. These are captured in

Definition 12.4.3. A nonempty set A in a PM space is

i. *bounded* if $l^- D_A(\infty) = 1$, i.e., if D_A is in $\mathcal{D}^+$;
ii. *semibounded* if $0 < l^- D_A(\infty) < 1$;
iii. *unbounded* if $l^- D_A(\infty) = 0$, i.e., if $D_A = \varepsilon_\infty$.

In a metric space a spherical neighborhood is a bounded set. Conversely, a set is bounded if and only if it is contained in such a neighborhood, or equivalently, if and only if its diameter is finite. In a PM space, these two characterizations of boundedness no longer coincide. To see this, let $S = \{p_n \mid n = 0, 1, 2, \ldots\}$ and define $\mathcal{F}$ via

$$F_{p_m p_n} = \tfrac{1}{2}(\varepsilon_0 + \varepsilon_{|n-m|}). \tag{12.4.3}$$

Then $F^{\wedge}_{p_0p_{n+m}} = F^{\wedge}_{p_0p_n} + F^{\wedge}_{p_0p_m}$, whence $(S, \mathfrak{F})$ is a Menger space under M. But $N_{p_0}(\frac{3}{4}) = S$ and $D_S = \frac{1}{2}(\varepsilon_0 + \varepsilon_\infty)$, whence $N_{p_0}(\frac{3}{4})$ is only semibounded.

Finally, we note that G. Bocşan and G. Constantin [1973, 1974] have used the probabilistic diameter to define a PM space analog of Kuratowski's measure of noncompactness.

12.5. Completion of Probabilistic Metric Spaces

A PM space $(S, \mathfrak{F})$ is *complete* in the strong topology if every strong Cauchy sequence in S is strongly convergent to a point in S; and a complete PM space $(S^*, \mathfrak{F}^*)$ is a *completion* of $(S, \mathfrak{F})$ if $(S, \mathfrak{F})$ is isometric to a dense subset of $(S^*, \mathfrak{F}^*)$. These definitions are the counterparts of those for metric spaces.

It is well known that every metric space has a completion that is unique up to isometry. The proof of this fact depends strongly on the completeness of the metric space of real numbers, the uniform continuity of the distance function, and the continuity of the binary operation of addition. In view of Theorem 12.2.2 and the fact that the metric space (Δ^+, d_L) is complete, the following result is to be expected.

Theorem 12.5.1. *Every* PM *space* $(S, \mathfrak{F}, \tau)$ *with a continuous triangle function* τ *has a completion* $(S^*, \mathfrak{F}^*, \tau)$ *that is unique up to isometry.*

H. Sherwood [1966] established this result for Menger spaces with a continuous t-norm, using the Lévy metric rather than d_L; and in [Sherwood 1971] he pointed out that with some minor modifications the same argument can be used to establish Theorem 12.5.1. His proof follows the lines of the standard metric space argument, with some changes required by the fact that one cannot subtract in the semigroup (Δ^+, τ) (see Lemma 12.2.1). We omit details.

In addition, Sherwood [1971] also established the following:

Theorem 12.5.2. *If* $(S, \mathfrak{F}, \tau)$ *is a complete* PM *space and* $\mathfrak{U}$ *is the strong uniformity on* S, *then* $(S, \mathfrak{U})$ *is a complete uniform space. If* d *is a metric that metrizes* $(S, \mathfrak{F}, \tau)$, *then* (S, d) *is a complete metric space.*

Theorem 12.5.3. *Let* $(S, \mathfrak{F}, \tau)$ *be a* PM *space and let* $(S^*, \mathfrak{F}^*, \tau)$ *be its completion. If* $\mathfrak{U}$ *is the strong uniformity on* S *and* $\mathfrak{U}^*$ *is the strong uniformity on* S^*, *then* $(S^*, \mathfrak{U}^*)$ *is the completion of* $(S, \mathfrak{U})$. *If* d *is a metric on* S *that metrizes* $(S, \mathfrak{F}, \tau)$ *and if* (S', d') *is the completion of* (S, d), *then there is a metric* d^* *on* S^* *that metrizes* $(S^*, \mathfrak{F}^*, \tau)$ *and is such that* (S^*, d^*) *and* (S', d') *are isometric.*

Furthermore, the analogs of Cantor's nested sets theorem and the Baire category theorem are valid in complete PM spaces.

The next lemma is an immediate consequence of Lemma 4.4.5 and Theorem 8.2.3 and the discussion following it.

Lemma 12.5.4. *Let $(S, \mathfrak{F}, \tau)$ be a* PM *space, and for any c in* (0, 1), *let d_c be defined by* (8.2.5). *Then $\{p_n\}$ strongly converges to p, or is a strong Cauchy sequence, if and only if $d_c(p_n, p) \to 0$, or $d_c(p_n, p_m) \to 0$, for each c in* (0, 1).

It does not follow from Lemma 12.5.4 that each of the spaces (S, d_c) is complete whenever $(S, \mathfrak{F}, \tau)$ is complete. This is illustrated by the following example:

Let S be the set of positive integers, let t be a fixed number in (0, 1), and let $\mathfrak{F}$ be defined via

$$F_{mn}(x) = \begin{cases} 0, & 0 < x \leqslant 1/\mathrm{Min}(m,n), \\ t, & 1/\mathrm{Min}(m,n) < x \leqslant 1, \\ 1, & 1 < x. \end{cases}$$

Then for any distinct m, n in S,

$$d_c(m,n) = \begin{cases} 1/\mathrm{Min}(m,n), & 0 < c \leqslant t, \\ 1, & t < c \leqslant 1. \end{cases}$$

Since $d_c(m,n) \leqslant d_c(m,k) + d_c(k,n)$ for any k, m, n in S and c in (0, 1], it follows that $(S, \mathfrak{F})$ is a Menger space under M. Furthermore, since the only Cauchy sequences in $(S, \mathfrak{F})$ are those that are eventually constant, $(S, \mathfrak{F})$ is complete. However, for any $c \leqslant t$, we have $\lim_{m,n\to\infty} d_c(m,n) = 0$, whence $\{n\}$ is a Cauchy sequence in (S, d_c); but since $\lim_{n\to\infty} d_c(m,n) = 1/m > 0$ for any m in S, the sequence $\{n\}$ does not converge. Thus for each $c \leqslant t$ the metric space (S, d_c) is not complete.

12.6. Contraction Maps

Contraction maps on PM spaces were first defined and studied by V. M. Sehgal [1966]. They were subsequently studied by Sherwood [1971], Sehgal and A. T. Bharucha-Reid [1972], and others (see [Bharucha-Reid 1976] p. 654).

Definition 12.6.1. Let $(S, \mathfrak{F}, \tau)$ be a PM space and φ a function from S into S. Then φ is a *contraction map* on $(S, \mathfrak{F}, \tau)$ if there is an α in (0, 1), the *contraction constant* of φ, such that

$$F_{\varphi p \varphi q}(x) \geqslant F_{pq}(x/\alpha) \tag{12.6.1}$$

for all p, q in S and all x in R^+; or equivalently, if for all p, q in S

$$F_{\varphi p \varphi q} \geqslant F_{pq}(j/\alpha) \tag{12.6.2}$$

where j is the identity function on R^+.

Lemma 12.6.2. *Let $(S, \mathcal{F}, \tau)$ be a* PM *space such that* $\operatorname{Ran} \mathcal{F} \subseteq \mathcal{D}^+$ *and let φ be a contraction map on $(S, \mathcal{F}, \tau)$. Then φ has at most one fixed point, i.e., there is at most one point p in S such that $\varphi(p) = p$.*

PROOF. Suppose that $\varphi p = p$ and $\varphi q = q$. Then for every $n \geqslant 1$ and every $x > 0$ we have $F_{pq}(x) = F_{\varphi^n p \varphi^n q}(x) \geqslant F_{pq}(x/\alpha^n)$. Since F_{pq} is in $\mathcal{D}^+$, letting $n \to \infty$ yields $F_{pq} = \varepsilon_0$, whence $p = q$. □

The hypothesis that $\operatorname{Ran} \mathcal{F} \subseteq \mathcal{D}^+$ is necessary. For example, if $S = \{p, q\}$ and $F_{pq} = \frac{1}{2}\varepsilon_0 + \frac{1}{2}\varepsilon_\infty$, then j_S is a contraction map with two fixed points. Consequently, in the remainder of this section, unless the contrary is stated, we assume that $\operatorname{Ran} \mathcal{F} \subseteq \mathcal{D}^+$.

Using (4.4.11), it follows that (12.6.1) is equivalent to

$$F_{pq}^{\wedge}(c) \leqslant \alpha F_{pq}^{\wedge}(c) \tag{12.6.3}$$

for all p, q in S and any c in $(0, 1)$; and (12.6.3) is, in turn, equivalent to

$$d_c(\varphi p, \varphi q) \leqslant \alpha d_c(p, q) \tag{12.6.4}$$

where d_c is the function from $S \times S$ into R^+ defined by $d_c(p, q) = F_{pq}^{\wedge}(c)$. Thus, if d_c is a pseudometric on S, then φ is a contraction map on the pseudometric space (S, d_c). Exploiting this observation yields the following alternate proof of a result first established by Sehgal and Bharucha-Reid [1972].

Theorem 12.6.3. *Suppose that the Menger space $(S, \mathcal{F}, \tau_M)$ is complete and that φ is a contraction map on $(S, \mathcal{F}, \tau_M)$. Then φ has a unique fixed point.*

PROOF. Note first that by Theorem 8.2.3, for each c in $(0, 1)$, the function d_c given by $d_c(p, q) = F_{pq}^{\wedge}(c)$ is a pseudometric on S; and that if $p \neq q$, then there is a c_{pq} in $(0, 1)$ such that $d_{c_{pq}}(p, q) \neq 0$. Consequently, if $d_c(p, q) = 0$ for all c in $(0, 1)$, then $p = q$. Let p be a point in S. Then in view of (12.6.4), the standard metric space argument (see [Apostol 1974, §4.21]) shows that the sequence of iterates $\{\varphi^n p\}$ is a Cauchy sequence in each of the pseudometric spaces (S, d_c). Thus, by Lemma 12.5.4, $\{\varphi^n p\}$ is a Cauchy sequence in $(S, \mathcal{F}, \tau_M)$. Since $(S, \mathcal{F}, \tau_M)$ is complete, there is a q in S such that $\varphi^n p \to q$ in the strong topology. Thus, again by Lemma 12.5.4, for each

c in $(0, 1)$, $d_c(\varphi^n p, q) \to 0$ whence, again using the standard metric space argument, we have $d_c(\varphi q, q) = 0$ for each c in $(0, 1)$, and by the remark above, $\varphi q = q$. □

A similar argument shows that if $(S, \mathfrak{F}, \tau)$ is complete and if for every pair of points p, q in S there is a number $x < \infty$ such that $F_{pq}(x) = 1$ (a stronger condition than $\operatorname{Ran} \mathfrak{F} \subseteq \mathfrak{D}^+$), then every contraction map on $(S, \mathfrak{F}, \tau)$ has a unique fixed point (see [Sherwood 1971]).

In general, however, the existence of fixed points of contraction maps is a delicate matter. Our discussion of this question is based on [Sherwood 1971].

Lemma 12.6.4. *Let $(S, \mathfrak{F}, \tau)$ be a* PM *space (with* Ran $\mathfrak{F} \subseteq \mathfrak{D}^+$*), and suppose that $\mathfrak{D}^+$ is closed under τ. Let φ be a contraction map on $(S, \mathfrak{F}, \tau)$. For any p in S and any $n \geqslant 0$, set $p_n = \varphi^n p$, and let*

$$\mathcal{O}(p, \varphi) = \{ p_n \mid n = 0, 1, 2, \ldots \} \tag{12.6.5}$$

be the trajectory of p under φ. Then for any p, q in S, $\mathcal{O}(p, \varphi)$ is bounded if and only if $\mathcal{O}(q, \varphi)$ is bounded.

PROOF. Suppose that $\mathcal{O}(p, \varphi)$ is bounded, i.e., that $D_{\mathcal{O}(p,\varphi)}$ is in $\mathfrak{D}^+$, and let p, q be distinct points of S. Then for any $m \geqslant 0$, repeated use of (12.6.2) yields

$$F_{p_m q_m} \geqslant F_{pq}(j/\alpha^m) \geqslant F_{pq}.$$

Hence for any m, $n \geqslant 0$,

$$\begin{aligned} F_{q_m q_n} &\geqslant \tau^2(F_{q_m p_m}, F_{p_m p_n}, F_{p_n q_n}) \\ &\geqslant \tau^2(F_{pq}, D_{\mathcal{O}(p,\varphi)}, F_{pq}), \end{aligned}$$

by virtue of property iii of Theorem 12.4.2, whence (12.4.1) and (12.4.2) yield $D_{\mathcal{O}(q,\varphi)} \geqslant \tau^2(F_{pq}, D_{\mathcal{O}(p,\varphi)}, F_{pq})$. Since F_{pq} and $D_{\mathcal{O}(p,\varphi)}$ are both in $\mathfrak{D}^+$, it follows that $D_{\mathcal{O}(q,\varphi)}$ is also in $\mathfrak{D}^+$, i.e., $\mathcal{O}(q, \varphi)$ is bounded. An interchange of p and q completes the proof. □

It follows that under the hypotheses of Lemma 12.6.4 the boundedness of the trajectories $\mathcal{O}(p, \varphi)$ is independent of the particular point p, hence is a property of the space $(S, \mathfrak{F}, \tau)$ and the map φ.

Theorem 12.6.5. *Let φ be a contraction map on the complete* PM *space $(S, \mathfrak{F}, \tau)$ and suppose that the trajectories under φ of all points in S are bounded. Then φ has a unique fixed point.*

PROOF. Let p be an arbitrary point in S and let α be the contraction constant of φ. Then for any $m, n > 0$

$$F_{p_n p_{m+n}} \geqslant F_{pp_m}(j/\alpha^n) \geqslant D_{\mathcal{O}(p,\varphi)}(j/\alpha^n).$$

Letting $n \to \infty$, and using the fact that $D_{\mathcal{O}(p,\varphi)}$ is in $\mathcal{D}^+$, it follows that $\{p_n\}$ is a Cauchy sequence. Since $(S, \mathfrak{F}, \tau)$ is complete, there is a point q in S such that $p_n \to q$. Next, for any $n > 0$ we have

$$\begin{aligned} F_{q\varphi q} &\geqslant \tau(F_{qp_n}, F_{p_n \varphi q}) \\ &\geqslant \tau(F_{qp_n}, F_{p_{n-1}q}(j/\alpha)). \end{aligned}$$

Letting $n \to \infty$ yields $F_{q\varphi q} = \varepsilon_0$, i.e., $\varphi q = q$. Thus q is a fixed point of φ that, by Lemma 12.6.2, is unique. □

Combining Lemma 12.6.4 and Theorem 12.6.5, we see that if φ is a contraction map on a complete PM space $(S, \mathfrak{F}, \tau)$ and if $\mathcal{D}^+$ is closed under τ, then either φ has a unique fixed point, or none of the trajectories $\mathcal{O}(p, \varphi)$ is bounded.

Definition 12.6.6. Let τ be a triangle function, α in $(0,1)$, and G in Δ^+. Then $G^{\tau,\alpha}$ is the function in Δ^+ given by

$$G^{\tau,\alpha} = l^-\left(\lim_{n\to\infty} \tau^n(G, G(j/\alpha), \ldots, G(j/\alpha^n))\right) \tag{12.6.6}$$

where τ^n is the nth serial iterate of τ (see Definition 6.3.4).

Note that the sequence $\{\tau^n(G, G(j/\alpha), \ldots, G(j/\alpha^n))\}$ is nondecreasing, so that its weak limit always exists.

Lemma 12.6.7. *Let φ be a contraction map with contraction constant α on the complete* PM *space $(S, \mathfrak{F}, \tau)$. If p is a point in S such that $F_{p\varphi p}^{\tau,\alpha}$ is in $\mathcal{D}^+$, then $\mathcal{O}(p, \varphi)$ is bounded.*

PROOF. For any $m > 0$, repeated use of (12.6.2) first yields

$$\begin{aligned} F_{pp_m} &\geqslant \tau^{m-1}(F_{pp_1}, F_{p_1 p_2}, \ldots, F_{p_{m-1}p_m}) \\ &\geqslant \tau^{m-1}\left(F_{pp_1}, F_{pp_1}\left(\frac{j}{\alpha}\right), \ldots, F_{pp_1}\left(\frac{j}{\alpha^{m-1}}\right)\right) \geqslant F_{p\varphi p}^{\tau,\alpha} \end{aligned}$$

and then, for any $m, n > 0$,

$$F_{p_n p_{m+n}} \geqslant F_{pp_m}\left(\frac{j}{\alpha^n}\right) \geqslant F_{p\varphi p}^{\tau,\alpha}\left(\frac{j}{\alpha^n}\right) \geqslant F_{p\varphi p}^{\tau,\alpha}.$$

Since $F^{\tau,\alpha}_{p\varphi p}$ is in $\mathcal{D}^+$, it follows that $D_{\Theta(p,\varphi)}$ is in $\mathcal{D}^+$, i.e., that $\Theta(p,\varphi)$ is bounded. □

Combining Lemmas 12.6.4, 12.6.7 and Theorem 12.6.5, we immediately have

Theorem 12.6.8. *Let τ be a triangle function such that $\mathcal{D}^+$ is closed under τ and $G^{\tau,\alpha}$ is in $\mathcal{D}^+$ for any G in $\mathcal{D}^+$ and any α in $(0,1)$. Then any contraction map on any complete* PM *space under τ has a unique fixed point.*

The hypotheses of Theorem 12.6.8 are rather severe. They do apply, however, to τ_M; for it is easy to show that $\mathcal{D}^+$ is closed under τ_M and that for any G in $\mathcal{D}^+$ and any α in $(0,1)$

$$G^{\tau_M,\alpha} = G((1-\alpha)j),$$

which is in $\mathcal{D}^+$. Thus Theorem 12.6.8 yields Theorem 12.6.3 as a corollary. Moreover, the following theorem shows that for a large class of triangle functions the hypotheses of Theorem 12.6.8 are not only sufficient, but also necessary.

Theorem 12.6.9. *Let τ be a triangle function that satisfies the following conditions:*

i. *$\mathcal{D}$ is closed under τ.*
ii. *τ is homogeneous on $\mathcal{D}^+$ in the sense that for any $\alpha > 0$ and all F, G in $\mathcal{D}^+$,*

$$\tau(F,G)(\alpha j) = \tau(F(\alpha j), G(\alpha j)). \tag{12.6.7}$$

iii. *For some $\alpha > 0$ and some G in $\mathcal{D}^+$ the function $G^{\tau,\alpha}$ is not in $\mathcal{D}^+$.*

Then there exists a complete PM *space $(S,\mathcal{F},\tau)$ and a contraction map φ on $(S,\mathcal{F},\tau)$ that has no fixed point.*

PROOF. Let S be the set of positive integers and let $\mathcal{F}$ be defined via

$$F_{n+m,n} = \tau^m\left(G\left(\frac{j}{\alpha^n}\right), G\left(\frac{j}{\alpha^{n+1}}\right), \ldots, G\left(\frac{j}{\alpha^{n+m-1}}\right)\right). \tag{12.6.8}$$

Then $(S,\mathcal{F})$ is a complete PM space (for details, see Theorem 3.5 in [Sherwood 1971]). The successor function s defined by $s(n) = n+1$ is—in view of (12.6.7) and (12.6.8)—a contraction map and, obviously, s has no fixed point. □

The class of homogeneous triangle functions includes convolution, the functions $\tau_{T,L}$ whenever L is homogeneous of degree 1, and hence all the functions τ_T. Moreover, if α is in (0, 1) and if G is the d.f. in $\mathcal{D}^+$ given by

$$G(x) = \begin{cases} 0, & x \leqslant \dfrac{1}{\alpha^2}, \\ 1 - \dfrac{1}{n}, & \dfrac{1}{\alpha^n} < x \leqslant \dfrac{1}{\alpha^{n+1}}, \quad n > 1, \end{cases}$$

then, using (7.7.6) and (7.7.7), it follows that, for any Archimedean t-norm T, $G^{\tau_T,\alpha}$ is not in $\mathcal{D}^+$. Thus, in the class of Menger spaces under Archimedean t-norms, the existence of fixed points of contraction maps appears to be the exception rather than the rule.

Sherwood has pointed out that the additional structure possessed by an E-space can be used to strengthen the definition of a contraction map—in effect, by requiring that the map be a contraction with respect to each of the pseudometrics given by (9.2.1)—and that every such *strict contraction map* on a complete E-space has a unique fixed point.

12.7. Product Spaces

In order to define a well-behaved product of a finite number of metric spaces, it is natural to take a function K from $R^+ \times R^+$ to R^+ and to define the K-product of two metric spaces (S_1, d_1) and (S_2, d_2) to be the space $(S_1 \times S_2, d_K)$ where

$$d_K(\mathbf{p}, \mathbf{q}) = K(d_1(p_1, q_1), d_2(p_2, q_2)) \tag{12.7.1}$$

for any $\mathbf{p} = (p_1, p_2)$ and $\mathbf{q} = (q_1, q_2)$ in $S_1 \times S_2$ (see [Motzkin 1936; Bohnenblust 1940; Tardiff 1980]). If K satisfies the conditions

$$K(a, 0) = a, \tag{12.7.2}$$

$$K(a, b) = K(b, a), \tag{12.7.3}$$

$$K(a, c) \leqslant K(b, c) \quad \text{whenever} \quad a \leqslant b, \tag{12.7.4}$$

then d_K is a semimetric. If in addition K satisfies the inequality

$$K(a + c, b + d) \leqslant K(a, b) + K(c, d), \tag{12.7.5}$$

then d_K satisfies the triangle inequality. If K is associative, then (12.7.1) can be used to define the product of three or more metric spaces in a consistent fashion. Lastly, to avoid pathologies, it is convenient to assume that K is continuous.

The set of continuous associative solutions of the system (12.7.2)–(12.7.4) is known (see Sections 5.3, 5.4, and 5.7). What is not known is precisely which functions in the set also satisfy (12.7.5). Among the ones that do are Max and the functions K_α given by (5.7.16). In particular, the function $K_1 = \mathrm{Sum}$ is a continuous associative solution of (12.7.2)–(12.7.5).

Similar considerations apply to the problem of defining the product of a finite number of PM spaces. Here, following R. M. Tardiff [1976] (see also [Urazov 1968]), it is natural to proceed as follows:

Definition 12.7.1. Let $(S_1, \mathfrak{F}_1)$ and $(S_2, \mathfrak{F}_2)$ be PSM spaces and let τ be a triangle function. The *τ-product* of $(S_1, \mathfrak{F}_1)$ and $(S_2, \mathfrak{F}_2)$ is the pair $(S_1 \times S_2, \mathfrak{F}_1 \tau \mathfrak{F}_2)$, where $\mathfrak{F}_1 \tau \mathfrak{F}_2$ is the function from $(S_1 \times S_2)^2$ into Δ^+ given by

$$(\mathfrak{F}_1 \tau \mathfrak{F}_2)(\mathbf{p}, \mathbf{q}) = \tau(\mathfrak{F}_1(p_1, q_1), \mathfrak{F}_2(p_2, q_2)) \tag{12.7.6}$$

for any $\mathbf{p} = (p_1, p_2)$ and $\mathbf{q} = (q_1, q_2)$ in $S_1 \times S_2$.

We shall frequently denote $(\mathfrak{F}_1 \tau \mathfrak{F}_2)(\mathbf{p}, \mathbf{q})$ by $F^{\tau}_{\mathbf{pq}}$ and, when there is no possible ambiguity, omit the superscript; in this case (12.7.6) assumes the form

$$F_{\mathbf{pq}} = \tau(F_{p_1 q_1}, F_{p_2 q_2}). \tag{12.7.7}$$

Convolution-products of Wald spaces, as well as several other types of products of PSM spaces, were first defined by V. Istrăţescu and I. Vaduva [1961]. If T is a t-norm and $\tau = \tau_T$, then the τ-product is the T-product as defined independently by R. J. Egbert [1968] and A. F. S. Xavier [1968]. It is immediate that the τ-product of two PSM spaces is a PSM space. However, to guarantee that the τ-product of two PM spaces is a PM space (i.e., to guarantee that the triangle inequality holds in the product space) a further condition is needed. This condition is a generalization of the relationship (12.7.5) between K and addition on R^+; and since we need the relevant concept in several different settings, we define it in some generality.

Definition 12.7.2. Let S be a partially ordered set and let f and g be associative binary operations on S with common identity e. Then f *dominates* g, and we write $f \gg g$, if for all x_1, x_2, y_1, y_2 in S,

$$f(g(x_1, y_1), g(x_2, y_2)) \geqslant g(f(x_1, x_2), f(y_1, y_2)). \tag{12.7.8}$$

Note that in this terminology (12.7.5) simply states that K is dominated by addition on R^+. Note also that when inequality is replaced by equality, (12.7.8) reduces to the well-known *bisymmetry equation* (see [Aczél 1966, §6.4]).

Setting $y_1 = x_2 = e$ in (12.7.8) yields $f(x_1, y_2) \geqslant g(x_1, y_2)$, i.e., $f \geqslant g$.

Thus if f dominates g, then f is stronger than g. The converse is false: Tardiff [1975] showed that if T_f and T_g are the t-norms additively generated by

$$f(x) = \frac{1}{x} - 1, \qquad 0 < x \leqslant 1,$$

and

$$g(x) = \begin{cases} \dfrac{1}{2x}, & 0 < x \leqslant \frac{1}{2}, \\ \left(\dfrac{1}{x} - 1\right)^{1/2}, & \frac{1}{2} \leqslant x \leqslant 1, \end{cases}$$

respectively, then T_f is stronger than T_g, but does not dominate T_g.

Since "stronger" is antisymmetric, it follows that "dominates" is likewise antisymmetric; and the associativity of f yields $f \gg f$, so that "dominates" is reflexive. By way of contrast, the transitivity of "dominates," particularly for the families of binary operations studied in Chapters 5 and 7, is still an open question.

Turning to triangle functions, we have the following lemmas, which we state without proof.

Lemma 12.7.3. *If T_1 and T_2 are continuous t-norms such that $T_1 \gg T_2$, then $\tau_{T_1} \gg \tau_{T_2}$, and conversely.*

Lemma 12.7.4. *If T is a continuous t-norm and $\mathbf{T}$ is the triangle function given by $\mathbf{T}(F,G)(x) = T(F(x), G(x))$ (see (7.1.2)), then*

i. *$\mathbf{T} \gg \tau_T$ for any t-norm T;*
ii. *$\mathbf{M} \gg \tau$ for any triangle function τ.*

Theorem 12.7.5. *Let $(S_1, \mathfrak{F}_1, \tau)$ and $(S_2, \mathfrak{F}_2, \tau)$ be PM spaces under the same triangle function τ. Suppose that $\tau_1 \gg \tau$. Then the τ_1-product $(S_1 \times S_2, \mathfrak{F}_1 \tau_1 \mathfrak{F}_2)$ is a PM space under τ.*

PROOF. We need only establish the triangle inequality in the product space. To this end, let $\mathbf{p} = (p_1, p_2)$, $\mathbf{q} = (q_1, q_2)$, and $\mathbf{r} = (r_1, r_2)$ be in $S_1 \times S_2$. Then

$$\begin{aligned} F_{\mathbf{pr}} &= \tau_1(F_{p_1r_1}, F_{p_2r_2}) \\ &\geqslant \tau_1(\tau(F_{p_1q_1}, F_{q_1r_1}), \tau(F_{p_2q_2}, F_{q_2r_2})) \\ &\geqslant \tau(\tau_1(F_{p_1q_1}, F_{p_2q_2}), \tau_1(F_{q_1r_1}, F_{q_2r_2})) \\ &= \tau(F_{\mathbf{pq}}, F_{\mathbf{qr}}). \end{aligned}$$

□

The result of Theorem 12.7.5 is best-possible in the sense that if τ_1 and τ are triangle functions and τ_1 does not dominate τ, then there exist PM spaces $(S_1, \mathfrak{F}_1, \tau)$ and $(S_2, \mathfrak{F}_2, \tau)$ whose τ_1-product is not a PM space under τ.

Corollary 12.7.6. *Under the hypotheses of Theorem* 12.7.5, *if* $\tau_1 = \tau$, *or* $\tau_1 = \mathbf{M}$, *or* $\tau = \tau_T$ *and* $\tau_1 = \mathbf{T}$ *for some continuous t-norm* T, *then* $(S_1 \times S_2, \mathfrak{F}_1 \tau_1 \mathfrak{F}_2)$ *is a* PM *space under* τ.

In certain special cases we can say more. Thus we have the following results, which are due to Egbert [1968].

Theorem 12.7.7. *If* $(S_1, \mathfrak{F}_1)$ *and* $(S_2, \mathfrak{F}_2)$ *are equilateral spaces with the same d.d.f.* F, *then their* **M**-*product is an equilateral space with* $\mathfrak{F}_{\mathbf{pq}} = F$.

Theorem 12.7.8. *Let* $(S_1, \mathfrak{F}_1)$ *and* $(S_2, \mathfrak{F}_2)$ *be the simple spaces* (S_1, d_1, G) *and* (S_2, d_2, G), *respectively. Let* d_{Max} *be the metric on* $S_1 \times S_2$ *defined by* (12.7.1) *with* $K =$ Max. *Then the* **M**-*product of* $(S_1, \mathfrak{F}_1)$ *and* $(S_2, \mathfrak{F}_2)$ *is the simple space* $(S_1 \times S_2, d_{\mathrm{Max}}, G)$.

Theorem 12.7.9. *For any* α *in* $(1, \infty)$, *let* $(S_1, \mathfrak{F}_1)$ *and* $(S_2, \mathfrak{F}_2)$ *be the* α-*simple spaces* $(S_1, d_1, G; \alpha)$ *and* $(S_2, d_2, G; \alpha)$, *respectively, where* G *is strict. For* $\beta \geqslant 1$, *let* T_β *be the strict t-norm whose additive generator is* $(G^{-1})^{-\beta/\alpha}$, *and let* d_β *be the metric defined on* $S_1 \times S_2$ *by* (12.7.1) *with* $K = K_\beta$ *(see* (5.7.16)). *Then the* $\mathbf{T}_\beta$-*product of* $(S_1, \mathfrak{F}_1)$ *and* $(S_2, \mathfrak{F}_2)$ *is the* α-*simple space* $(S_1 \times S_2, d_\beta, G; \alpha)$.

It follows from Theorem 8.6.5 that the α-simple spaces in Theorem 12.7.9 are all Menger spaces under the t-norm T_G whose additive generator is $(G^{-1})^{1/1-\alpha}$. If we want to have $T_\beta = T_G$, then we must have $1/(1-\alpha) = -\beta/\alpha$, which is equivalent to

$$\frac{1}{\alpha} + \frac{1}{\beta} = 1.$$

Since triangle functions are associative, the foregoing constructions and results extend in an obvious fashion to the product of any finite number of PM spaces. Furthermore, it is easy to see that the strong topology in the product space is equivalent to the product topology of the strong topologies on the factor spaces.

Finally we have the following result (see [Sherwood and Taylor 1974] and [Höhle 1978]).

Theorem 12.7.10. *If $\mathcal{F}$ and τ are triangle functions such that $\mathcal{F} \gg \tau$, then $(\Delta^+, \mathcal{F}, \tau)$ is a* PM *space.*

12.8. Countable Products

Products of a countably infinite number of PM spaces were first considered by V. Radu [1977], whose work we summarize in the following definition and theorem.

Definition 12.8.1. Let $\{(S_n, \mathcal{F}_n)\}$ be a sequence of PSM spaces and $\{\lambda_n\}$ a sequence of positive numbers with $\sum_{n=1}^{\infty} \lambda_n = 1$. Then the $\{\lambda_n\}$*-product* of the spaces $(S_n, \mathcal{F}_n)$ is the space $(S, \mathcal{F})$, where $S = \times_{n=1}^{\infty} S_n$ *and $\mathcal{F}$ is the function from $S \times S$ into Δ^+ given by*

$$\mathcal{F}(p, q) = \sum_{n=1}^{\infty} \lambda_n \mathcal{F}_n(p_n, q_n) \tag{12.8.1}$$

where $p = \{p_n\}$, $q = \{q_n\}$.

Theorem 12.8.2. *Let $\{(S_n, \mathcal{F}_n)\}$ be a sequence of* PM *spaces. Then*

i. *If each $(S_n, \mathcal{F}_n)$ is a proper* PM *space, then their $\{\lambda_n\}$-product is a proper* PM *space.*
ii. *If each $(S_n, \mathcal{F}_n)$ is a Menger space under W, then so is their $\{\lambda_n\}$-product.*
iii. *The strong topology on the $\{\lambda_n\}$-product is the product topology of the strong topologies on the spaces $(S_n, \mathcal{F}_n)$.*

These and other results were independently found by C. Alsina [1978a], who also obtained the following result, which shows that the t-norm W in Part ii of Theorem 12.8.2 cannot be replaced by any stronger t-norm.

Theorem 12.8.3. *Let T be a t-norm such that the $\{2^{-n}\}$-product of any sequence of Menger spaces under T is also a Menger space under T. Then $T \leqslant W$.*

This theorem is a consequence of the following interesting property of t-norms [Alsina 1980]:

Theorem 12.8.4. *Let T be a t-norm and suppose that*

$$T\left(\sum_{n=1}^{\infty} \frac{a_n}{2^n}, \sum_{n=1}^{\infty} \frac{b_n}{2^n}\right) \leqslant \sum_{n=1}^{\infty} \frac{1}{2^n} T(a_n, b_n) \tag{12.8.2}$$

for all sequences $\{a_n\}$, $\{b_n\}$ of numbers in I. Then $T \leqslant W$.

Motivated by the conclusion of Theorem 12.8.3, Alsina constructed examples that show that the $\{2^{-n}\}$-product of a sequence of simple spaces, even a sequence of identical simple spaces, need not be a simple space, or even a Menger space under M, and that the $\{2^{-n}\}$-product of Wald spaces need not be a Wald space. Since it seems desirable that the product of identical spaces be a space of the same type, Alsina defined a different product, related to the τ-product of the preceding section.

Definition 12.8.5. Let $\{\tau_n\}$ be a sequence of triangle functions. Let the operations $\tau^{(n)}$ be recursively defined by $\tau^{(1)} = \tau_1$, and for $n > 1$, $\tau^{(n)}$ is the function from $(\Delta^+)^{n+1}$ into Δ^+ given by

$$\tau^{(n)}(F_1, \ldots, F_n, F_{n+1}) = \tau_n\left(\tau^{(n-1)}(F_1, \ldots, F_n), F_{n+1}\right). \quad (12.8.3)$$

For any sequence $\{F_n\}$ of functions in Δ^+ let $\tau^{(\infty)}\{F_n\}$ be the weak limit in Δ^+ (which always exists) of the sequence $\{\tau^{(n)}(F_1, \ldots, F_{n+1})\}$. The *$\tau^{(\infty)}$-product* of the sequence $\{(S_n, \mathfrak{F}_n, \tau_n)\}$ of PM spaces is the space $(S, \mathfrak{F})$, where $S = \times_{n=1}^{\infty} S_n$ and $\mathfrak{F}$ is the function from $S \times S$ into Δ^+ given by

$$\mathfrak{F}(p, q) = \tau^{(\infty)}\{\mathfrak{F}_n(p_n, q_n)\} \quad (12.8.4)$$

for $p = \{p_n\}$, $q = \{q_n\}$.

Theorem 12.8.6. *Let $\{(S_n, \mathfrak{F}_n, \tau_n)\}$ be a sequence of* PM *spaces and suppose that there is a continuous triangle function τ such that $\tau_n \geqslant \tau$ for all n. Then the $\tau^{(\infty)}$-product of the spaces $(S_n, \mathfrak{F}_n, \tau_n)$ is a* PM *space under τ.*

Thus $\tau^{(\infty)}$-products avoid one of the drawbacks of $\{\lambda_n\}$-products. On the other hand, since the infinite product in (12.8.4) has a strong tendency to diverge to ε_∞, $\tau^{(\infty)}$-products have drawbacks of their own. In particular, we have

Theorem 12.8.7. *Let $(S, \mathfrak{F})$ be the $\tau^{(\infty)}$-product of the spaces $(S_n, \mathfrak{F}_n, \tau_n)$. Then the strong topology on $(S, \mathfrak{F})$ is finer than the product topology of the strong topologies on the spaces $(S_n, \mathfrak{F}_n, \tau_n)$, i.e., every neighborhood in the product topology on $(S, \mathfrak{F})$, contains a strong neighborhood. The converse may fail, whence for $\tau^{(\infty)}$-products the product topology and the strong topology are in general not identical.*

Results such as Theorem 12.8.7 led Alsina to modify the definition of $\tau^{(\infty)}$-products as follows: Replace the space $(S_n, \mathfrak{F}_n, \tau_n)$ by the space

$(S_n, \mathcal{F}'_n, \tau_n)$, where $\mathcal{F}'_n$ is given in terms of $\mathcal{F}_n$ by

$$\mathcal{F}'_n(p,q)(x) = \begin{cases} \mathcal{F}_n(p,q)(x), & x \text{ in } [0, 2^{-n}], \\ 1, & x \text{ in } (2^{-n}, \infty] \end{cases} \tag{12.8.5}$$

(i.e., subject each d.d.f. of $(S_n, \mathcal{F}_n, \tau_n)$ to a drastic truncation) and define the modified $\tau^{(\infty)}$-product of the spaces $(S_n, \mathcal{F}_n, \tau_n)$ to be the $\tau^{(\infty)}$-product of the spaces $(S_n, \mathcal{F}'_n, \tau_n)$. Theorem 12.8.6 remains valid for these modified $\tau^{(\infty)}$-products, and at the same time, the product and strong topologies coincide in every modified $\tau^{(\infty)}$-product.[1]

12.9. The Probabilistic Hausdorff Distance

The *Hausdorff distance* between two nonempty subsets A and B of a metric space (S, d) is the number

$$d(A,B) = \operatorname{Max}\Big(\sup_{p \text{ in } A} \inf_{q \text{ in } B} d(p,q), \sup_{q \text{ in } B} \inf_{p \text{ in } A} d(p,q) \Big). \tag{12.9.1}$$

This quantity admits an immediate probabilistic generalization.

Definition 12.9.1. Let $(S, \mathcal{F})$ be a PSM space and, for any two nonempty subsets A and B of S, let Γ_{AB} and G_{AB} be the functions on R^+ defined by

$$\Gamma_{AB}(x) = \inf_{p \text{ in } A} \sup_{q \text{ in } B} F_{pq}(x), \tag{12.9.2}$$

$$G_{AB} = l^- \Gamma_{AB}. \tag{12.9.3}$$

Then the *probabilistic Hausdorff distance* between A and B is the d.f.

$$F_{AB} = \mathbf{M}(G_{AB}, G_{BA}) \tag{12.9.4}$$

where $\mathbf{M}$ is the maximal triangle function (see Theorem 7.1.4).

Theorem 12.9.2. *The probabilistic Hausdorff distance has the following properties:*

i. *If $\{p\}$ and $\{q\}$ are singleton subsets of S, then $F_{\{p\}\{q\}} = F_{pq}$.*

---

[1] *Note added in proof*: Recently C. Alsina and B. Schweizer have defined a countable product that avoids truncations and has the following properties: (i) The product of any sequence of PM spaces under a given triangle function τ is a PM space under τ. (ii) The strong topology on the product space is the product topology. The key to its construction is the PM space analog of a metric transform (*Houston J. Math.*, to appear).

ii. *Let A and B be nonempty subsets of S with closures $k(A)$ and $k(B)$, respectively. Then $F_{AB} = F_{BA}$, $F_{AB} = F_{k(A)k(B)}$, and $F_{AB} = \varepsilon_0$ if and only if $k(A) = k(B)$. In particular, $F_{AA} = \varepsilon_0$.*

It follows that if $(S, \mathcal{F})$ is a PM space, then the probabilistic Hausdorff distance is a probabilistic semimetric on the set of all closed subsets of S. In order to be able to specify when this distance is a probabilistic metric, we need an additional concept.

Definition 12.9.3. A triangle function τ is *sup continuous* if for every family $\{F_\lambda | \lambda \text{ in } \Lambda\}$ of d.f.'s in Δ^+ and every G in Δ^+

$$\sup_{\lambda \text{ in } \Lambda} \tau(F_\lambda, G) = \tau\left(\sup_{\lambda \text{ in } \Lambda} F_\lambda, G\right). \tag{12.9.5}$$

Note that in view of Lemma 4.3.5, the suprema in (12.9.5) are in Δ^+.

Thus sup continuity is upper semicontinuity with respect to the partial order on Δ^+. If T is a continuous t-norm, then τ_T is sup continuous. On the other hand, convolution is not sup continuous.

Lemma 12.9.4. *Let $(S, \mathcal{F}, \tau)$ be a* PM *space. If τ is sup continuous, then for any nonempty subsets A, B, C of S*

$$G_{AC} \geqslant \tau(G_{AB}, G_{BC}). \tag{12.9.6}$$

Using Lemma 12.7.4 and Theorem 12.9.2, it is easy to establish

Theorem 12.9.5. *Let $(S, \mathcal{F}, \tau)$ be a* PM *space and suppose that τ is sup continuous. Then the probabilistic Hausdorff distance is a probabilistic pseudometric on $\mathcal{P}(S)$ and a probabilistic metric on any collection of closed subsets of S.*

For details concerning these results, see [Tardiff 1976]. An earlier formulation, couched in the language of t-norms rather than triangle functions, may be found in [Egbert 1968].

12.10. Discernibility Relations

In his note on probabilistic geometry, Menger [1951b] introduced three relations for pairs of points of a PM space. These were reconsidered in [Schweizer 1964], [Egbert 1968], and [Menger 1968, 1979]. They are defined as follows.

Definition 12.10.1. Let $(S, \mathcal{F})$ be a PSM space. For any p, q in S, note that $\sup\{x \mid F_{pq}(x) = 0\} = F_{pq}^{\wedge}(0+)$. Then p and q are

A. *certainly discernible* if $F_{pq}^{\wedge}(0+) > 0$;
B. *barely discernible* if $F_{pq}^{\wedge}(0+) = 0$ and $F_{pq}(0+) = 0$;
C. *perhaps indiscernible* if $F_{pq}(0+) > 0$.

For example, let p and q be elements of an E-space. Then p and q are certainly discernible if there is a positive number b such that $P\{\omega \mid d(p(\omega), q(\omega)) < b\} = 0$, barely discernible if $P\{\omega \mid d(p(\omega), q(\omega)) < b\} > 0$ for all $b > 0$ but $P\{\omega \mid d(p(\omega, q(\omega)) = 0\} = 0$, and perhaps indiscernible if $P\{\omega \mid d(p(\omega), q(\omega)) = 0\} > 0$.

Since $F_{pp}(0+) = 1$ and $F_{pq} = F_{qp}$, the relation C of being perhaps indiscernible is reflexive and symmetric, and thus a *tolerance relation* in the sense of Zeeman [1962]. Furthermore, we have

Theorem 12.10.2. *Let $(S, \mathcal{F}, \tau)$ be a* PM *space and suppose that the set of d.d.f.'s F for which $F(0+) > 0$ is closed under τ, i.e., that*

$$\tau(F, G)(0+) > 0 \quad \textit{whenever} \quad F(0+) > 0 \quad \textit{and} \quad G(0+) > 0. \tag{12.10.1}$$

Then the relation C is an equivalence relation.

Triangle functions having the property (12.10.1) include convolution and the functions τ_T for positive t-norms T (see Definition 5.5.10).

Next, let (A') be the relation on S defined by $p(A')q$ if and only if p and q are not certainly discernible, i.e., if and only if $F_{pq}^{\wedge}(0+) = 0$. It is immediate that (A') is reflexive and symmetric; and if the function $t: S \times S \to R^+$ defined by $t_{pq} = F_{pq}^{\wedge}(0+)$ is a pseudometric, then (A') is also transitive and hence an equivalence relation. A sufficient condition for transitivity is provided by

Theorem 12.10.3. *Let $(S, \mathcal{F}, \tau)$ be a* PM *space and suppose that for any F, G in Δ^+ and any $x, y > 0$*

$$\tau(F, G)(x+y) > 0 \quad \textit{whenever} \quad F(x) > 0 \quad \textit{and} \quad G(y) > 0. \tag{12.10.2}$$

Then (A') is an equivalence relation.

Since (12.10.1) implies (12.10.2), it follows that convolution and the functions τ_T for positive T satisfy (12.10.2). Next, we have

Theorem 12.10.4. *Under the hypotheses of Theorem* 12.10.3, *the equivalence classes in S determined by (A') are closed subsets of S in the strong topology.*

PROOF. Let p be a point in S and let $A'(p)$ be the equivalence class containing p, i.e., $A'(p) = \{q \mid F_{pq}^{\wedge}(0+) = 0\}$. We show that the complement of $A'(p)$ is open. To this end, suppose that r is not in $A'(p)$. Then $F_{pr}^{\wedge}(0+) > 0$, whence $F_{pr}(F_{pr}^{\wedge}(0+)) = 0$. Now choose x in $(0, F_{pr}^{\wedge}(0+))$ and $y < 1$ in $(0, F_{pr}^{\wedge}(0+) - x)$. Let q be any point in $N_r(y)$, so that $F_{qr}(y) > 1 - y > 0$. Suppose that $F_{pq}(x) > 0$. Then since $x + y < F_{pr}^{\wedge}(0+)$, (12.10.2) yields

$$0 = F_{pr}\big(F_{pr}^{\wedge}(0+)\big) \geqslant F_{pr}(x+y) \geqslant \tau(F_{pq}, F_{qr})(x+y) > 0,$$

which is a contradiction. Thus $F_{pq}(x) = 0$, whence $F_{pq}^{\wedge}(0+) \geqslant x > 0$ and q is not in $A'(p)$. It follows that $N_r(y)$ belongs to the complement of $A'(p)$, which is therefore open. Hence $A'(p)$ is closed. □

Combining Theorems 12.10.4 and 12.9.5 yields

Theorem 12.10.5. *Let* $(S, \mathfrak{F}, \tau)$ *be a* PM *space. Suppose that* τ *is sup continuous and satisfies* (12.10.2). *Let* $\mathfrak{A}' = \{A'(p) \mid p \text{ in } S\}$ *be the set of equivalence classes of* (A') *and let* $\mathfrak{F}_H$ *be the probabilistic Hausdorff distance on* $\mathfrak{A}'$. *Then* $(\mathfrak{A}', \mathfrak{F}_H, \tau)$ *is a* PM *space.*

Note that since the triangle function τ_W fails to satisfy (12.10.2)—and a fortiori (12.10.1)—these results may not apply to all E-spaces.

Egbert [1968] introduced the following relations on S.

Definition 12.10.6. Let $(S, \mathfrak{F})$ be a PSM space. For any p, q in S, note that $\inf\{x \mid F_{pq}(x) = 1\} = F_{pq}^{\wedge}(1-)$. Then p and q are

A*. *certainly accessible* if $F_{pq}^{\wedge}(1-) < \infty$;
B*. *barely accessible* if $F_{pq}^{\wedge}(1-) = \infty$ and F_{pq} is in $\mathfrak{D}^+$;
C*. *perhaps inaccessible* if F_{pq} is not in $\mathfrak{D}^+$.

The conditions A*, B*, and C* are clearly the duals of A, B, and C, respectively. Likewise, it can be shown that under appropriate restrictions on τ, Theorems 12.10.2–12.10.5 have their dual counterparts.

12.11. Open Problems

As noted at the end of Section 12.6, every strict contraction map on a complete E-space has a unique fixed point. This still leaves

Problem 12.11.1. Does every contraction map on a complete E-space have a unique fixed point?

Problem 12.11.2. Determine all continuous associative solutions of the system (12.7.2)–(12.7.5). Note that the statement that K_α, for $\alpha \geqslant 1$, is such a solution is equivalent to Minkowski's inequality.

Problem 12.11.3. Is the relation "dominates" defined by (12.7.8) always transitive? If not, under what conditions is it transitive?

Problem 12.11.4. Find all t-norms that dominate a given t-norm. Find all t-norms that are dominated by a given t-norm. What does convolution dominate and what dominates convolution?

13

Profile Functions

13.1. Profile Closures

Recall from Section 1.8 that a *profile function* for a PSM space is a fixed d.f. φ in Δ^+ whose value for any $x > 0$ is interpreted as the maximum degree of confidence that can be assigned to statements about distances less than x between points of S. With this interpretation, if p and q are points of S such that $F_{pq} \geqslant \varphi$, then p and q are indistinguishable relative to φ. Pursuing this observation leads naturally to a family of topological structures for PSM spaces. Such structures were first considered by E. O. Thorp [1962] and subsequently by R. Fritsche [1971] and R. M. Tardiff [1976]. In our presentation we follow Tardiff.

Definition 13.1.1. Let $(S, \mathcal{F})$ be a PSM space and let φ be a profile function. For any p in S and any h in $(0, 1]$ the *(φ, h)-neighborhood of p* is the set

$$N_p(\varphi, h) = \{q \mid F_{pq}(x+h) + h \geqslant \varphi(x) \quad \text{for all } x \text{ in } (0, 1/h)\}. \tag{13.1.1}$$

The *φ-neighborhood system at p* is the family

$$\mathfrak{N}_p^{\varphi} = \{N_p(\varphi, h) \mid h \text{ in } (0, 1]\}; \tag{13.1.2}$$

and the *φ-neighborhood system for S* is the union

$$\mathfrak{N}^{\varphi} = \bigcup_{p \text{ in } S} \mathfrak{N}_p^{\varphi}. \tag{13.1.3}$$

Note that q is in $N_p(\varphi, h)$ if and only if $[F_{pq}, \varphi; h]$ holds (see (4.3.1)), and that $[F_{pq}, \varphi; h]$ is one of the two conditions for $d_L(F_{pq}, \varphi) \leqslant h$ (see (4.3.2)).

Since $N_p(\varphi, h_1) \subseteq N_p(\varphi, h_2)$ whenever $0 < h_1 \leqslant h_2$, an appeal to Section 3.5 yields

Theorem 13.1.2. *If $(S, \mathcal{F})$ is a* PSM *space with profile function φ, then the neighborhood system $\mathcal{N}^\varphi$ satisfies* N1 *and* N2, *whence $(S, \mathcal{N}^\varphi)$ is a V_D-space. Furthermore, the function c^φ from $\mathcal{P}(S)$ into $\mathcal{P}(S)$ defined by*

$$c^\varphi(A) = \{p \text{ in } S \mid p \text{ is } \mathcal{N}^\varphi\text{-contiguous to } A\} \tag{13.1.4}$$

is a Čech closure operation for S—the profile closure determined by φ.

Similarly, if φ and ψ are profile functions such that $\varphi \geqslant \psi$, then $N_p(\varphi, h) \subseteq N_p(\psi, h)$ for all p in S and all h in $(0, 1]$. Hence we have

Theorem 13.1.3. *If $(S, \mathcal{F})$ is a* PSM *space and φ, ψ are profile functions such that $\varphi \geqslant \psi$, then $c^\varphi(A) \subseteq c^\psi(A)$ for all A in $\mathcal{P}(S)$.*

Definition 13.1.4. A triangle function τ is *lower semicontinuous* at the point (φ, ψ) in $\Delta^+ \times \Delta^+$ if, for every t in $(0, 1]$, there is an h_t in $(0, 1]$ such that $[\tau(F, G), \tau(\varphi, \psi); t]$ holds whenever both $[F, \varphi; h_t]$ and $[G, \psi; h_t]$ hold.

Clearly, if τ is continuous at (φ, ψ), then τ is lower semicontinuous at (φ, ψ); and if τ is lower semicontinuous at $(\varepsilon_0, \varepsilon_0)$, then τ is continuous at $(\varepsilon_0, \varepsilon_0)$.

The principal result of this section relates several profile closures. It is

Theorem 13.1.5. *Let $(S, \mathcal{F}, \tau)$ be a* PM *space and let φ, ψ be profile functions. If τ is lower semicontinuous at (φ, ψ), then for all $A \subseteq S$,*

$$c^\varphi(c^\psi(A)) \subseteq c^{\tau(\varphi, \psi)}(A). \tag{13.1.5}$$

PROOF. Let t be in $(0, 1]$ and choose h_t in accordance with Definition 13.1.4. Let $A \subseteq S$ and suppose that p is in $c^\varphi(c^\psi(A))$. Then there is a q in $c^\psi(A)$ such that $[F_{pq}, \varphi; h_t]$ holds; and similarly, there is an r in A such that $[F_{qr}, \psi; h_t]$ holds. Since τ is lower semicontinuous at (φ, ψ), it follows that $[\tau(F_{pq}, F_{qr}), \tau(\varphi, \psi); t]$ holds. But $F_{pr} \geqslant \tau(F_{pq}, F_{qr})$. Hence $[F_{pr}, \tau(\varphi, \psi); t]$ holds, whence p is in $c^{\tau(\varphi, \psi)}(A)$. □

The inclusion in (13.1.5) can be proper. For example, consider the space $\mathcal{E}(I)$ constructed in Section 9.1. This space is a Menger space under τ_W. Let $\varphi = \frac{1}{2}(\varepsilon_0 + \varepsilon_\infty)$. Then $\tau_W(\varphi, \varphi) = \varepsilon_\infty$, whence it follows that the $c^{\tau_W(\varphi, \varphi)}$-

closure of any nonempty subset of $\mathcal{E}(I)$ is the whole space $\mathcal{E}(I)$. On the other hand, for any function p in $\mathcal{E}(I)$, q is in $c^{\varphi}\{p\}$ if and only if $F_{pq}(0+) \geqslant \frac{1}{2}$, i.e., if and only if $\lambda\{t \text{ in } I \,|\, p(t) = q(t)\} \geqslant \frac{1}{2}$; and from this it follows that $c^{\varphi}(c^{\varphi}\{p\})$ is not all of $\mathcal{E}(I)$.

Theorem 13.1.6. *Let $(S, \mathfrak{F}, \tau)$ be a* PM *space and φ a profile function. If τ is lower semicontinuous at (φ, φ) and $\tau(\varphi, \varphi) = \varphi$, then c^{φ} is a Kuratowski closure operation.*

PROOF. By (13.5.6), (13.1.5), and the idempotence of φ, for any $A \subseteq S$ we have

$$c^{\varphi}(A) \subseteq c^{\varphi}(c^{\varphi}(A)) \subseteq c^{\tau(\varphi,\varphi)}(A) = c^{\varphi}(A). \qquad \square$$

Corollary 13.1.7. *If $(S, \mathfrak{F}, \tau)$ is a* PM *space and τ is continuous at $(\varepsilon_0, \varepsilon_0)$, then c^{ε_0} is a Kuratowski closure operation.*

Lemma 13.1.8. *In any* PSM *space $(S, \mathfrak{F})$ the following neighborhood systems for S are equivalent:*

i. *the strong neighborhood system $\mathfrak{N}$;*
ii. *the ε_0-neighborhood system $\mathfrak{N}^{\varepsilon_0}$;*
iii. *the c^{ε_0}-neighborhood system.*

PROOF. On comparing (12.1.1) and (13.1.1), it is readily seen that for any p in S and any h, t satisfying $0 < h < t \leqslant 1$ we have

$$N_p(\varepsilon_0, h) \subseteq N_p(t) \subseteq N_p(\varepsilon_0, t).$$

On the other hand, if $t > 1$, then it is immediate that

$$N_p(t) = N_p(\varepsilon_0, 1) = S.$$

Thus $\mathfrak{N}_p$ and $\mathfrak{N}_p^{\varepsilon_0}$ are equivalent for any p in S, whence $\mathfrak{N}$ and $\mathfrak{N}^{\varepsilon_0}$ are equivalent. The equivalence of $\mathfrak{N}^{\varepsilon_0}$ and the c^{ε_0}-neighborhood system follows from Theorem 3.5.9. $\square$

Combining Corollary 13.1.7 and Lemma 13.1.8 with Lemma 3.5.14, we have

Theorem 13.1.9. *Let $(S, \mathfrak{F}, \tau)$ be a* PM *space. If τ is continuous at $(\varepsilon_0, \varepsilon_0)$, then the strong neighborhood system for S, or equivalently the closure operation c^{ε_0}, determines the strong topology for S.*

Theorem 13.1.9 is a slight improvement of the corresponding result obtained in Chapter 12—see the remarks following Theorem 12.1.2.

We conclude this section with several results about separation and metrizability properties. For details, see [Tardiff 1976].

A straightforward consequence of (13.1.1)–(13.1.4) is

Lemma 13.1.10. *For any q in S, p is in $c^{\varphi}\{q\}$ if and only if $F_{pq} \geqslant \varphi$.*

From this lemma, we immediately obtain

Theorem 13.1.11. *Let $(S, \mathfrak{F})$ be a* PSM *space and φ a profile function. Then the neighborhood system $\mathfrak{N}^{\varphi}$ for S has the Fréchet separation property—i.e., if $p \neq q$, then there is a neighborhood of p that does not contain q and a neighborhood of q that does not contain p—if and only if for any distinct p, q in S there is an x_{pq} in $(0, \infty)$ such that $F_{pq}(x_{pq}) < \varphi(x_{pq})$.*

Theorem 13.1.12. *Let $(S, \mathfrak{F}, \tau)$ be a* PM *space and φ a profile function. If τ is continuous and if the neighborhood system $\mathfrak{N}^{\tau(\varphi,\varphi)}$ has the Fréchet separation property, then the neighborhood system $\mathfrak{N}^{\varphi}$ has the Hausdorff separation property* N4.

Lemma 13.1.13. *If $(S, \mathfrak{F})$ is a* PSM *space and φ is a profile function, then the neighborhood system $\mathfrak{N}^{\varphi}$ is first-countable.*

Theorem 13.1.14. *Let $(S, \mathfrak{F}, \tau)$ be a* PM *space and φ a profile function. If τ is continuous and φ is an idempotent of τ, then the topology induced by the neighborhood system $\mathfrak{N}^{\varphi}$ is pseudo-metrizable. If in addition, $c^{\varphi}\{p\} = \{p\}$ for every p in S, then this topology is metrizable.*

13.2. Distinguishability

If A is a subset of a metric space and p is a point in the closure of A, then $\inf\{d(p,q) \mid q \text{ in } A\} = 0$ so that p is at zero distance from A and, in this sense, indistinguishable from points in A. Similarly, if A is a subset of a PM space and p a point in the strong closure of A, then by (12.1.8), $\sup\{F_{pq} \mid q \text{ in } A\} = F_{pA} = \varepsilon_0$ and p is again indistinguishable from A. However, if $\varphi \neq \varepsilon_0$ and p is in the c^{φ}-closure of A, then as the example following Theorem 13.1.5 shows, this need no longer be so, i.e., we may have $F_{pA} \neq \varepsilon_0$. In this case, however, we may regard the c^{φ}-closure of A as a *penumbra* or *haze* about A, within which it is hard to distinguish between members and nonmembers of A. These considerations lead to the concept of profile-distinguishability, introduced in [Schweizer 1975b].

Definition 13.2.1. Let $(S, \mathfrak{F})$ be a PSM space and φ a profile function. For any $A \subseteq S$, the *φ-haze of A* is the set $c^{\varphi}(A)$. For any A, $B \subseteq S$, we say that A and B are *indistinguishable modulo* φ if $A \subseteq c^{\varphi}(B)$ and $B \subseteq c^{\varphi}(A)$; and we write $A(\operatorname{ind}\varphi)B$ or, if $\{p\}$, $\{q\}$ are singleton sets, simply $p(\operatorname{ind}\varphi)q$.

Clearly, indistinguishability modulo φ is reflexive and symmetric. But since we can have $A(\operatorname{ind}\varphi)B$ and $B(\operatorname{ind}\varphi)C$ without having $A(\operatorname{ind}\varphi)C$, indistinguishability modulo φ is in general not transitive and so not an equivalence relation. This situation is an example of the "Poincaré paradox."[1] From Lemma 13.1.10 we have at once

Lemma 13.2.2. *If p and q are points of a* PSM *space, then $p(\operatorname{ind}\varphi)q$ if and only if $F_{pq} \geqslant \varphi$.*

(Note that p and q are perhaps indiscernible in the sense of Definition 12.10.1 if and only if there is an a in $(0, 1]$ such that $p(\operatorname{ind}\langle a\rangle)q$, where $\langle a\rangle = a\varepsilon_0 + (1 - a)\varepsilon_\infty$.)

Lemma 13.2.2 admits the following partial extension to arbitrary subsets of S:

Lemma 13.2.3. *For any profile function φ, let $\alpha = l^-\varphi(\infty)$ and set*

$$\langle\alpha\rangle = \alpha\varepsilon_0 + (1 - \alpha)\varepsilon_\infty. \tag{13.2.1}$$

Let $(S, \mathfrak{F})$ be a PSM *space and, for A, $B \subseteq S$, let F_{AB} be the probabilistic Hausdorff distance given by Definition* 12.9.1. *If $F_{AB} \geqslant \langle\alpha\rangle$, then $A(\operatorname{ind}\varphi)B$.*

Given a PM space $(S, \mathfrak{F}, \tau)$ and a profile function φ, we can define hazes of higher order by considering the τ-powers of φ. Recall that, as in (5.1.3),

---

[1] So called because H. Poincaré referred to it in his popular writings as the typical, but to him unacceptable, property of the "physical continuum." Thus, in Chapter II of [Poincaré 1905] he writes:

> It has, for instance, been observed that a weight A of 10 grammes and a weight B of 11 grammes produced identical sensations, that the weight B could no longer be distinguished from a weight C of 12 grammes, but that the weight A was readily distinguished from the weight C. Thus the rough result of the experiments may be expressed by the following relations: A = B, B = C, A < C which may be regarded as the formula of the physical continuum. But here is an intolerable disagreement with the law of contradiction, and the necessity of banishing this disagreement has compelled us to invent the mathematical continuum . . . . We cannot believe that two quantities which are equal to a third are not equal to one another.

these are defined recursively via

$$\varphi^1 = \varphi \quad \text{and} \quad \varphi^{n+1} = \tau(\varphi^n, \varphi). \tag{13.2.2}$$

The φ^n-haze of any subset A of S is then the φ^n-closure of A, and we have $A(\text{ind}\,\varphi^n)B$ if each of the sets A, B is contained in the φ^n-haze of the other.

We can also introduce a measure of the proximity of points in a PM space as follows.

Definition 13.2.4. Let $(S, \mathfrak{F}, \tau)$ be a PM space and φ a profile function. For any subsets A, B of S, the *degree of proximity* (*mod* φ) of A and B, written $\delta_\varphi(A, B)$, is the least positive integer n, if any, such that $A(\text{ind}\,\varphi^n)B$; if there is no such integer, then $\delta_\varphi(A, B) = \infty$.

Note that $\delta_\varphi(A, B) = 1$ if and only if A and B are indistinguishable modulo φ.

For any two points p, q in S, Lemma 13.2.2 implies that $\delta_\varphi(p, q)$ is finite if and only if there is an n such that $F_{pq} \geqslant \varphi^n$, in which case we have

$$\delta_\varphi(p, q) = \inf\{n \mid F_{pq} \geqslant \varphi^n\}. \tag{13.2.3}$$

Theorem 13.2.5. *For any p, q, r in S we have*

$$\delta_\varphi(p, r) \leqslant \delta_\varphi(p, q) + \delta_\varphi(q, r). \tag{13.2.4}$$

PROOF. If either $\delta_\varphi(p, q)$ or $\delta_\varphi(q, r)$ is infinite, then (13.2.4) is immediate. Otherwise, suppose that $\delta_\varphi(p, q) = m$ and $\delta_\varphi(q, r) = n$. Then, using (13.2.3) and (5.1.4), we have

$$F_{pr} \geqslant \tau(F_{pq}, F_{qr}) \geqslant \tau(\varphi^m, \varphi^n) = \varphi^{m+n},$$

whence $\delta_\varphi(p, r) \leqslant m + n$. □

If we define p and q to be *weakly indistinguishable modulo* φ if $\delta_\varphi(p, q) < \infty$, then (13.2.4) shows that this relation is transitive. Since it is clearly reflexive and symmetric, it is an equivalence relation. On the other hand, if φ is an *Archimedean element* of the semigroup (Δ^+, τ), i.e., if for every $F \neq \varepsilon_\infty$ in Δ^+ there is an n such that $\varphi^n \leqslant F$, then in any PM space $(S, \mathfrak{F}, \tau)$, the relation of weak indistinguishability modulo φ is trivial. Such Archimedean elements exist in many semigroups (Δ^+, τ): e.g., for any a in $(0, 1)$, any $b > 0$, and any Archimedean t-norm T, the function $a\varepsilon_b + (1 - a)\varepsilon_\infty$ is an Archimedean element of (Δ^+, τ_T).

In certain cases, the notion of degree of proximity can be refined to include nonintegral degrees. In particular, if T is a strict t-norm and φ is T-log-concave, then we can use Theorem 7.8.11 to define $\varphi^{[\alpha]}$ for all nonnegative α; and given a PM space $(S, \mathfrak{F}, \tau_T)$, we can define a function δ_φ^* on $S \times S$ via

$$\delta_\varphi^*(p,q) = \begin{cases} \infty, & \text{if } \delta_\varphi(p,q) = \infty, \\ \inf\{\alpha \mid F_{pq} \geqslant \varphi^{[\alpha]}\}, & \text{if } \delta_\varphi(p,q) < \infty. \end{cases} \tag{13.2.5}$$

Then, when combined with (7.8.22), the argument used to establish (13.2.4) yields

$$\delta_\varphi^*(p,r) \leqslant \delta_\varphi^*(p,q) + \delta_\varphi^*(q,r) \tag{13.2.6}$$

for all p, q, r in S.

Note that $\delta_\varphi^*(p,q)$ is finite if and only if $\delta_\varphi(p,q)$ is finite; indeed, if $\delta_\varphi^*(p,q)$ is finite, then $\delta_\varphi(p,q)$ is the least integer $\geqslant \delta_\varphi^*(p,q)$.

If a PM space $(S, \mathfrak{F}, \tau)$ admits a function δ_φ^*, then two points p, q in S are indistinguishable modulo φ if and only if $\delta_\varphi^*(p,q) \leqslant 1$. In this case, we can use the value of $\delta_\varphi^*(p,q)$ to specify an exact *level of indistinguishability modulo* φ. This can also be done in a simpler but coarser fashion without the function δ_φ^*, as follows:

For any p, q in S, if there is a positive integer n such that $F_{pq}^n < \varphi$, let $\iota_\varphi(p,q)$ be the least such positive integer; otherwise, set $\iota_\varphi(p,q) = \infty$. Clearly, if $p(\operatorname{ind}\varphi)q$, then $\iota_\varphi(p,q) \geqslant 2$; and the exact value of $\iota_\varphi(p,q)$ will serve to specify various levels of indistinguishability modulo φ.

14

Betweenness

14.1. Wald Betweenness

In a metric space there is just one natural way to introduce a notion of betweenness (see Section 3.3). On the other hand, as shown in [Moynihan and Schweizer 1979], in a PM space there are a number of distinct possibilities. In this chapter we consider several of these. We begin with a straightforward generalization of A. Wald's [1943] definition of betweenness for Wald spaces.

For distinct points p, q, r in a Wald space, Wald postulated that "q lies between p and r" if and only if

$$F_{pr} = F_{pq} * F_{qr} \tag{14.1.1}$$

where $*$ denotes convolution. He showed that betweenness in this sense has the standard properties (3.3.2)–(3.3.4) of metric betweenness.

If we replace convolution in (14.1.1) by an arbitrary triangle function τ, we obtain an expression that is meaningful in any PM space $(S, \mathfrak{F}, \tau)$. This observation motivates

Definition 14.1.1. Let $(S, \mathfrak{F}, \tau)$ be a PM space. For any p, q, r in S, we say that *q is Wald-between p and r*, and we write $\mathfrak{W}(pqr)$, if p, q, r are distinct, $F_{pr} \neq \varepsilon_\infty$, and

$$F_{pr} = \tau(F_{pq}, F_{qr}). \tag{14.1.2}$$

It follows from Theorem 8.4.2 and its proof that in any simple space (S, d, G) Wald betweenness coincides with betweenness with respect to the generating metric d. Similarly, (8.6.7) yields the same conclusion for any α-simple space $(S,d,G;\alpha)$ with G strict and $\alpha>1$.

If $\mathfrak{W}(pqr)$, then each of F_{pq}, F_{qr}, F_{pr} is distinct from both ε_0 and ε_∞. Furthermore, $\mathfrak{W}(pqr)$ implies $\mathfrak{W}(rqp)$, whence Wald betweenness satisfies (3.3.2). On the other hand, since a particular triangle function may not have all the pleasant properties of convolution, it is to be expected that Wald betweenness may sometimes lack one or more of the properties of metric betweenness. This is in fact the case. For example, suppose that τ has a nontrivial idempotent, i.e., that there is a function G in Δ^+ distinct from both ε_0 and ε_∞ such that $\tau(G,G)=G$. Let $(S,\mathfrak{F},\tau)$ be an equilateral space in which $F_{pq}=G$ for all distinct p, q in S. Then it is immediate from (14.1.2) that each one of any three distinct points of S is Wald-between the other two, whence (3.3.3) does not hold for Wald betweenness in $(S,\mathfrak{F},\tau)$.

To study the connection between properties of triangle functions and properties of Wald betweenness, we begin with

Definition 14.1.2. Let τ be a triangle function and Γ a subset of Δ^+. Then Γ_τ, the *τ-closure* of Γ, is the smallest subset of Δ^+ that contains Γ and is closed under τ. Furthermore:

i. τ is *cancellative on* Γ if $\tau(F,G)=\tau(F,H)$ implies that $G=H$ for any F, G, H in Γ_τ such that $F\neq\varepsilon_\infty$.
ii. τ is *strictly increasing on* Γ if $G<H$ implies that $\tau(F,G)<\tau(F,H)$ for any F, G, H in Γ_τ such that $F\neq\varepsilon_\infty$.
iii. τ is *Archimedean on* Γ if $\tau(F,G)<F$ for any F, G in Γ_τ such that $F\neq\varepsilon_\infty$ and $G\neq\varepsilon_0$.

Lemma 14.1.3. *Let τ be a triangle function. Then:*

i. *If τ is cancellative on Γ, then τ is strictly increasing on Γ.*
ii. *If τ is strictly increasing on Γ, then τ is Archimedean on Γ.*
iii. *If τ is Archimedean on Γ, then there are no nontrivial idempotents of τ in Γ_τ.*
iv. *If τ is continuous and has no nontrivial idempotents in Δ^+, then τ is Archimedean on Δ^+.*

PROOF. We omit the straightforward proofs of statements i, ii, and iii. As for statement iv, for any G in Δ^+ let G^n denote the nth τ-power of G (see (5.1.3)). Then for any $G\neq\varepsilon_0$ we have $\varepsilon_0>G\geqslant G^2\geqslant\cdots\geqslant G^n\geqslant\cdots$, whence the sequence $\{G^n\}$ has a weak limit H in Δ^+ distinct from ε_0. Since τ is continuous,

$$\tau(H,H)=\tau(\lim G^n,\lim G^n)=\lim\tau(G^n,G^n)=\lim G^{2n}=H,$$

so that H is idempotent; and since τ has no nontrivial idempotents, necessarily $H = \varepsilon_\infty$. Now suppose that τ is not Archimedean on Δ^+. Then there exist $F \neq \varepsilon_\infty$, $G \neq \varepsilon_0$ such that $\tau(F,G) = F$. Hence

$$\begin{aligned} F &= \tau(F,G) = \tau(\tau(F,G),G) = \tau(F,G^2) = \cdots \\ &= \tau(F,G^n) = \cdots = \tau(F,\lim G^n) = \tau(F,\varepsilon_\infty) = \varepsilon_\infty, \end{aligned}$$

which is a contradiction. □

Theorem 14.1.4. *Let* $(S,\mathfrak{F},\tau)$ *be a* PM *space and let* p, q, r, s *be in* S. *Then*:

i. *If* τ *is Archimedean on* $\operatorname{Ran}\mathfrak{F}$, *then* $\mathfrak{W}(pqr)$ *implies that neither* $\mathfrak{W}(prq)$ *nor* $\mathfrak{W}(qpr)$ *hold, i.e., Wald betweenness satisfies* (3.3.3).
ii. $\mathfrak{W}(pqr)$ *and* $\mathfrak{W}(prs)$ *together imply* $\mathfrak{W}(pqs)$, *i.e., Wald betweenness satisfies half of* (3.3.4). *If* τ *is strictly increasing on* $\operatorname{Ran}\mathfrak{F}$, *then* $\mathfrak{W}(pqr)$ *and* $\mathfrak{W}(prs)$ *together imply* $\mathfrak{W}(qrs)$, *i.e., Wald betweenness satisfies the other half of* (3.3.4).
iii. *If* τ *is continuous*, p *and* r *are distinct, and* $F_{pr} \neq \varepsilon_\infty$, *then the set* $A(p,r) = \{q \mid \mathfrak{W}(pqr)\} \cup \{p,r\}$ *is closed in the strong topology, i.e., Wald betweenness satisfies* (3.3.5).

PROOF. Suppose that τ is Archimedean on $\operatorname{Ran}\mathfrak{F}$ and $\mathfrak{W}(pqr)$. Then

$$\tau(F_{pr},F_{qr}) = \tau(\tau(F_{pq},F_{qr}),F_{qr}) = \tau(F_{pq},\tau(F_{qr},F_{qr})) < F_{pq},$$

whence $\mathfrak{W}(prq)$ does not hold. Similarly, $\mathfrak{W}(qpr)$ does not hold. This proves Part i.

Now suppose that $\mathfrak{W}(pqr)$ and $\mathfrak{W}(prs)$. Then

$$F_{ps} = \tau(F_{pr},F_{rs}) = \tau(\tau(F_{pq},F_{qr}),F_{rs}) = \tau(F_{pq},\tau(F_{qr},F_{rs})),$$

whence, since $\tau(F_{qr},F_{rs}) \leqslant F_{qs}$, we have $F_{ps} \leqslant \tau(F_{pq},F_{qs})$. But F_{ps} is greater than or equal to $\tau(F_{pq},F_{qs})$. Therefore $F_{ps} = \tau(F_{pq},F_{qs})$, whence $\mathfrak{W}(pqs)$. Next, if τ is strictly increasing on $\operatorname{Ran}\mathfrak{F}$, then $F_{qs} > \tau(F_{qr},F_{rs})$ implies

$$\begin{aligned} F_{ps} &\geqslant \tau(F_{pq},F_{qs}) > \tau(F_{pq},\tau(F_{qr},F_{rs})) \\ &= \tau(\tau(F_{pq},F_{qr}),F_{rs}) = \tau(F_{pr},F_{rs}) = F_{ps}, \end{aligned}$$

which is impossible. Thus $\mathfrak{W}(qrs)$, and Part ii is established.

Turning to Part iii, suppose p and r are distinct points with $F_{pr} \neq \varepsilon_\infty$, and let q be an accumulation point of the set $A(p,r)$. Then there is a sequence $\{q_n\}$ of points in $A(p,r)$ that converges to q in the strong topology. For each q_n in the sequence, we have $F_{pr} = \tau(F_{pq_n},F_{q_nr})$. Therefore, since τ is continuous, $F_{pr} = \tau(F_{pq},F_{qr})$, whence either $q = p$ or $q = r$ or $\mathfrak{W}(pqr)$. In each case, q is in $A(p,r)$, which establishes Part iii and completes the proof. □

Using known properties of particular triangle functions and applying Lemma 14.1.3 and Theorem 14.1.4, we have the following special cases.

1. Since convolution is continuous [Schweizer 1975a] and cancellative (see [Feller 1966, §13.1]) on Δ^+, the conclusions of Theorem 14.1.4 are valid in any Wald space (cf. [Wald 1943] and [Rhodes 1964]). (Recall that convolution is not cancellative on $\mathfrak{D}$ [Feller 1966, §15.2].)
2. The triangle function τ_M is cancellative on $\mathfrak{D}^+$, though not on Δ^+ (see [Moynihan 1975]), and continuous on Δ^+ (Theorem 7.2.8). Hence the conclusions of Theorem 14.1.4 are valid in any space $(S, \mathfrak{F}, \tau_M)$ in which $\operatorname{Ran} \mathfrak{F} \subseteq \mathfrak{D}^+$. On the other hand, there exist spaces $(S, \mathfrak{F}, \tau_M)$ in which $\operatorname{Ran} \mathfrak{F}$ is not a subset of $\mathfrak{D}^+$ and (3.3.4) fails for Wald betweenness.
3. Let T be a strict t-norm. Then τ_T is continuous on Δ^+ (Theorem 7.2.8) and cancellative on the set Δ_T^+ of all T-log-concave functions in Δ^+ (Definition 7.8.7 and Theorem 7.8.10). Hence the conclusions of Theorem 14.1.4 are valid in any space $(S, \mathfrak{F}, \tau_T)$ in which $\operatorname{Ran} \mathfrak{F} \subseteq \Delta_T^+$. But τ_T is not cancellative on Δ^+ (see [Moynihan 1975]) and there exist spaces $(S, \mathfrak{F}, \tau_T)$ in which $\operatorname{Ran} \mathfrak{F}$ is not a subset of Δ_T^+ and (3.3.4) fails for Wald betweenness (see [Moynihan and Schweizer 1979]).

To discuss Wald betweenness in E-spaces, we need

Lemma 14.1.5. *Let $(S, \mathfrak{F})$ be a canonical E-space with base $(\Omega, \mathfrak{A}, P)$ and target (M, d) such that d never assumes the value ∞. Given p, q, r in S, let C be a copula of the random variables $d(p,q)$ and $d(q,r)$. Then*

$$d(p,r) = d(p,q) + d(q,r) \quad \text{a.s.} \tag{14.1.3}$$

if and only if

$$F_{pr} = \sigma_C(F_{pq}, F_{qr}). \tag{14.1.4}$$

PROOF. Since d never assumes the value infinity, it follows that F_{pr} and $df(d(p,q) + d(q,r))$ are in $\mathfrak{D}^+$. By Lemma 7.6.2 and Definition 7.4.1,

$$df(d(p,q) + d(q,r)) = \sigma_C(F_{pq}, F_{qr}).$$

Since $d(p,r) \leqslant d(p,q) + d(q,r)$ everywhere, the equivalence of (14.1.3) and (14.1.4) follows immediately from Lemma 4.1.5. □

Now every canonical E-space is a Menger space under W (Theorem 9.1.2), and if T is a t-norm stronger than W, then there are canonical E-spaces that are not Menger spaces under T (Theorem 9.2.5). Thus in considering E-spaces in general, it is natural to restrict our attention to

Wald betweenness relative to the triangle function τ_W. We then have

Theorem 14.1.6. *Let $(S, \mathcal{F}, \tau_W)$ be a canonical E-space with base $(\Omega, \mathcal{Q}, P)$ and target (M, d), and suppose that d never assumes the value ∞. If p, q, r are distinct points in S, then $\mathfrak{W}(pqr)$ if and only if* (14.1.3) *holds and there is an a in $(0, \infty)$ such that either $F_{pq} = \varepsilon_a$ or $F_{qr} = \varepsilon_a$.*

PROOF. Suppose $\mathfrak{W}(pqr)$, i.e., that $F_{pr} = \tau_W(F_{pq}, F_{qr})$. Let C be a copula of $d(p,q)$ and $d(q,r)$. Since d is a metric, for any x in $(0,\infty)$ we have

$$\begin{aligned}\sigma_C(F_{pq}, F_{qr})(x) &= P\{\omega \text{ in } \Omega \mid (d(p,q) + d(q,r))(\omega) < x\} \\ &\leqslant P\{\omega \text{ in } \Omega \mid d(p,r)(\omega) < x\} \\ &= F_{pr}(x) = \tau_W(F_{pq}, F_{qr})(x),\end{aligned}$$

whence $\sigma_C(F_{pq}, F_{qr}) \leqslant \tau_W(F_{pq}, F_{qr})$. But since by (7.5.2), $\tau_W \leqslant \sigma_C$, we have

$$F_{pr} = \tau_W(F_{pq}, F_{qr}) = \sigma_C(F_{pq}, F_{qr}), \tag{14.1.5}$$

which is (14.1.4) and implies (14.1.3). Next, by Corollary 2 of Theorem 9 of [Moynihan, Schweizer, and Sklar 1978], the second equality in (14.1.5) holds if and only if there is an a in $(0, \infty)$ such that either $F_{pq} = \varepsilon_a$ or $F_{qr} = \varepsilon_a$.

In the other direction, if (14.1.3) holds and either $F_{pq} = \varepsilon_a$ or $F_{qr} = \varepsilon_a$, then Lemma 14.1.5 and Corollary 2 of Theorem 9 of [Moynihan, Schweizer, and Sklar 1978] immediately yield $\mathfrak{W}(pqr)$. □

To illustrate, consider the space $\mathfrak{S}(I)$ of Section 9.1. For distinct p, q, r in this space, we have $\mathfrak{W}(pqr)$ if and only if the graph of q lies between the graphs of p and r almost everywhere, and either $p - q$ or $q - r$ is constant (and nonzero) almost everywhere. In particular, if $p = r$ on a set of positive Lebesgue measure, then there is no q such that $\mathfrak{W}(pqr)$. Thus in $\mathfrak{S}(I)$, Wald betweenness is considerably stronger than betweenness with respect to either the L_∞ (ess sup) metric or the L_1 metric (in which betweenness is simply pointwise almost everywhere betweenness). For $1 < \beta < \infty$, Wald betweenness is not comparable to betweenness with respect to the L_β metric.

14.2. Transform Betweenness

Any Menger space under a continuous Archimedean t-norm admits an interesting variation on Wald betweenness, whose definition is motivated by Theorem 7.8.10.

Definition 14.2.1. Let $(S, \mathcal{F}, \tau_T)$ be a Menger space for which T is continuous and Archimedean. Let p, q, r be in S. Then q is *transform-between p*

and r, and we write $\mathfrak{T}(pqr)$, if p, q, r are distinct, $F_{pr} \neq \varepsilon_\infty$, and

$$F_{prT} = \tau_T(F_{pqT}, F_{qrT}) \tag{14.2.1}$$

or equivalently,

$$c_T F_{pr} = c_T F_{pq} \cdot c_T F_{qr}, \tag{14.2.2}$$

where F_T is given by (7.8.15) and c_T is the T-conjugate transform.

Using (14.2.2) and Theorem 8.2.9 we immediately have

Theorem 14.2.2. *Let* $(S, \mathfrak{F}, \tau_T)$ *be as in Definition* 14.2.1. *For any* x *in* $(0, \infty)$, *let* $d_{T,x}$ *be the metric on* S *defined by* (8.2.14). *Then for any* p, q, r *in* S, $\mathfrak{T}(pqr)$ *if and only if* $\langle pqr \rangle$ *in* $(S, d_{T,x})$ *for every* x *in* $(0, \infty)$.

Corollary 14.2.3. *Transform betweenness satisfies* (3.3.2)–(3.3.4), *and for every distinct* p, r *in* S *with* $F_{pr} \neq \varepsilon_\infty$ *the set* $\{q \mid \mathfrak{T}(pqr)\} \cup \{p, r\}$ *is closed in the strong topology.*

If the t-norm T is strict, then (7.8.12) yields the following simple relation between Wald betweenness and transform betweenness.

Theorem 14.2.4. *If* $(S, \mathfrak{F}, \tau_T)$ *is a Menger space under a strict t-norm* T, *then* $\mathfrak{W}(pqr)$ *implies* $\mathfrak{T}(pqr)$.

In PM spaces $(S, \mathfrak{F}, \tau_T)$ in which T is Archimedean but not strict, the relations of Wald betweenness and transform betweenness are generally not comparable. However, we do have

Theorem 14.2.5. *Let* $(S, \mathfrak{F}, \tau)$ *be as in Definition* 14.2.1, *and let the function* $\mathfrak{F}_T$ *be defined on* $S \times S$ *by*

$$\mathfrak{F}_T(p, q) = (\mathfrak{F}(p, q))_T. \tag{14.2.3}$$

Then $(S, \mathfrak{F}_T)$ *is a Menger space under* T *in which Wald betweenness and transform betweenness coincide.*

14.3. Menger Betweenness

The first definition of betweenness for PM spaces was the one proposed by Menger [1942]. Within his framework (see Section 1.2), Menger said that q lies between p and r if

$$1 - F_{pr}(x + y) \geqslant T(1 - F_{pq}(x), 1 - F_{qr}(y)) \tag{14.3.1}$$

for all x, y in $(0, \infty)$.

As in Section 1.1, the inequality (14.3.1) has the following natural interpretation: "The probability that the distance between p and r is at least $x+y$ is at least as great as a function of the probability that the distance between p and q is at least x and the probability that the distance between q and r is at least y."

In a Wald space, where the distances of different pairs of points can be regarded as independent r.v.'s (see Section 1.3), it is natural to rephrase the interpretation of (14.3.1) as follows: "The probability that the distance between p and r is at least $x+y$ is at least as great as the probability that the sum of the distance between p and q and the distance between q and r is at least $x+y$." Since this latter probability is $1-(F_{pq}*F_{qr})(x+y)$, (14.3.1) becomes

$$1-F_{pr}(x+y) \geqslant 1-(F_{pq}*F_{qr})(x+y)$$

for all x, y in $(0,\infty)$, or equivalently,

$$F_{pr}(x) \leqslant (F_{pq}*F_{qr})(x) \tag{14.3.2}$$

for all x in $(0,\infty)$. Combining (14.3.2) with (1.3.1) yields $F_{pr}=F_{pq}*F_{qr}$, which is Wald's definition of betweenness.

Returning to (14.3.1), observe that it is equivalent to

$$F_{pr}(x+y) \leqslant T^*(F_{pq}(x), F_{qr}(y))$$

for all x, y in $(0,\infty)$, where T^* is given by (5.7.1). Since T^* belongs to $\mathfrak{T}^*$ (see Definition 7.1.5), this yields $F_{pr} \leqslant \tau_{T^*}(F_{pq}, F_{qr})$ (see Definition 7.3.1), and motivates

Definition 14.3.1. Let $(S, \mathfrak{F}, \tau_T)$ be a Menger space. For any p, q, r in S we say that *q is Menger-between p and r*, and we write $\mathfrak{M}(pqr)$, if p, q, r are distinct and

$$F_{pr} \leqslant \tau_{T^*}(F_{pq}, F_{qr}). \tag{14.3.3}$$

Since $\tau_T \leqslant \tau_{T^*}$ for any t-norm T (by (7.3.4)), and since $\tau_M = \tau_{M^*}$ (by (7.5.11)), (14.3.3) is consistent with the triangle inequality (8.1.5). Furthermore, we have

Theorem 14.3.2. *In any Menger space, Wald betweenness implies Menger betweenness. In a Menger space under M, Wald betweenness and Menger betweenness coincide.*

In general, however, Menger betweenness is much weaker than Wald betweenness. To illustrate this, consider an α-simple space $(S, d, G; \alpha)$ with

G strict and $\alpha > 1$. Some calculations using (14.3.1), (8.6.1), and (8.6.5) yield the fact that $\mathfrak{M}(pqr)$ is equivalent to the validity of the inequality

$$(d(p,q))^{\alpha}H(u) + (d(q,r))^{\alpha}H(v) \leqslant (d(p,r))^{\alpha}H(u+v) \quad (14.3.4)$$

for all u, v in R^+, where H is the continuous and strictly increasing function defined on R^+ by

$$H(u) = G^{-1}(1 - G(u^{1-\alpha})). \quad (14.3.5)$$

Since $H(0) = 0$, it follows from (14.3.4) that a necessary condition for $\mathfrak{M}(pqr)$ is

$$d(p,r) \geqslant \mathrm{Max}(d(p,q), d(q,r)); \quad (14.3.6)$$

and when H is superadditive, it is easily seen that (14.3.6) is a sufficient condition for $\mathfrak{M}(pqr)$ as well. There are many strict d.f.'s G for which H is superadditive. In particular, any G such that $G(1) = \frac{1}{2}$ and $G(x) = 1 - G(1/x)$ for $x \geqslant 1$ yields $H(x) = x$ for all x. For such a d.f. G, it follows that if p, q, r are distinct points in an α-simple space $(S, d, G; \alpha)$ with $\alpha > 1$, then at least one of them is Menger-between the other two; and if the points form an equilateral triangle in (S, d), then each of the three points is Menger-between the other two. In particular, when (S, d) is the Euclidean plane, the set $\{q \mid \mathfrak{M}(pqr)\} \cup \{p, r\}$ is the closed convex region bounded by two circular arcs of radius $d(p,r)$, one with center at p, the other with center at r. Further details concerning Menger-betweenness in α-simple spaces are given in [Alsina and Schweizer 1982].

14.4. Probabilistic Betweenness

The betweenness relations considered in the preceding sections are deterministic: Given three distinct points p, q, r, either q is between p and r or it is not. It is of course more in the spirit of our theory to have a probabilistic notion of betweenness. H. Sherwood [1970] showed that such a notion is definable in E-spaces, where one can speak meaningfully of the probability that one point is between two others.

Definition 14.4.1. Let $(S, \mathfrak{F})$ be a canonical E-space with base $(\Omega, \mathcal{Q}, P)$ and target (M, d). For any three points p, q, r in S, the *probability that q is between p and r* is the number $B(p,q,r)$ given by

$$\begin{aligned} B(p,q,r) = P\{\omega \text{ in } \Omega \mid\, & d(p(\omega), r(\omega)) \\ & = d(p(\omega), q(\omega)) + d(q(\omega), r(\omega))\}. \end{aligned} \quad (14.4.1)$$

Note that an immediate consequence of Lemma 14.1.5 is

Theorem 14.4.2. *Let $(S, \mathfrak{F})$ be a canonical E-space with target (M, d) and such that d never assumes the value ∞. Then for any p, q, r in S, $B(p,q,r) = 1$ if and only if* (14.1.4) *holds, i.e., if and only if $F_{pr} = \sigma_C(F_{pq}, F_{qr})$.*

Note further that we do not require the points p, q, r in Definition 14.4.1 to be distinct. This allows us to consider expressions such as $B(p,q,p)$, which is easily seen to be $F_{pq}(0+)$; and $B(p,p,q)$, which is immediately seen to be 1. Another immediate consequence of the definition is $B(p,q,r) = B(r,q,p)$ for all p, q, r. This is the probabilistic analog of (3.3.2). By using arguments similar to those in the proof of Theorem 9.1.2, Sherwood obtained analogs of (3.3.3) and (3.3.4) as well. These are contained in

Theorem 14.4.3. *For any three points p, q, r in a canonical E-space, we have*

$$B(p,q,r) + B(q,p,r) + B(p,r,q) \leqslant 1 + B(p,q,p) + B(q,r,q) + B(r,p,r). \qquad (14.4.2)$$

For any four points p, q, r, s in a canonical E-space, we have

$$B(p,q,s) \geqslant W(B(p,q,r), B(p,r,s)), \qquad (14.4.3)$$

$$B(q,r,s) \geqslant W(B(p,q,r), B(p,r,s)). \qquad (14.4.4)$$

Corollary 14.4.4. *If p, q, r are points in a canonical E-space such that F_{pq}, F_{qr}, and F_{pr} are continuous at 0, then*

$$B(p,q,r) + B(q,p,r) + B(p,r,q) \leqslant 1, \qquad (14.4.5)$$

whence $B(p,q,r) = 1$ implies $B(q,p,r) = B(p,r,q) = 0$.

Sherwood also proved the following analog of Part iii of Theorem 14.1.3:

Theorem 14.4.5. *For any two points p, r in a canonical E-space and any t in I, the set $\{q \mid B(p,q,r) \geqslant t\}$ is closed in the strong topology.*

As a corollary to Theorem 14.1.6, we have

Theorem 14.4.6. *Under the hypotheses of Theorem 14.1.6, if p, q, r are distinct points in S, then $\mathfrak{W}(pqr)$ if and only if $B(p,q,r) = 1$ and there is an a in $(0, \infty)$ such that either $F_{pq} = \varepsilon_a$ or $F_{qr} = \varepsilon_a$.*

If p, q, r are points in an E-space, and C is any 2-copula, then (7.5.2) and (7.5.10) yield

$$\sigma_C(F_{pq}, F_{qr}) \leqslant \tau_{W^*}(F_{pq}, F_{qr}).$$

Hence an immediate consequence of Theorem 14.4.2 is

Theorem 14.4.7. *Let $(S, \mathcal{F}, \tau_W)$ be a canonical E-space with target (M, d) and such that d never assumes the value ∞. If p, q, r are distinct points in S, then $B(p,q,r) = 1$ implies $\mathfrak{M}(pqr)$.*

The converse is false. In [Moynihan and Schweizer 1979] there is an example, using the space $\mathcal{E}(I)$, with $\mathfrak{M}(pqr)$ but $B(p,q,r) = 7/8$. A more elaborate example in [Sherwood 1970] has $\mathfrak{M}(pqr)$ but $B(p,q,r) = 0$.

14.5. Open Problems

The outstanding open problem is a long-term project: Elaborate one or more of the theories of betweenness and develop the corresponding theories of segments, arc length, and curvature in PM spaces.

A key step in such development would be the demonstration that certain sets are convex, either in the sense of Menger, according to which a set A is *metrically convex* if for any distinct points p and r in A there is a point q in A between p and r; or in the classical Minkowski sense, according to which A is convex if whenever p and r are in A and q is between p and r, then q is in A. These considerations motivate

Problem 14.5.1. Let $(S, \mathcal{F}, \tau)$ be a PM space and let p and r be distinct points in S. Under what conditions is the set $\{q \mid \mathfrak{W}(pqr)\}$ convex? If τ is of the form τ_T for some t-norm T, then under what conditions is the set $\{q \mid \mathfrak{M}(pqr)\}$ convex? If, in addition, T is continuous and Archimedean, then under what conditions is the set $\{q \mid \mathfrak{T}(pqr)\}$ convex?

Problem 14.5.2. Let $(S, \mathcal{F})$ be a canonical E-space with base $(\Omega, \mathcal{Q}, P)$ and target (M, d). Let p, r be points in S, and α a number in $(0, 1]$. Let $A_{pr}(\alpha) = \{q \mid B(p,q,r) \geqslant \alpha\}$. Under what conditions does $A_{pr}(\alpha)$ have the property that if q_1 and q_3 are in $A_{pr}(\alpha)$ and q_2 is in $A_{q_1q_3}(\alpha)$, then q_2 is in $A_{pr}(\alpha)$? More generally, if q_1 and q_3 are in $A_{pr}(\alpha)$ and q_2 is in $A_{q_1q_3}(\beta)$, is q_2 in $A_{pr}(W(\alpha, \beta))$? If so, can W be replaced by a stronger t-norm, e.g., a positive one?

15
Supplements

15.1. Probabilistic Normed Spaces

Recall that a *normed linear space* (or *Banach space*, if complete) is a pair $(V, \|\cdot\|)$, where V is a vector space over the field K of real or complex numbers and $\|\cdot\|$ is a *norm*, i.e., a function from V into $[0, \infty)$ such that for any p, q in V and a in K the following conditions hold.

i. $\|\theta\| = 0$ where θ is the null-vector in V. (15.1.1)
ii. $\|p\| \neq 0$ if $p \neq \theta$. (15.1.2)
iii. $\|ap\| = |a|\|p\|$. (15.1.3)
iv. $\|p + q\| \leqslant \|p\| + \|q\|$. (15.1.4)

A normed linear space is a metric space under the metric d defined by

$$d(p,q) = \|p - q\|. \tag{15.1.5}$$

If $\|\cdot\|$ is a function from V into $[0, \infty)$ satisfying (15.1.1), (15.1.3), and (15.1.4), but not necessarily (15.1.2), then $\|\cdot\|$ is a *pseudonorm* on V, and the function d defined by (15.1.5) is a pseudometric on V.

Following the procedure used in going from metric to PM spaces, we probabilize these notions by replacing the norm by a function from V into Δ^+ and the operation of addition by a suitable triangle function.

Definition 15.1.1. A *probabilistic normed space* (briefly, a *PN space*) is a triple (V, ν, τ) where V is a vector space over the field K of real or complex numbers, ν is a function from V into Δ^+, τ is a continuous triangle function, and for any p, q in V and any $a \neq 0$ in K the following conditions hold.

i. $\nu(\theta) = \varepsilon_0$. (15.1.6)
ii. $\nu(p) \neq \varepsilon_0$ if $p \neq \theta$. (15.1.7)
iii. $\nu(ap)(x) = \nu(p)(x/|a|)$ for all x in R^+. (15.1.8)
iv. $\nu(p+q) \geqslant \tau(\nu(p), \nu(q))$. (15.1.9)

If ν satisfies (15.1.6), (15.1.8), and (15.1.9), then (V, ν, τ) is a *probabilistic pseudonormed space* (briefly, a *PPN space*).

As in the case of PM spaces, when convenient we write ν_p rather than $\nu(p)$ and express (15.1.8) in the equivalent form

$$\nu_{ap} = \nu_p(j/|a|). \tag{15.1.10}$$

An immediate consequence of Definition 15.1.1 is

Theorem 15.1.2. *Let* (V, ν, τ) *be a* PN *space and let* $\mathfrak{F}$ *be the function from* $V \times V$ *into* Δ^+ *defined by*

$$\mathfrak{F}(p, q) = \nu(p - q). \tag{15.1.11}$$

Then $(V, \mathfrak{F}, \tau)$ *is a* PM *space.*

Just as every ordinary metric space can be viewed as a PM space, so every ordinary normed linear space can be viewed as a probabilistic normed space, one in which each function $\nu(p)$ is of the form ε_a for some a in $[0, \infty)$.

Probabilistic normed spaces were introduced by A. N. Šerstnev [1962, 1963ab, 1964b]. His principal concern in these papers was with problems of best approximation. Such problems can often be formulated in general PM spaces, a typical example being the following: Let $(S, \mathfrak{F}, \tau)$ be a PM space, p a point in S, and A a subset of S. Find a point q in A that is "closest" to p in the sense that $F_{pr} \ngtr F_{pq}$ for any r in A. (Note that this is much weaker than requiring that $F_{pq} = F_{pA}$, where F_{pA} is the probabilistic distance from p to A as defined in (12.1.7).) Šerstnev [1963b, Th. 4; 1964b, Th. 5], showed that if (V, ν, τ) is a PN space and A is a finite-dimensional subspace of V, then such a point q always exists for any p in V. He also considered the uniqueness question and other related problems of best approximation in PN spaces.

If $(V, \|\cdot\|)$ is a normed linear space and G a d.f. in Δ^+ distinct from ε_0 and ε_∞, then the *simple PN space* generated by $(V, \|\cdot\|)$ and G is the PN space in which ν is defined by

$$\nu_p = G(j/\|p\|). \tag{15.1.12}$$

It is immediate that the associated PM space given by Theorem 15.1.2 is a simple PM space, that a simple PN space is a PN space under τ_M, etc. Furthermore, comparison of (15.1.12) and (15.1.8) shows that any one-dimensional subspace of a PN space is a simple PN space. However, in view of the homogeneity condition (15.1.8), there is no PN space analog of an α-simple PM space for any $\alpha \neq 1$.

For any p, q in V and any $a > 0$, we can estimate the quantity $\nu_{a^{-1}p+a^{-1}q}$ in two ways: First, by

$$\nu_{a^{-1}p+a^{-1}q} \geqslant \tau(\nu_{a^{-1}p}, \nu_{a^{-1}q}) = \tau(\nu_p(aj), \nu_q(aj));$$

and second, by

$$\nu_{a^{-1}p+a^{-1}q} = \nu_{a^{-1}(p+q)} = \nu_{p+q}(aj) \geqslant \tau(\nu_p, \nu_q)(aj).$$

If we want these estimates to be consistent, then it is reasonable to require that τ be homogeneous in the sense of (12.6.7). This is supported by the fact that convolution and all the functions τ_T are homogeneous, and by the following results of Muštari and Šerstnev [1965, 1978]:

Lemma 15.1.3. *If τ is a triangle function, then for any $a > 0$ the function τ_a defined on $\Delta^+ \times \Delta^+$ by*

$$\tau_a(F, G) = \tau(F(aj), G(aj))(j/a) \tag{15.1.13}$$

is also a triangle function.

Note that τ is homogeneous if and only if $\tau = \tau_a$ for all $a > 0$.

Lemma 15.1.4. *If (V, ν, τ) is a* PN *space, then (V, ν, τ_a) is also a* PN *space for any $a > 0$.*

Theorem 15.1.5. *If τ is the strongest triangle function under which the triple (V, ν, τ) is a* PN *space, then τ is homogeneous on* Ran ν.

PROOF. For any p, q in V and any $a > 0$ we have

$$\begin{aligned} \nu_{p+q} &\geqslant \tau(\nu_p, \nu_q) \geqslant \tau_a(\nu_p, \nu_q) \\ &= \tau(\nu_p(aj), \nu_q(aj))(j/a) \\ &\geqslant \tau_{a^{-1}}(\nu_p(aj), \nu_q(aj))(j/a) = \tau(\nu_p, \nu_q), \end{aligned}$$

whence $\tau = \tau_a$ on Ran ν. □

It follows that in the study of PN spaces we can, without significant loss of generality, restrict our attention to homogeneous triangle functions.

The PN space analogs of E-spaces are readily defined as follows.

Definition 15.1.6. Let $(\Omega, \mathscr{A}, P)$ be a probability space, $(V, \|\cdot\|)$ a normed linear space over the field K of real or complex numbers, and S a set of functions from Ω into V. Let ν be a function from S into Δ^+. Then (S, ν) is an *E-norm space* with *base* $(\Omega, \mathscr{A}, P)$ and *target* $(V, \|\cdot\|)$ if the following conditions hold.

i. S, under pointwise addition and scalar multiplication, is a linear space over K. The zero element in S is the constant function $\boldsymbol{\theta}$ given by $\boldsymbol{\theta}(\omega) = \theta$ for all ω in Ω.
ii. For all p in S and all x in R^+ the set $\{\omega \text{ in } \Omega \mid \|p(\omega)\| < x\}$ belongs to $\mathscr{A}$, i.e., the composite function $\|p\|$ is P-measurable.
iii. For all p in S, $\nu_p = df(\|p\|)$, i.e.,

$$\nu_p(x) = P\{\omega \text{ in } \Omega \mid \|p(\omega)\| < x\}. \tag{15.1.14}$$

If for any p in S, $p = \boldsymbol{\theta}$ a.s. only if $p = \boldsymbol{\theta}$, then we call (S, ν) a *canonical* E-norm space.

Theorem 15.1.7. *If (S, ν) is an E-norm space, then (S, ν, τ_W) is a* PPN *space. If (S, ν) is a canonical E-norm space, then (S, ν, τ_W) is a* PN *space.*

E-norm spaces were first considered in depth by D. H. Muštari [1970], who studied the relationship between almost sure convergence and convergence in measure in canonical E-norm spaces in which the elements of S are themselves random variables. The principal result is that these modes of convergence agree if and only if S is locally convex and nuclear in the topology of convergence in measure. These results were subsequently used by R. L. Taylor [1975] to obtain various criteria for convergence in the probabilistic norm topology, and then to derive various laws of large numbers for Banach space valued random variables; and F. F. Sultanbekov [1976] showed that certain canonical E-norm spaces can be represented as combinations of Banach spaces. This work was extended in [Sultanbekov 1979ab].

Now let (S, ν) be an E-norm space with base $(\Omega, \mathscr{A}, P)$ and target $(V, \|\cdot\|)$. Then for any ω in Ω the function n_ω from V into K defined by $n_\omega(p) = \|p(\omega)\|$ is a pseudonorm on V. Thus every E-norm space is a *pseudonorm-generated space* (à la Definition 9.2.1) that is *separated* whenever the E-norm space is canonical. In the other direction, H. Sherwood [1979] has shown that every (separated) pseudonorm-generated PN space is isomorphically isometric to a (canonical) E-norm space. Thus, as in the case of PM spaces, these two classes of PN spaces are again coextensive. The proof in the PN space case, however, is considerably more involved than the corresponding proof of Theorem 9.2.2. This is because in addition to a metric structure, a linear structure has to be preserved.

The countably normed PN spaces introduced and studied by Šerstnev [1963ab, 1964b] are norm-generated spaces. Since Šerstnev assumes that the generating collection of norms is linearly ordered, these spaces are PN spaces under τ_M (cf. Theorem 9.2.6).

If (S, ν) is an E-norm space with base $(\Omega, \mathscr{A}, P)$, then for any p in S the function X_p defined by

$$X_p(\omega) = \|p(\omega)\| \quad \text{for all } \omega \text{ in } \Omega \tag{15.1.15}$$

is in $L_1^+(\Omega)$ and satisfies the following conditions.

i. $X_p(\omega) = 0$ for all ω in Ω iff $p = \theta$. (15.1.16)

ii. $X_{ap} = |a| X_p$ for all a in K and p in S. (15.1.17)

iii. $X_{p+q} \leqslant X_p + X_q$ for all p, q in S. (15.1.18)

Thus an E-norm space may be viewed as a collection of nonnegative random variables satisfying (15.1.16)–(15.1.18) and, following Section 9.3, we may define and study *random normed spaces*.

Other papers on PN spaces include [Prochaska 1969], where various properties of the strong topology are investigated; [Prochaska 1967], where boundedness and continuity of linear transformations are considered; [Muštari 1968], where the Mazur–Ulam theorem on the linearity of isometric mappings of normed linear spaces is extended to PN spaces; [Muštari 1969], where convergence in Wiener measure of linear operators in Hilbert space is shown to be equivalent to convergence with respect to a suitably chosen probabilistic norm; [Matveičuk 1971], where various properties of von Neumann algebras are expressed in the terminology of PN spaces; and [Matveičuk 1974], where random norms are studied. Further work on PN spaces has been done by G. Bocşan [1976], O. Hadžić [1977, 1978], and V. Radu [1975ab, 1976].

15.2. Probabilistic Inner Product Spaces

Hilbert spaces are complete normed linear spaces in which the norm is derivable from an *inner product* satisfying appropriate conditions. Correspondingly, one would expect a theory of probabilistic inner product or Hilbert spaces. There are complications, however, arising in part from the fact that the value of an inner product can be negative (or even complex). Thus at the very least, any theory of probabilistic inner products must contend with functions in Δ rather than just in Δ^+.

The first, and to date the only, attempt to develop a theory of probabilistic inner product spaces was made by M. L. Senechal [1965]. The basic features are the following.

Let S be a vector space over the reals. Let $\mathcal{G}$ be a function from $S \times S$ into Δ, and for any p, q in S denote the d.f. $\mathcal{G}(p,q)$ by G_{pq}. Given any x in R, the value $G_{pq}(x)$ is interpreted as the probability that the inner product of p and q is less than x. With this interpretation, the following conditions are direct translations of some of the elementary properties of ordinary inner products:

i. For every p in S

$$G_{pp} \text{ is in } \Delta^+; \tag{15.2.1}$$

$$G_{pp} = \varepsilon_0 \quad \text{iff} \quad p = \theta; \tag{15.2.2}$$

$$G_{\theta p} = \varepsilon_0. \tag{15.2.3}$$

ii. For every p, q in S, $G_{pq} = G_{qp}$. (15.2.4)

iii. For every p, q in S and every x in R

$$G_{pq}(x) = G_{\alpha p, \beta q}(\alpha\beta x) \tag{15.2.5}$$

for all α, β in $(-\infty, \infty)$ such that $\alpha\beta > 0$; and

$$G_{-p,q}(x) = 1 - l^+ G_{pq}(-x). \tag{15.2.6}$$

Translation of the distributivity property and Schwarz inequality leads to some ambiguity. Here Senechal makes the following definitions:

A *probabilistic inner product space* is a pair $(S, \mathcal{G})$ in which, in addition to (15.2.1)–(15.2.6), the following conditions are satisfied:

iv. For every p, q, r in S and every x, y in R

$$\text{if } G_{pr}(x) = 1 \text{ and } G_{qr}(y) = 1, \text{ then } G_{p+q,r}(x+y) = 1. \tag{15.2.7}$$

v. For every p, q in S and every x, y in R^+

$$\text{if } G_{pp}(x) = 1 \text{ and } G_{qq}(y) = 1, \text{ then } G_{pq}\left(\sqrt{xy}\right) = 1. \tag{15.2.8}$$

A *weak Menger inner product space* is a probabilistic inner product space in which (15.2.8) is strengthened to the following.

vi. For every p, q in S and every x, y in R^+

$$G_{pq}\left(\sqrt{xy}\right) \geqslant T(G_{pp}(x), G_{qq}(y)) \tag{15.2.9}$$

where T is a t-norm; while in addition to (15.2.7), we have

vii. For every p, q, r in S and every x, y in R such that $xy \geqslant 0$

$$G_{p+q,r}(x+y) \geqslant M(G_{pr}(x), G_{qr}(y)). \tag{15.2.10}$$

A *Menger inner product space* is a weak Menger inner product space in which (15.2.10) holds for all x, y in R.

In any probabilistic inner product space $(S, \mathcal{G})$, let ν be the function from S into Δ^+ defined by $\nu_\theta = \varepsilon_0$, and for $p \neq \theta$ by

$$\nu_p(x) = G_{pp}(x^2) \quad \text{for all } x \text{ in } R^+. \tag{15.2.11}$$

Then (S, ν) is always a PN space in the weak sense, i.e., if $\nu_p(x) = 1$ and $\nu_q(y) = 1$, then $\nu_{p+q}(x+y) = 1$, while if $(S, \mathcal{G})$ is a weak Menger inner product space under T, then (S, ν, τ_T) is a PN space.

Senechal also defined *simple* probabilistic inner product spaces and showed that every such space is a weak Menger inner product space under any choice of T, but generally not a Menger inner product space under any T.

15.3. Probabilistic Topologies

The discussion in this section is based on [Frank 1971] and [Wagner 1965]. Most of the results are due to Frank. To begin, let $(S, \mathcal{F})$ be a PSM space and, for any a in I, let c^a be the profile closure determined as in (13.1.4) by the d.f.

$$\langle a \rangle = a\varepsilon_0 + (1-a)\varepsilon_\infty. \tag{15.3.1}$$

For any a, b in I and A, $B \subseteq S$ these profile closures satisfy:

i. $c^a(\varnothing) = \varnothing$. (15.3.2)
ii. $A \subseteq c^a(A)$. (15.3.3)
iii. $c^a(A) \cup c^a(B) = c^a(A \cup B)$. (15.3.4)
iv. If $b \leqslant a$, then $c^a(A) \subseteq c^b(A)$. (15.3.5)
v. If $a > 0$ and $A \neq \varnothing$, then $c^a(A) = \bigcap_{b<a} c^b(A)$. (15.3.6)

Of these properties, i, ii, and iii follow from Theorem 13.1.2; iv from Theorem 13.1.3; and v from the fact that for any $A \neq \varnothing$, p is in $c^a(A)$ if and only if $F_{pA} \geqslant \langle a \rangle$, where F_{pA} is given by (12.1.7). Now let μ be the function from $S \times \mathcal{P}(S)$ into I defined by

$$\mu(p, \varnothing) = 0 \tag{15.3.7}$$

and

$$\mu(p, A) = F_{pA}(0+) \quad \text{for } A \neq \varnothing. \tag{15.3.8}$$

The number $\mu(p, A)$ is naturally interpreted as "the probability that p is in the closure of A." The conditions (15.3.3)–(15.3.6) yield the following:

i. $\mu(p, A) = 1$ if p is in A. (15.3.9)
ii. $\mu(p, A \cup B) = \text{Max}(\mu(p, A), \mu(p, B))$. (15.3.10)
iii. If $A \neq \varnothing$, then $\mu(p, A) = \sup\{a \mid p \text{ is in } c^a(A)\}$. (15.3.11)

Next, suppose that $(S, \mathcal{F}, \tau)$ is a PM space in which:

a. τ is lower semicontinuous at $(\langle a\rangle, \langle b\rangle)$ for all a, b in I (see Definition 13.1.4).
b. The set of profile functions $\{\langle a\rangle \mid a \text{ in } I\}$ is closed under τ.

Then the function T defined on $I \times I$ via

$$\langle T(a,b)\rangle = \tau(\langle a\rangle, \langle b\rangle) \tag{15.3.12}$$

is a t-norm, and Theorem 13.1.5 yields

$$c^a\big(c^b(A)\big) \subseteq c^{T(a,b)}(A) \tag{15.3.13}$$

for all a, b in I and all $A \subseteq S$. (For triangle functions of the form $\tau_{T,L}$, the t-norm in (15.3.12) is T itself; for convolution, it is Π.) Since p is in $c^a(A)$ if and only if $\mu(p,A) \geqslant a$, (15.3.13) is equivalent to the condition:

$$\text{If } \mu\big(p, c^b(A)\big) \geqslant a, \text{ then } \mu(p,A) \geqslant T(a,b). \tag{15.3.14}$$

The preceding discussion motivates the following generalization.

Definition 15.3.1. A *probabilistic semitopological space* (briefly, a *PST space*) is a pair (S, μ), where S is a set and μ a function from $S \times \mathcal{P}(S)$ into I satisfying (15.3.7), (15.3.9), and

$$\mu(p, A \cup B) \geqslant \operatorname{Max}(\mu(p,A), \mu(p,B)). \tag{15.3.15}$$

Note that (15.3.15) is equivalent to

$$\mu(p,A) \leqslant \mu(p,B) \quad \text{whenever} \quad A \subseteq B. \tag{15.3.16}$$

If (S, μ) is a PST space, then for any a in I we define a function c^a from $\mathcal{P}(S)$ into $\mathcal{P}(S)$ via

$$c^a(\varnothing) = \varnothing \tag{15.3.17}$$

and

$$c^a(A) = \{p \mid \mu(p,A) \geqslant a\} \quad \text{for } A \neq \varnothing. \tag{15.3.18}$$

The following is then immediate:

Theorem 15.3.2. *If* (S, μ) *is a* PST *space, then the family of functions* $\{c^a \mid a \text{ in } I\}$ *satisfies the conditions* (15.3.2), (15.3.3), (15.3.5), (15.3.6), *and*

$$c^a(A) \cup c^a(B) \subseteq c^a(A \cup B). \tag{15.3.19}$$

Furthermore, $\mu(p,A) = \sup\{a \mid p \text{ is in } c^a(A)\}$ *for any* $A \neq \varnothing$. *Finally, if* μ *satisfies* (15.3.10), *then the functions* c^a *satisfy* (15.3.4).

Thus a PST space determines a collection of Fréchet closure operations (see Definition 3.5.7) that is ordered and continuous from above in the sense of (15.3.5) and (15.3.6). In the other direction, we have

Theorem 15.3.3. *Let S be a set and $\{c^a \mid a \text{ in } I\}$ a collection of Fréchet closure operations on S. Let $\mu(p,A)$ be the function from $S \times \mathcal{P}(S)$ into I defined by* (15.3.7) *and* (15.3.11). *Then (S, μ) is a* PST *space. If, furthermore, the collection $\{c^a \mid a \text{ in } I\}$ satisfies* (15.3.5) *and* (15.3.6), *then $c^a(A) = \{p \mid \mu(p,A) \geqslant a\}$ for any $A \neq \emptyset$. Finally, if the functions c^a satisfy* (15.3.4), *then μ satisfies* (15.3.10).

Definition 15.3.4. A *probabilistic topological space* (briefly, a *PT space*) is a triple (S, μ, T) where (S, μ) is a PST space, T is a t-norm, and (15.3.14) holds, with the functions c^a given (15.3.17) and (15.3.18).

Theorem 15.3.5. *If (S, μ, T) is a* PT *space, then in addition to the conclusions of Theorem* 15.3.2, *the functions c^a also satisfy* (15.3.13).

Theorem 15.3.6. *Suppose that S is a set, $\{c^a \mid a \text{ in } I\}$ is a collection of Fréchet closures satisfying* (15.3.5) *and* (15.3.6), *and T is a t-norm such that* (15.3.13) *holds. Let μ be defined as in Theorem* 15.3.3. *Then (S, μ, T) is a* PT *space.*

Definition 15.3.7. Let S be a set and μ a function from $S \times \mathcal{P}(S)$ into I. Then (S, μ) is a *V-generated space* if there is a probability space $(\Omega, \mathcal{A}, P)$ such that:

i. For each ω in Ω there is a Fréchet closure c_ω on S.
ii. For each p in S and $A \subseteq S$, the set

$$\{\omega \mid p \text{ is in } c_\omega(A)\} \text{ is in } \mathcal{A}.$$

iii. $\mu(p,A) = P\{\omega \mid p \text{ is in } c_\omega(A)\}$.

It is immediate that every V-generated space is a PST space. Note that equality in (15.3.15) will generally fail, even when all the c_ω are Čech closures. This is the reason for using (15.3.15), rather than (15.3.10), in Definition 15.3.1.

If $(S, \mathcal{F})$ is a pseudometrically generated PM space with base $(\Omega, \mathcal{A}, P)$ and if, for each ω in Ω, c_ω is the closure operation determined by the d_ω-pseudometric topology, then it follows from Definition 15.3.7 that the quantity $\mu_V(p,A)$ given by

$$\mu_V(p,A) = P\{\omega \mid d_\omega(p,A) = 0\} \tag{15.3.20}$$

is well defined. Thus (S, μ_V) is a V-generated space, which we call the *associated V-generated space*.

Theorem 15.3.8. *Let $(S, \mathcal{F})$ be a pseudometrically generated* PM *space and let (S, μ_V) be the associated V-generated space. Then (S, μ_V) is a* PT *space under the t-norm W.*

Theorem 15.3.8 is a consequence of

Lemma 15.3.9. *Let (S, μ) be a V-generated space and, for every $A \neq \emptyset$ and any a in I, let E_a be the set given by*

$$E_a = \{\omega \mid c_\omega(c^a(A)) = c^a(A) \cup c_\omega(A)\} \tag{15.3.21}$$

where $c^a(A)$ is given by (15.3.18). *If $P(E_a) \geqslant a$ for all a in I, then (S, μ, W) is a* PT *space. If $P(E_a) = 1$ for all a in I, then (S, μ, M) is a* PT *space.*

Given a pseudometrically generated PM space $(S, \mathcal{F})$, we can also use the family of profile closures determined by (15.3.1) to define a probabilistic topological structure on S. The discussion at the beginning of this section shows that if μ^* is defined on $S \times \mathcal{P}(S)$ by

$$\mu^*(p, A) = F_{pA}(0+) \quad \text{for } A \neq \emptyset, \tag{15.3.22}$$

then (S, μ^*, W) is a PT space. The functions μ_V and μ^* are simply related by

$$\mu_V(p, A) \geqslant \mu^*(p, A), \tag{15.3.23}$$

and there exist pseudometrically generated spaces for which this inequality is strict.

In conclusion, we note the following: Let S be a set. Then a *fuzzy subset* of S, as defined by L. Zadeh [1965], is a function f_A from S into I. For any p in S, $f_A(p)$ is interpreted as the *grade of membership* of p in f_A (cf. [Menger 1951c]). The collection of these functions $\{f_A \mid A \subseteq S\}$ is assumed to satisfy the following conditions:

i. $f_\emptyset(p) = 0$ for all p in S.
ii. $f_A(p) = 1$ for all p in A.
iii. $f_A(p) \leqslant f_B(p)$ for all p in S whenever $A \subseteq B$.

On setting $\mu(p, A) = f_A(p)$, it follows at once from Definition 15.3.1 and the equivalence of (15.3.15) and (15.3.16) that a collection of fuzzy subsets of a set S is a probabilistic semitopology for S, and conversely. Thus the theories of probabilistic semitopological spaces and fuzzy sets are coextensive (see [Frank 1971; Nguyen 1975]).

In a series of papers, U. Höhle has exploited this connection and extensively developed both theories (see, e.g., [Höhle 1976, 1978, 1979]).

15.4. Probabilistic Information Spaces

In classical information theory, an event A is a measurable set in a probability space $(\Omega, \mathcal{Q}, P)$ and its information content $J(A)$ is given by

$$J(A) = \frac{1}{c} \log P(A) \tag{15.4.1}$$

where c is a positive constant, usually $\log 2$. It follows that $J(\emptyset) = \infty$, that $J(\Omega) = 0$, and that J is order inverting, i.e., $J(A) \leqslant J(B)$ whenever $B \subseteq A$. Furthermore, if A and B are independent, i.e., if $P(A \cap B) = P(A)P(B)$, then

$$J(A \cap B) = J(A) + J(B); \tag{15.4.2}$$

while if $A \cap B = \emptyset$, then

$$J(A \cup B) = \mathrm{Max}\Big(0, -\frac{1}{c} \log(e^{-cJ(A)} + e^{-cJ(B)})\Big). \tag{15.4.3}$$

In 1967, J. Kampé de Fériet and B. Forte began the development of a generalized theory of information (see [Kampé de Fériet and Forte 1967; Kampé de Fériet, Forte, and Benvenuti 1969; Kampé de Fériet 1974]). In discussing this theory in [S² 1978ab], we introduced the notion of an information space as follows:

Definition 15.4.1. An *information space* is a pair $(\mathfrak{S}, J)$ where $\mathfrak{S}$ is a partially ordered set with minimal element $\emptyset$ and maximal element Ω, and J is an order-inverting function from $\mathfrak{S}$ into R^+ with $J(\emptyset) = \infty$ and $J(\Omega) = 0$. The elements of $\mathfrak{S}$ are called *events*. Two events A, B are *independent* if they have a (necessarily unique) greatest lower bound $A \cap B$ in $\mathfrak{S}$ and (15.4.2) holds. If L is a *composition law* (see Definition 5.7.5), then $(\mathfrak{S}, J)$ is *compositive* with respect to L if whenever the events A, B are such that $A \cap B = \emptyset$, then they have a least upper bound $A \cup B$ in $\mathfrak{S}$ and

$$J(A \cup B) = L(J(A), J(B)). \tag{15.4.4}$$

Note that (15.4.3) expresses compositivity with respect to the Wiener–Shannon law L_W given by (5.7.12). For our purposes it is convenient to introduce two additional notions. We say that A and B are *concordant* if

$A \cap B$ exists in $\mathfrak{S}$ and

$$J(A \cap B) \leqslant J(A) + J(B); \tag{15.4.5}$$

and that $(\mathfrak{S}, J)$ is *weakly compositive* with respect to a composition law L if, whenever $A \cap B = \emptyset$, then $A \cup B$ exists in $\mathfrak{S}$ and

$$J(A \cup B) \geqslant L(J(A), J(B)). \tag{15.4.6}$$

Thus independence and compositivity are special cases of concordance and weak compositivity, respectively.

For example, if (S, μ) is a PST space (see Definition 15.3.1), and if for any p in S, J_p is the function defined on $\mathcal{P}(S)$ by

$$J_p(A) = -\log \mu(p, A),$$

then $(\mathcal{P}(S), J_p)$ is an information space that is weakly compositive with respect to the Inf-law (see Section 5.7).

A probabilistic generalization of the notion of an information space is obtained in the customary manner by replacing the range of the function J by a subset of Δ^+. This yields

Definition 15.4.2. A *probabilistic information space* (briefly, a *PI space*) is a pair $(\mathfrak{S}, K)$ where $\mathfrak{S}$ is as in Definition 15.4.1 and K is an order-preserving function from $\mathfrak{S}$ into Δ^+, with $K(\emptyset) = \varepsilon_\infty$ and $K(\Omega) = \varepsilon_0$.

For any event A in a PI space and any x in R^+ we interpret $K(A)(x)$ as the probability that the information content of A is less than x.

As usual, information spaces are the special cases of PI spaces in which each d.f. $K(A)$ is of the form ε_a for some a in R^+. Similarly, if $(\mathfrak{S}, J)$ is an information space and G is in $\mathcal{D}^+$, then the *simple PI space* generated by $(\mathfrak{S}, J)$ and G is obtained by setting $K(A) = G(j/J(A))$.

Information spaces can also be obtained as special cases of random information spaces.

Definition 15.4.3. A *random information space* (briefly, an *RI space*) is a triple $(\mathfrak{S}, (\Theta, \mathcal{Q}, P), \mathfrak{X})$ (briefly, $(\mathfrak{S}, \Theta, \mathfrak{X})$), where $\mathfrak{S}$ is as in Definition 15.4.1, $(\Theta, \mathcal{Q}, P)$ is a probability space, and $\mathfrak{X}$ is a function from $\mathfrak{S} \times \Theta$ into R^+ such that:

i. For each A in $\mathfrak{S}$ the function X_A from Θ into R^+ defined by $X_A(\omega) = \mathfrak{X}(A, \omega)$ is in $L_1^+(\Theta)$.
ii. For each ω in Θ, J_ω is the function from $\mathfrak{S}$ into R^+ defined by $J_\omega(A) = \mathfrak{X}(A, \omega)$, and the pair $(\mathfrak{S}, J_\omega)$ is an information space for almost all ω in Θ.

Two events A, B are *concordant* if they have a greatest lower bound $A \cap B$ in $\mathfrak{E}$ and

$$X_{A \cap B} \leqslant X_A + X_B \quad \text{a.s.}; \tag{15.4.7}$$

and A, B are *independent* if they are concordant and (15.4.7) holds with equality a.s. The RI space is *weakly compositive* (*compositive*) with respect to the composition law L if the spaces $(\mathfrak{E}, J_\omega)$ are weakly compositive (compositive) with respect to L for almost all ω in Θ.

Following Kampé de Fériet [1969, 1970, 1971, 1973], we regard the elements of Θ as *observers*. Each observer sees the same set $\mathfrak{E}$ of events, but different observers generally assign different information contents to a given event. Note that the information content assigned to the event A by the observer ω can be written indifferently as $X_A(\omega)$ or $J_\omega(A)$. The set Θ itself is a (*weighted*) *set of observers*.

Theorem 15.4.4. *Let* $(\mathfrak{E}, \Theta, \mathfrak{X})$ *be an* RI *space. Let K be the function from* $\mathfrak{E}$ *into* Δ^+ *defined by*

$$K(A) = df(X_A). \tag{15.4.8}$$

Then $(\mathfrak{E}, K)$ *is a* PI *space.*

We call $(\mathfrak{E}, K)$ the *associated PI space* of $(\mathfrak{E}, \Theta, \mathfrak{X})$.

Theorem 15.4.5. *Let* $(\mathfrak{E}, \Theta, \mathfrak{X})$ *be an* RI *space and* $(\mathfrak{E}, K)$ *its associated* PI *space. If A and B are concordant events in* $(\mathfrak{E}, \Theta, \mathfrak{X})$ *and if C is a copula of* X_A *and* X_B, *then*

$$K(A \cap B) \geqslant \sigma_C(K(A), K(B)). \tag{15.4.9}$$

If A and B are independent, then (15.4.9) *holds with equality. If* $(\mathfrak{E}, \Theta, \mathfrak{X})$ *is weakly compositive with respect to the composition law L and* $A \cap B = \emptyset$, *then*

$$K(A \cup B) \leqslant \sigma_{C,L}(K(A), K(B)). \tag{15.4.10}$$

If $(\mathfrak{E}, \Theta, \mathfrak{X})$ *is compositive with respect to L, then* (15.4.10) *holds with equality.*

One could use (15.4.9) and (15.4.10) to define concordance and weak compositivity in arbitrary PI spaces. This would be highly unnatural, since the copula C in (15.4.9) and (15.4.10) depends, not on the distribution functions $K(A)$ and $K(B)$, but on random variables that are in general extraneous to the space. On the other hand, the inequalities (7.5.2) motivate:

Definition 15.4.6. Let $(\mathfrak{E}, K)$ be a PI space. Then the events A and B are *concordant* if $A \cap B$ exists in $\mathfrak{E}$ and

$$K(A \cap B) \geqslant \tau_W(K(A), K(B)). \tag{15.4.11}$$

The space $(\mathfrak{E}, K)$ is *weakly compositive* with respect to the composition law L if, whenever $A \cap B = \emptyset$, then $A \cup B$ exists in $\mathfrak{E}$ and

$$K(A \cup B) \leqslant \rho_{W,L}(K(A), K(B)). \tag{15.4.12}$$

Theorem 15.4.7. *Let $(\mathfrak{E}, \mathfrak{O}, \mathfrak{X})$ be an* RI *space and $(\mathfrak{E}, K)$ its associated* PI *space. If A, B are concordant events in $(\mathfrak{E}, \mathfrak{O}, \mathfrak{X})$, then they are concordant events in $(\mathfrak{E}, K)$. If $(\mathfrak{E}, \mathfrak{O}, \mathfrak{X})$ is weakly compositive with respect to the composition law L, then $(\mathfrak{E}, K)$ is weakly compositive with respect to L.*

The question whether Theorems 15.4.4 and 15.4.7 admit valid converses, i.e., whether for a given PI space $(\mathfrak{E}, K)$ there always exists an RI space whose associated space is $(\mathfrak{E}, K)$, and for which the converse of Theorem 15.4.7 holds, is the *problème réciproque* as formulated by Kampé de Fériet [1969]. In view of the negative results of Sections 7.6 and 9.3, it is to be expected that this problem is generally unsolvable. This is indeed the case. Specifically, we have:

Theorem 15.4.8. *There exists a* PI *space $(\mathfrak{E}, K)$ (indeed, a five-event space) for which there does not exist any random information space $(\mathfrak{E}, \mathfrak{O}, \mathfrak{X})$ whose associated space is $(\mathfrak{E}, K)$, and which is such that two events are concordant in $(\mathfrak{E}, \mathfrak{O}, \mathfrak{X})$ if they are concordant in $(\mathfrak{E}, K)$. For any composition law L belonging to a large class of composition laws, there exists a* PI *space $(\mathfrak{E}, K)$ (indeed, a six-event space) that is weakly compositive with respect to L and for which there does not exist any* RI *space $(\mathfrak{E}, \mathfrak{O}, \mathfrak{X})$ whose associated space is $(\mathfrak{E}, K)$ and which is also weakly compositive with respect to L.*

Proofs of the theorems in this section may be found in [S^2 1978ab]. Some of the results go back to [S^2 1969, 1971].

15.5. Generalized Metric Spaces

As we have pointed out on several occasions, the generalization from ordinary to probabilistic metric spaces may be viewed as the replacement of an R^+-valued metric by a Δ^+-valued one. In 1967, E. Trillas, motivated by the work on PM spaces, by Menger's metrics for groups, and by Blumenthal's Boolean metrics, introduced a common abstraction of these concepts, the notion of a generalized metric space.

Definition 15.5.1. A *T-semigroup* is a quadruple $(G, +, e, \leqslant)$, where $(G, +)$ is a commutative semigroup with identity e, and $\leqslant$ is a partial order relation on G such that:

i. $e \leqslant a$ for all a in G.
ii. $a \leqslant b$ implies that $a + c \leqslant b + c$ for all a, b, c in G.

Definition 15.5.2. A *generalized metric space* is a triple $(S, \mathcal{G}, d)$, where S is a nonempty set, $\mathcal{G} = (G, +, e, \leqslant)$ is a T-semigroup, and d is a mapping from $S \times S$ into G such that for all p, q, r in S:

i. $d(p,q) = e$ iff $p = q$. (15.5.1)
ii. $d(p,q) = d(q,p)$. (15.5.2)
iii. $d(p,r) \leqslant d(p,q) + d(q,r)$. (15.5.3)

If $\mathcal{G} = (R^+, +, 0, \leqslant)$, then $(S, \mathcal{G}, d)$ is an ordinary metric space. If τ is a triangle function and $\mathcal{G} = (\Delta^+, \tau, \varepsilon_0, \geqslant)$ (note the reversal of the order relation), then $(S, \mathcal{G}, d)$ is a PM space. Other examples are obtained as follows:

Let $(G, +, e, \leqslant, \cup, \cap)$ be a commutative lattice-ordered group. This means that $(G, +)$ is an Abelian group with identity e, $(G, \cup, \cap)$ is a lattice, $\leqslant$ is defined by

$$a \leqslant b \quad \text{iff} \quad a \cap b = a \quad \text{iff} \quad a \cup b = b, \tag{15.5.4}$$

and

$$a + (b \cup c) = (a + b) \cup (a + c), \qquad a + (b \cap c) = (a + b) \cap (a + c) \tag{15.5.5}$$

for all a, b, c in G. For any a, b in G let $a - b$ denote the element $a + (-b)$, where $-b$ is the group inverse of b. It then follows from (15.5.4) and (15.5.5) that for any a, b, c in G we have

$$e \leqslant (a \cup b) - (a \cap b) \tag{15.5.6}$$

and

$$(a \cup c) - (a \cap c) \leqslant ((a \cup b) - (a \cap b)) + ((b \cup c) - (b \cap c)). \tag{15.5.7}$$

Now let $G^+ = \{a \text{ in } G \mid e \leqslant a\}$, let f be a one-to-one function from a set S into G, and define d on $S \times S$ by

$$d(p,q) = (f(p) \cup f(q)) - (f(p) \cap f(q)). \tag{15.5.8}$$

Then $\mathcal{G} = (G^+, +, \leqslant, e)$ is a T-semigroup and $(S, \mathcal{G}, d)$ is a generalized metric space.

Since the symmetric difference is a commutative group operation on any Boolean algebra, it follows that the class of generalized metric spaces also includes the autometrized Boolean algebras and Boolean metric spaces (see [Blumenthal 1953, Ch. 15; Blumenthal and Menger 1970, Ch. 3]).

Another class of generalized metric spaces derives from consideration of the set-valued semimetrics for groups introduced by Menger [1931]. These are defined as follows: Given an Abelian group $(G, +)$ with identity e, let m be the function from $G \times G$ into $\mathcal{P}(G)$ given by

$$m(a,b) = \{a - b, b - a\}.$$

Let S be the family of nonempty subsets of G and S^+ the family of subsets of G that contain e. Define $+$ on $S \times S$ by

$$p + q = \{a + b \mid a \text{ in } p, b \text{ in } q\}.$$

Then S^+ is closed under $+$, and, with $\subseteq$ denoting inclusion, $\mathbb{S} = (S^+, +, \{e\}, \subseteq)$ is a T-semigroup. Define d on $S \times S$ by

$$d(p,q) = \{e\} \cup A(p,q)$$

where $A(p,q)$ is the (possibly empty) union of the sets $m(a,b)$ over all a in p, b in q such that either a is not in q or b is not in p. Then $(S, \mathbb{S}, d)$ is a generalized metric space.

For further details, examples and references, see [Trillas 1967; Trillas and Alsina 1978; Alsina 1978b].

References

N. H. Abel, 1826. Untersuchungen der Functionen zweier unabhängig veränderlichen Grössen x und y wie $f(x, y)$, welche die Eigenschaft haben, dass $f(z, f(x, y))$ eine symmetrische Function von x, y und z ist. *J. Reine Angew. Math.* 1, 11–15 (*Oeuvres Complètes de N.H. Abel*, Vol. 1, Christiana, 1881, pp. 61–65).

J. Aczél, 1949. Sur les opérations définies pour nombres réels. *Bull. Soc. Math. France* 76, 59–64.

———, 1965. Quasigroups, nets and nomograms. *Adv. in Math.* 1, 383–450.

———, 1966. *Lectures on Functional Equations and Their Applications*. New York: Academic Press.

R. L. Adler and T. J. Rivlin, 1964. Ergodic and mixing properties of Chebyshev polynomials. *Proc. Amer. Math. Soc.* 15, 794–796.

C. Alsina, 1978a. On countable products and algebraic convexifications of probabilistic metric spaces. *Pacific J. Math.* 76, 291–300.

———, 1978b. Producto, convexificación y completación de espacios métricos generalizados y probabilisticos. *Stochastica* 2, 35–49.

———, 1980. On a family of functional inequalities, in *General Inequalities* 2 (E. F. Beckenbach, ed.) pp. 419–427. Basel: Birkhäuser Verlag.

C. Alsina and E. Bonet, 1979. On sums of dependent uniformly distributed random variables. *Stochastica* 3, 33–43.

C. Alsina and B. Schweizer, 1982. Menger-betweenness in α-simple spaces, in *General Inequalities* 3 (E. F. Beckenbach, ed.). Basel: Birkhäuser Verlag.

D. E. Amos, 1978. Evaluation of some cumulative distribution functions by numerical quadrature. *SIAM Rev.* 20, 778–800.

J. M. Anthony, H. Sherwood, and M. Taylor, 1974. Sequential convergence and probabilistic metric spaces. *Rev. Roumaine Math. Pures Appl.* 19, 1177–1187.

References

T. Apostol, 1974. *Mathematical Analysis*, 2nd ed. Reading, MA: Addison-Wesley.

A. Appert and Ky-Fan, 1951. Espaces topologiques intermédiaires. *Actualités Sci. Ind.* 1121. Paris: Hermann et Cie.

V. I. Arnold, 1957. Concerning the representability of functions of two variables in the form $\chi[\phi(x) + \psi(y)]$. *Uspehi Mat. Nauk* 12, 119–121.

R. Bellman and W. Karush, 1961. On a new functional transform in analysis: the maximum transform. *Bull. Amer. Math. Soc.* 67, 501–503.

———, 1962. Mathematical programming and the maximum transform, *J. Soc. Ind. Appl. Math.* 10, 550–567.

———, 1963. On the maximum transform. *J. Math. Anal. Appl.* 6, 67–74.

A. T. Bharucha-Reid, 1976. Fixed point theorems in probabilistic analysis. *Bull. Amer. Math. Soc.* 82, 641–657.

P. Billingsley, 1968. *Convergence of Probability Measures*. New York: Wiley.

N. H. Bingham, 1971. Factorization theory and domains of attraction for generalized convolution algebras. *Proc. London Math. Soc.* (3) 23, 16–30.

M. Black, 1937. Vagueness. *Philosophy of Science* 4, 427–455.

W. Blaschke and G. Bol, 1938. *Geometrie der Gewebe*. Springer-Verlag.

D. I. Blokhintsev, 1971. Stochastic spaces. *Communs. Joint Inst. Nuclear Research Dubna*, No. P2-6094.

———, 1973. *Space and Time in the Microworld*. Dordrecht: Reidel. (First published by Nauka Publishers, Moscow, 1970.)

L. M. Blumenthal, 1953. *Theory and Applications of Distance Geometry*. New York and London: Oxford Univ. Press.

———, 1961. *A Modern View of Geometry*. San Francisco: W. H. Freeman.

L. M. Blumenthal and K. Menger, 1970. *Studies in Geometry*. San Francisco: W. H. Freeman.

R. P. Boas, 1972. *A Primer of Real Functions*, 2nd ed. Carus Mathematical Monograph No. 13, Math. Assn. of Amer.

G. Bocşan, 1976. Sur certaines semi-normes aléatoires et leurs applications. *C. R. Acad. Sci. Paris* 282A, 1319–1321.

G. Bocşan and G. Constantin, 1973. The Kuratowski function and some applications to probabilistic metric spaces. *Atti Accad. Naz. Lincei, Rend.* (8) 55, 236–240.

———, 1974. On some measures of noncompactness in probabilistic metric spaces. *Proc. Fifth Conf. Probability Theory* (Braşov 1974), pp. 163–168. Bucharest: Editura Acad. R.S.R., 1977.

H. F. Bohnenblust, 1940. An axiomatic characterization of L_p-spaces. *Duke Math. J.* 6, 627–640.

L. Boltzmann, 1896. Entgegnung auf die wärmetheoretischen Betrachtungen des Hrn. E. Zermelo. *Ann. Physik* 57, 773–784.

M. Born, 1955. Continuity, determinism and reality. *Kgl. Danske Videnskab. Selskab. Mat. Fys. Medd.* 30, No. 2. (Reprinted in *Max Born, Ausgewählte Abhandlungen*, Erster Band, Göttingen, pp. 196–219.)

R. C. Bose, 1935. On the exact distribution and moment-coefficients of the D^2-statistic. *Sankhyā* 2, 143–154.

L. Brillouin, 1956. *Science and Information Theory*, 2nd ed. New York: Academic Press.

L. de Broglie, 1935. Une remarque sur l'interaction entre la matière et le champ électromagnétique. *C. R. Acad. Sci. Paris* 200, 361–363.

L. E. J. Brouwer, 1909. Die Theorie der endlichen kontinuierlichen Gruppen unabhängig von den Axiomen von Lie. *Math. Ann.* 67, 127–136.

J. B. Brown, 1972. Stochastic metrics, *Z. Wahrsch. verw. Geb.* 24, 49–62.

———, 1973. On the relationship between Menger spaces and Wald spaces. *Colloq. Math.* 27, 323–330.

G. L. Cain and R. H. Kasriel, 1976. Fixed and periodic points of local contraction mappings on probabilistic metric spaces. *Math. Systems Theory* 9, 289–297.

P. Calabrese, 1978. The probability of the triangle inequality in probabilistic metric spaces. *Aequationes Math.* 18, 187–205.

É. Cartan, 1930. La théorie des groupes finis et continus et l'analyse situs. *Mem. Sci. Math.* 42, Paris: Gauthier-Villars.

E. Čech, 1966. *Topological Spaces*. New York: Wiley (Interscience).

K. L. Chung, 1967. *Markov Chains with Stationary Transition Probabilities*, 2nd ed. New York: Springer-Verlag.

A. H. Clifford, 1954. Naturally totally ordered commutative semigroups. *Amer. J. Math.* 76, 631–646.

A. C. Climescu, 1946. Sur l'équation fonctionnelle de l'associativité. *Bul. Inst. Politehn. Iaşi* 1, 1–16.

G. Dall'Aglio, 1959. Sulla compatibilità delle funzioni di ripartizione doppia. *Rend. Mat.* (5) 18, 385–413.

———, 1972. Fréchet classes and compatibility of distribution functions. *Symposia Math.* 9, 131–150.

G. Darboux, 1915. *Lecons sur la théorie générale des surfaces*, Deuxième Partie. Deuxième éd., Paris: Gauthier-Villars, (republished by Bronx, NY: Chelsea, 1972.)

C. A. Drossos, 1977. Stochastic Menger spaces and convergence in probability. *Rev. Roumaine Math. Pures Appl.* 22, 1069–1076.

R. O. Duda and P. E. Hart, 1973. *Pattern Classification and Scene Analysis*. New York: Wiley.

A. S. Eddington, 1953. *Fundamental Theory*. New York and London: Cambridge Univ. Press.

R. J. Egbert, 1968. Products and quotients of probabilistic metric spaces. *Pacific J. Math.* 24, 437–455.

T. Erber, P. Everett, and P. W. Johnson, 1979. The simulation of random processes on digital computers with Čebyšev mixing transformations. *J. Computational Phys.* 32, 168–211.

T. Erber, B. N. Harmon, and H. G. Latal, 1971. The origin of hysteresis in simple magnetic systems. *Advances in Chemical Physics* (I. Prigogine and S. Rice, eds.) Vol. 20, pp. 71–134. New York: Wiley (Interscience).

T. Erber, S. A. Guralnick, and H. G. Latal, 1972. A general phenomenology of hysteresis. *Ann. Physics* 69, 116–192.

T. Erber, B. Schweizer, and A. Sklar, 1971. Probabilistic metric spaces and hysteresis systems. *Comm. Math. Phys.* 20, 205–219.

———, 1973. Mixing transformations on metric spaces. *Comm. Math. Phys.* 29, 311–317.

T. Erber and A. Sklar, 1974. Macroscopic irreversibility as a manifestation of micro-instabilities, in *Modern Developments in Thermodynamics* (B. Gal Or, ed.) pp. 281–301. Jerusalem: Israel Univ. Press.

W. Feller, 1966. *An Introduction to Probability Theory and Its Applications*, Vol. 2. New York: Wiley.

W. Fenchel, 1953. Convex cones, sets, and functions. *Lecture Notes*, Princeton Univ.

R. Féron, 1956. Sur les tableaux de corrélation dont les marges sont données. *Publ. Inst. Statist. Univ. Paris* 5, 3–12.

D. J. Foulis, 1969. *Fundamental Concepts of Mathematics*. Boston: Prindle, Weber & Schmidt.

M. J. Frank, 1971. Probabilistic topological spaces. *J. Math. Anal. Appl.* 34, 67–81.

———, 1975. Associativity in a class of operations on a space of distribution functions. *Aequationes Math.* 12, 121–144.

———, 1979. On the simultaneous associativity of $F(x, y)$ and $x + y - F(x, y)$. *Aequationes Math.* 19, 194–226.

———, 1980. Diagonals and sections determine associative functions. *Proc. 18th Int. Symposium on Functional Equations* (Univ. of Waterloo), p. 13.

M. J. Frank and B. Schweizer, 1979. On the duality of generalized infimal and supremal convolutions, *Rend. Mat.* (6) 12, 1–23.

M. Fréchet, 1906. Sur quelques points du calcul fonctionnel. *Rend. Circ. Mat. Palermo* 22, 1–74.

———, 1917. Sur la notion de voisinage dans les ensembles abstraits. *C. R. Acad. Sci. Paris* 165, 359–360. (Reprinted on pp. 277–278 of *Les espaces abstraits*.)

———, 1928. *Les espaces abstraits*. Paris: Gauthier-Villars.

———, 1936. *Recherches théoriques modernes sur le calcul des probabilités*. Premier livre: *Généralités sur les probabilités*; *Eléments aléatoires* (deuxième éd. 1950). Paris: Gauthier-Villars.

———, 1948. Les éléments aléatoires de nature quelconque dans un espace distancié. *Ann. Inst. H. Poincaré* 10, 215–310.

———, 1951. Sur les tableaux de corrélation dont les marges sont données. *Ann. Univ. Lyon* 9, Sec. A, 53–77.

———, 1956. Les inégalités de Minkowski dégénérées et leurs applications en calcul de probabilitiés. *Ann. Inst. H. Poincaré* 15, 1–33.

A. Frenkel, 1977. On the possible connection between quantum mechanics and gravitation, in *Quantum Mechanics: A Half Century Later*, pp. 19–38. Dordrecht: Reidel.

R. Fritsche, 1971. Topologies for probabilistic metric spaces. *Fund. Math.* 72, 7–16.

J. Gilewski, 1972. Generalized convolutions and Delphic semigroups. *Colloq. Math.* 25, 281–289.

B. V. Gnedenko and A. N. Kolmogorov, 1954. *Limit Distributions for Sums of Independent Random Variables*. Reading, MA: Addison-Wesley.

S. Gudder, 1968. Elementary length topologies in physics. *SIAM J. Appl. Math.* 16, 1011–1019.

O. Hadžić, 1977. A fixed point theorm in random normed spaces. *Zbornik Radov Prirod-Mat. Fakult. Novi Sad* 7, 23–27.

———, 1978. A fixed point theorem in probabilistic locally convex spaces. *Rev. Roumaine Math. Pures Appl.* 23, 735–744.

P. Halmos, 1960. *Naive Set Theory*. New York: Van Nostrand Reinhold.

J. M. Hammersley, 1974. Postulates for subadditive processes. *Ann. Probab.* 2, 652–680.

G. H. Hardy, J. E. Littlewood, and G. Pólya, 1952. *Inequalities*, 2nd ed. New York and London: Cambridge Univ. Press.

F. Hausdorff, 1914. *Grundzüge der Mengenlehre*. Leipzig: Veit und Comp.

W. Hengartner and R. Theodorescu, 1973. *Concentration Functions*. New York: Academic Press.

———, 1979. A characterization of strictly unimodal distribution functions by their concentration functions. *Publ. Inst. Statist. Univ. Paris* 24, 1–10.

E. Hille, 1962. *Analytic Function Theory*, Vol. 2. Waltham, MA: Blaisdell.

———, 1965. Topics in classical analysis, in *Lectures on Modern Mathematics* (T. L. Saaty, ed.) Vol. 3, pp. 1–57. New York: Wiley.

E. Hille and R. S. Philips, 1957. Functional Analysis and Semigroups, *Amer. Math. Soc. Colloq. Publ.* 31.

J. Hjelmslev, 1923. Die natürliche Geometrie. *Abh. Math. Sem. Univ. Hamburg* 2, 1–36.

U. Höhle, 1976. Masse auf unscharfen Mengen. *Z. Wahrsch. Verw. Geb.* 36, 179–188.

———, 1978. Probabilistische Topologien. *Manuscripta Math.* 26, 223–245.

———, 1979. Probabilistisch kompakte L-unscharfe Mengen. *Manuscripta Math.* 26, 345–356.

J. A. R. Holbrook, 1981. Stochastic independence and space-filling curves, *Amer. Math. Monthly* 88, 426–432.

H. Hotelling, 1928. Spaces of statistics and their metrization. *Science* 67, 149–150.

I. A. Ibragimov, 1956. On the composition of unimodal distributions. *Theory Prob. Appl.* 1, 255–260.

I. Istrăţescu, 1976. Some remarks on nonarchimedean probabilistic metric spaces. *Glas. Mat.* 11 (31), 155–161.

I. Istrăţescu and I. Vaduva, 1961. Products of statistical metric spaces. *Stud. Cerc. Mat.* 12, 567–574.

V. I. Istrăţescu, 1974. *Introducere in teoria spatiilor metrice probabiliste cu aplicatii.* Bucharest: Editura Technica.

V. I. Istrăţescu and I. Săcuiu, 1973. Fixed point theorems for contraction mappings on probabilistic metric spaces, *Rev. Roumaine Math. Pures Appl.* 18, 1375–1380.

K. Jacobs, 1978. *Measure and Integral.* New York: Academic Press.

M. F. Janowitz, 1978. An order theoretic model for cluster analysis. *SIAM J. Appl. Math.* 34, 55–72.

N. L. Johnson and S. Kotz, 1970. *Continuous Univariate Distributions—2.* Boston: Houghton Mifflin.

———, 1977. Propriétés de dépendance des distributions itérées, généralisées à deux variables Farlie-Gumbel-Morgenstern. *C. R. Acad. Sci. Paris* 285A, 277–281.

P. Johnson and A. Sklar, 1976. Recurrence and dispersion under iteration of Čebyšev polynomials. *J. Math. Anal. Appl.* 54, 752–771.

J. Kampé de Fériet, 1969. Measures de l'information par un ensemble d'observateurs. *C. R. Acad. Sci. Paris* 269A, 1081–1085.

———, 1970. Mesures de l'information par un ensemble d'observateurs. *C. R. Acad. Sci. Paris* 271A, 1017–1021.

———, 1971. Applications of functional equations and inequalities to information theory—measure of information by a set of observers: a functional equation, in *Functional Equations and Inequalities* (B. Forte, ed.), pp. 163–193. Rome: Edizioni Cremonese.

———, 1973. Mesure de l'information par un ensemble d'observateurs indépendants. *Trans. Sixth Prague Conf. Information Theory, Statist. Decision Functions, Random Processes*, pp. 315–329. Prague: Publishing House of the Czechoslovak Academy of Sciences.

———, 1974. *La théorie généralisée de l'information et la mesure subjective de l'information* (Lecture Notes in Math., Vol 398), pp. 1–35. New York: Springer-Verlag.

J. Kampé de Fériet and B. Forte, 1967. Information et probabilité. *C. R. Acad. Sci. Paris* 265A, 110–114; 142–146; 350–353.

J. Kampé de Fériet, B. Forte, and P. Benvenuti, 1969. Forme générale de l'opération de composition continue d'une information. *C. R. Acad. Sci. Paris* 269A, 529–534.

J. L. Kelley, 1955. *General Topology.* New York: Van Nostrand Reinhold.

M. G. Kendall, 1961. *A Course in Multivariate Analysis.* New York: Hafner.

C. H. Kimberling, 1973a. On a class of associative functions. *Publ. Math. Debrecen* 20, 21–39.

———, 1973b. Exchangeable events and completely monotonic sequences. *Rocky Mountain J. Math.* 3, 565–574.

———, 1974. A probabilistic interpretation of complete monotonicity. *Aequationes Math.* 10, 152–164.

J. F. C. Kingman, 1964. Metrics for Wald spaces. *J. London Math. Soc.* 39, 129–130.

J. F. C. Kingman and S. J. Taylor, 1966. *Introduction to Measure and Probability*. New York and London: Cambridge Univ. Press.

A. N. Kolmogorov, 1933. *Grundbegriffe der Wahrscheinlichkeitsrechnung*. Springer-Verlag. (English translation: *Foundations of the Theory of Probability*, 1950. Bronx, NY: Chelsea.)

G. Krause, 1981. A strengthened form of Ling's theorem on associative functions. Ph.D. thesis, Illinois Institute of Technology.

I. Kurosaki, 1964. A note on statistical metric spaces. *Proc. Japan Acad.* 40, 396–399.

N. S. Landkof, 1972. *Foundations of Modern Potential Theory*, New York: Springer-Verlag.

P. Lévy, 1937. *Théorie de l'addition des variables aléatoires*. Paris: Gauthier-Villars (2nd ed. 1954).

C. H. Ling, 1965. Representation of associative functions. *Publ. Math. Debrecen* 12, 189–212.

Ju. V. Linnik, 1964. *Decomposition of Probability Distributions*. Edinburgh: Oliver & Boyd.

M. Loève, 1965. On stochastic processes, in *Lectures on Modern Mathematics* (T. L. Saaty, ed.), Vol. 3, pp. 245–276. New York: Wiley.

———, 1977. *Probability Theory*, 4th ed., Vol. 1 (Vol. 2, 1978). New York: Springer-Verlag.

E. Lukacs, 1970. *Characteristic Functions*, 2nd ed. London: C. Griffin Co.

P. C. Mahalanobis, 1936. On the generalized distance in statistics. *Proc. Nat. Inst. Sci. India* 2, 49–55.

———, 1949. Historical note on the D^2-statistic. *Sankhyā* 9, 237–240.

A. A. J. Marley, 1971. An observable property of a generalization of Kinchla's diffusion model of perceptual memory. *J. Math. Psych.* 8, 491–499.

M. S. Matveičuk, 1971. Random norms and characteristic probabilities on orthoprojections associated with factors. *Probabilistic Methods and Cybernetics, Kazan Univ.* 9, 73–77.

———, 1974. On rings with random norms. *Probabilistic Methods and Cybernetics, Kazan Univ.* 10–11, 43–50.

K. Menger, 1928. Untersuchungen über allgemeine Metrik. *Math. Ann.* 100, 75–163. Erste (76–113), Theorie der Konvexität; Zweite (113–141), Die Euklidische Metrik; Dritte (142–163), Entwurf einer Theorie der *n*-dimensionalen Metrik.

———, 1930. Zur metrik der Kurven; Vierte Untersuchung über allgemeine Metrik. *Math. Ann.* 103, 466–501.

———, 1931. Beiträge zur Gruppentheorie I: Über einen Abstand in Gruppen. *Math. Z.* 33, 396–418.

———, 1932. *Kurventheorie*. Leipzig: Teubner. (Reprinted by Chelsea, Bronx, NY, 1967.)

———, 1942. Statistical metrics. *Proc. Nat. Acad. Sci. U.S.A.* 28, 535–537.

———, 1949. The theory of relativity and geometry, in *Albert Einstein: Philosopher-Scientist* (Library of Living Philosophers, Vol. VII; pp. 459–474. P.A. Schilpp, ed., Evanston, Ill.)

———, 1951a. Probabilistic theories of relations. *Proc. Nat. Acad. Sci. U.S.A.* 37, 178–180.

———, 1951b. Probabilistic geometry. *Proc. Nat. Acad. Sci. U.S.A.* 37, 226–229.

———, 1951c. Ensembles flous et fonctions aléatoires. *C. R. Acad. Sci. Paris* 232, 2001–2003.

———, 1954. Géométrie générale. *Mem. Sci. Math.* 124, Chapter 7. Paris: Gauthier-Villars.

———, 1956. Random variables from the point of view of a general theory of variables, in *Proc. Third Berkeley Symp. Math. Statist. Prob.* 2, 215–229.

———, 1968. Mathematical implications of Mach's ideas: Positivistic geometry, the clarification of functional connections, in *Ernst Mach, Physicist and Philosopher* (Boston Studies in Philosophy of Science, R. S. Cohen and R. J. Seeger, eds., Vol. 6), pp. 107–125. Dordrecht: Reidel.

———, 1979. *Selected Papers in Logic and Foundations, Didactics, Economics.* Dordrecht: Reidel.

H. Minkowski, 1910. *Geometrie der Zahlen.* Leipzig: Teubner.

D. S. Moore and M. C. Spruill, 1975. Unified large-sample theory of general chi-squared statistics for tests of fit. *Ann. Statist.* 3, 599–616.

R. E. Moore, 1979. *Methods and Applications of Interval Analysis.* Philadelphia: SIAM.

B. Morrel and J. Nagata, 1978. Statistical metric spaces as related to topological spaces. *General Topology Appl.* 9, 233–237.

P. S. Mostert and A. L. Shields, 1957. On the structure of semigroups on a compact manifold with boundary. *Ann. of Math.* 65, 117–143.

T. S. Motzkin, 1936. Sur le produit des espaces métriques. *C. R. Congrès Int. Mathématiciens* (Oslo), Vol. 2, pp. 137–138.

R. Moynihan, 1975. On the class of τ_T-semigroups of probability distribution functions. *Aequationes Math.* 12, 249–261.

———, 1976. Probabilistic metric spaces induced by Markov chains. *Z. Wahrsch. Verw. Geb.* 35, 177–187.

———, 1978a. On τ_T-semigroups of probability distribution functions, II. *Aequationes Math.* 17, 19–40.

———, 1978b. Infinite τ_T products of probability distribution functions. *J. Austral. Math. Soc.* (Ser. A), 26, 227–240.

———, 1980a. Conjugate transforms for τ_T-semigroups of probability distribution functions. *J. Math. Anal. Appl.* 74, 15–30.

———, 1980b. Conjugate transforms and limit theorems for τ_T-semigroups. *Studia Math.* 69, 1–18.

R. Moynihan and B. Schweizer, 1979. Betweenness relations in probabilistic metric spaces. *Pacific J. Math.* 81, 175–196.

R. Moynihan, B. Schweizer, and A. Sklar, 1978. Inequalities among operations on probability distribution functions, in *General Inequalities* 1, (E. F. Beckenbach, ed.), pp. 133–149. Basel: Birkhäuser Verlag.

D. H. Muštari, 1967. On the completion of random metric spaces. *Kazan Gos. Univ. Učen. Zap.* 127, 109–119.

———, 1968. On the linearity of isometric mappings of random normed spaces. *Kazan. Gos. Univ. Učen. Zap.* 128, 86–90.

———, 1969. Convergence in Wiener measure of linear operations in Hilbert space. *Kazan. Gos. Univ. Učen. Zap.* 129, 80–84.

———, 1970. On almost sure convergence in linear spaces of random variables. *Theory Prob. Appl.* 15, 337–342.

D. H. Muštari and A. N. Šerstnev, 1965. A problem about triangle inequalities for random normed spaces. *Kazan. Gos. Univ. Učen. Zap.* 125, 102–113.

———, 1966. On methods of introducing a topology into random metric spaces. *Izv. Vysh. Mat.* 6(55), 99–106.

———, 1978. Les fonctions du triangle pour les espaces normés aléatoires, in *General Inequalities* 1, (E. F. Beckenbach, ed.), pp. 255–260. Basel: Birkhäuser Verlag.

J. Neveu, 1965. *Mathematical Foundations of the Calculus of Probabilities*. San Francisco: Holden-Day.

Hung T. Nguyen, 1975. Mesures d'information, ensembles flous et espaces topologiques aléatoires. Thèse, Univ. des Sciences et Techniques de Lille.

E. Nishiura, 1970. Constructive methods in probabilistic metric spaces, *Fund. Math.* 67, 115–124.

O. Onicescu, 1964. *Nombres et systèmes aléatoires*. Bucharest: Éditions de l'Académie de la R. P. Roumaine.

J. R. Oppenheimer, 1962. *The Flying Trapeze: Three Crises for Physicists*. New York and London: Oxford Univ. Press.

J. C. Oxtoby, 1971. *Measure and Category*. New York: Springer-Verlag.

O. Penrose, 1979. Foundations of statistical mechanics. *Rep. Prog. Phys.* 42, 1937–2006.

R. Penrose, 1975. Twistor theory, its aims and achievements, in *Quantum Gravity*, pp. 268–407. New York and London: Oxford Univ. Press.

R. Penrose and M. A. H. MacCullum, 1973. Twistor theory: an approach to the quantisation of fields and space-time. *Phys. Rep.* 6C (4), 241–315.

H. Poincaré, 1905. *Science and Hypothesis*. Lancaster Pa.: The Science Press (Republished by Dover, New York, 1952.)

———, 1913. *The Foundations of Science*. Lancaster, Pa.: The Science Press.

J. L. Porteseil, 1978. Two general features of the hysteretic behavior of ferromagnets: anisotrophy of demagnetization and "reptation" (creep). *Physica* 93B, 201–211.

J. M. Prager, 1979. Segmentation of static and dynamic schemes. Univ. of Massachusetts Dept. of Computer and Information Sciences Tech. Rep. 79-7.

B. J. Prochaska, 1967. On random normed spaces. Ph.D. thesis, Clemson University.

———, 1969. A note on the ε, δ-topology in random normed spaces. *Metron* 27, 82–87.

M. H. Quenouille, 1949. The evaluation of probabilities in a normal multivariate distribution, with special reference to the correlation ratio. *Proc. Edinburgh Math. Soc.* (Ser. 2), 8, 95–100.

V. Radu, 1975a. Linear operators in random normed spaces. *Bull. Math. Soc. Sci. Math. R. S. Roumanie* (N.S.), 17(65), 217–220.

———, 1975b. Sur une norme aléatoire et la continuité des opérateurs linéaires dans des espaces normés aléatoires. *C. R. Acad. Sci. Paris* 280A, 1303–1305.

———, 1976. Equicontinuity, affine mean ergodic theorem and linear equations in random normed spaces. *Proc. Amer. Math. Soc.* 57, 299–303.

———, 1977. On some probabilistic structures on product spaces. *Proc. Fifth Conf. Probability Theory* (Braşov, 1974), pp. 277–282. Bucharest: Editura Acad. R.S.R.

C. R. Rao, 1949. On the distance between two populations. *Sankhyā* 9, 246–248.

A. Rényi, 1959. On measures of dependence. *Acta Math. Acad. Sci. Hungar.* 10, 441–451.

———, 1970. *Probability Theory*. Amsterdam: North-Holland.

F. Rhodes, 1964. Convexity in Wald's statistical metric spaces. *J. London Math. Soc.* 39, 117–128.

R. Rice, 1978. On mixing transformations. *Aequationes Math.* 17, 104–108.

F. Riesz, 1907. Die Genesis des Raumbegriffes. *Math. Naturwiss. Ber. Ungarn* 24, 309–353. (Reprinted in Friedrich Riesz, *Gesammelte Arbeiten* I. pp. 110–154. Budapest, 1960.)

B. A. Rogozin, 1965. On the maximum of the density of the sum of random variables with a unimodal distribution. *Litovsk. Mat. Sbornik* 5, 499–503. (*Selected Translations in Math. Statist. Prob.* 9, 69–74 (1970).)

N. Rosen, 1947. Statistical geometry and fundamental particles. *Phys. Rev.* 72, 298–303.

———, 1962. Quantum geometry. *Ann. Physics* 19, 165–172.

A. Saleski, 1974. Entropy of self-homeomorphisms of statistical pseudometric spaces. *Pacific J. Math.* 51, 537–542.

———, 1975. A conditional entropy for the space of pseudo-Menger maps. *Pacific J. Math.* 59, 525–533.

B. Schweizer, 1964. Equivalence relations in probabilistic metric spaces. *Bul. Inst. Politehn. Iaşi* 10, 67–70.

———, 1966. On the uniform continuity of the probabilistic distance. *Z. Wahrsch. Verw. Geb.* 5, 357–360.

———, 1967. Probabilistic metric spaces—the first 25 years. *New York Statistician* 19, 3–6.

———, 1975a. Multiplications on the space of probability distribution functions. *Aequationes Math.* 12, 156–183.

———, 1975b. Sur la possibilité de distinguer les points dans un espace métrique aléatoire. *C. R. Acad. Sci. Paris* 280A, 459–461.

B. Schweizer and A. Sklar, 1958. Espaces métriques aléatoires, *C. R. Acad. Sci. Paris* 247, 2092–2094.

———, 1960. Statistical metric spaces. *Pacific J. Math.* 10, 313–334.

———, 1961a. Associative functions and statistical triangle inequalities. *Publ. Math. Debrecen* 8, 169–186.

———, 1961b. Topology and Tchebycheff. *Amer. Math. Monthly* 68, 760–762.

———, 1962. Statistical metric spaces arising from sets of random variables in Euclidean *n*-space. *Theory Prob. Appl.* 7, 447–456.

———, 1963a. Inequalities for the confluent hypergeometric function. *J. Math Phys.* 42, 329–330.

———, 1963b. Triangle inequalities in a class of statistical metric spaces. *J. London Math. Soc.* 38, 401–406.

———, 1963c. Associative functions and abstract semigroups. *Publ. Math. Debrecen* 10, 69–81.

———, 1965. The algebra of functions, III. *Math. Ann.* 161, 171–196.

———, 1967. Function systems. *Math. Ann.* 172, 1–16.

———, 1968. A grammar of functions, Part I. *Aequationes Math.* 2, 62–85.

———, 1969. Mesures aléatoires de l'information. *C. R. Acad. Sci. Paris* 269A, 721–723.

———, 1971. Mesure aléatoire de l'information et mesure de l'information par un ensemble d'observateurs. *C. R. Acad. Sci. Paris* 272A, 149–152.

———, 1973. Probabilistic metric spaces determined by measure-preserving transformations. *Z. Wahrsch. Verw. Geb.* 26, 235–239.

———, 1974. Operations on distribution functions not derivable from operations on random variables. *Studia Math.* 52, 43–52.

———, 1978a. Probabilistic information spaces, in *Théorie de l'information; Développements récents et applications* (Colloques Internationaux du Centre National de la Recherche Scientifique, No. 276, C-F. Picard, ed.), pp. 485–494. Paris: C.N.R.S.

———, 1978b. Unsolvability of the general 'problème réciproque' for probabilistic and random information spaces, in *Théorie de l'information; Développements récents et applications* (Colloques Internationaux du Centre National de la Recherche Scientifique, No. 276, C-F. Picard, ed.), pp. 495–501. Paris: C.N.R.S.

———, 1980. How to derive all L_p metrics from a single probabilistic metric, in *General Inequalities* 2 (E. F. Beckenbach, ed.), pp. 429–434. Basel: Birkhäuser Verlag.

B. Schweizer, A. Sklar, and E. Thorp, 1960. The metrization of statistical metric spaces. *Pacific J. Math.* 10, 673–675.

B. Schweizer and E. F. Wolff, 1976. Sur une mesure de dépendance pour les variables aléatoires. *C. R. Acad. Sci. Paris* 283A, 659–661.

———, 1981. On nonparametric measures of dependence for random variables. *Ann. Statist.* 9, 879–885.

V. M. Sehgal, 1966. Some fixed point theorems in functional analysis and probability. Ph.D. thesis, Wayne State Univ.

V. M. Sehgal and A. T. Bharucha-Reid, 1972. Fixed points of contraction mappings on probabilistic metric spaces. *Math. Systems Theory* 6, 97–102.

M. L. Senechal, 1965. Approximate functional equations and probabilistic inner product spaces. Ph.D. thesis, Illinois Inst. of Technology.

A. N. Šerstnev, 1962. Random normed spaces: problems of completeness. *Kazan. Gos. Univ. Učen. Zap.* 122, 3–20.

———, 1963a. On the notion of a random normed space, *Dokl. Akad. Nauk SSSR* 149(2), 280–283.

———, 1963b. Best approximation problems in random normed spaces. *Dokl. Akad. Nauk SSSR* 149(3), 539–542.

———, 1964a. On a probabilistic generalization of metric spaces. *Kazan. Gos. Univ. Učen. Zap.* 124, 3–11.

———, 1964b. Some best approximation problems in random normed spaces. *Rev. Roumaine Math. Pures Appl.* 9, 771–789.

———, 1965. Triangle inequalities for random metric spaces. *Kazan. Gos. Univ. Učen. Zap.* 125, 90–93. (*Selected Translations in Math. Statist. Prob.* 11, 27–30 (1973).)

———, 1967. On the connection between probabilistic metric spaces and Špaček spaces. *Kazan. Gos. Univ. Učen. Zap.* 127, 120–123.

———, 1968. On Boolean logics. *Kazan. Gos. Univ. Učen. Zap.* 128, 48–62.

C. E. Shannon, 1948. The mathematical theory of communication. *Bell System Tech. J.* 27, 379–423, 623–656. (Reprinted in *The Mathematical Theory of Communication*, Univ. of Illinois Press, 1949.)

R. N. Shepard, 1980. Multidimensional scaling, tree-fitting, and clustering. *Science* 210, 390–398.

H. Sherwood, 1966. On the completion of probabilistic metric spaces. *Z. Wahrsch. Verw. Geb.* 6, 62–64.

———, 1969. On *E*-spaces and their relation to other classes of probabilistic metric spaces. *J. London Math. Soc.* 44, 441–448.

———, 1970. Betweenness in probabilistic metric spaces. *Rev. Roumaine Math. Pures Appl.* 15, 1061–1068.

———, 1971. Complete probabilistic metric spaces. *Z. Wahrsch. Verw. Geb.* 20, 117–128.

———, 1976. A note on PM spaces determined by measure preserving transformations. *Z. Wahrsch. Verw. Geb.* 33, 353–354.

———, 1979. Isomorphically isometric probabilistic normed linear spaces. *Stochastica* 3, 71–77.

H. Sherwood and M. D. Taylor, 1974. Some PM structures on the set of distribution functions. *Rev. Roumaine Math. Pures Appl.* 19, 1251–1260.

D. A. Sibley, 1971. A metric for weak convergence of distribution functions, *Rocky Mountain J. Math.* 1, 427–430.

W. Sierpiński, 1956. *General Topology* 2nd ed. Toronto: Univ. Toronto Press.

G. Sîmboan and R. Theodorescu, 1962. Statistical spaces. *Rev. Roumaine Math. Pures Appl.* 7, 699–703.

A. Sklar, 1959. Fonctions de répartition à *n* dimensions et leurs marges. *Publ. Inst. Statist. Univ. Paris* 8, 229–231.

———, 1973. Random variables, joint distribution functions, and copulas. *Kybernetika* 9, 449–460.

A. Špaček, 1956. Note on K. Menger's probabilistic geometry, *Czechoslovak Math. J.* 6(81), 72–74.

———, 1960. Random metric spaces. *Trans. Second Prague Conf. Information Theory, Statist. Decision Functions and Random Processes*, pp. 627–638.

M. S. Srivastava and C. G. Khatri, 1979. *An Introduction to Multivariate Statistics*. New York: Elsevier North Holland.

R. R. Stevens, 1968. Metrically generated probabilistic metric spaces. *Fund. Math.* 61, 259–269.

S. S. Stevens, 1959. Measurement, psychophysics, and utility, in *Measurement, Definitions, and Theories* (C. West Churchman and P. Ratoosh, eds.), pp. 18–63. New York: Wiley.

F. F. Sultanbekov, 1976. On random normed spaces of strongly measurable functions. *Mat. Meh. Kazan. Univ.*, pp. 1–13.

———, 1977. On regular triangle functions. *Konstruktivnaja Teor. Funkcij i Funkcionalnij Analiz Kazan*, pp. 54–57.

———, 1979a. On random functionals in spaces of strongly measurable functions. *Issled. Priklad. Mat.* 6, Izd-vo KGU, 74–82.

———, 1979b. On weak random functionals in spaces of strongly measurable functions, *Konstruktivnaja Teor. Funkcij i Funkcionalnij Analiz* 2, Izd-vo KGU, 88–92.

R. M. Tardiff, 1975. Topologies for probabilistic metric spaces. Ph.D. thesis, Univ. of Massachusetts.

———, 1976. Topologies for probabilistic metric spaces. *Pacific J. Math.* 65, 233–251.

———, 1980. On a functional inequality arising in the construction of the product of several metric spaces. *Aequationes Math.* 20, 51–58.

R. L. Taylor, 1975. Convergence of elements in random normed spaces. *Bull. Austral. Math. Soc.* 12, 156–183.

G. Thomsen, 1927. Un teorema topologico sulle schiere di curve e una caratterizzazione geometrica delle superficie isotermoasintotiche. *Bol. Un. Mat. Ital.* 6, 80–85.

E. Thorp, 1960. Best-possible triangle inequalities for statistical metric spaces. *Proc. Amer. Math. Soc.*, 11, 734–740.

———, 1962. Generalized topologies for statistical metric spaces. *Fund. Math.* 51, 9–21.

E. Trillas, 1967. Sobre distancias aleatorias. *Actas R.A.M.E.* (C.S.I.C.), Santiago de Compostela.

E. Trillas and C. Alsina, 1978. Introducción a los espacios métricos generalizados. *Fundacion Juan March Ser. Univ.* 49, Madrid.

V. K. Urazov, 1968. On products of probabilistic metric spaces. *Vestnik Kazan. Univ. Mat.* 4, 15–18.

K. Urbanik, 1964. Generalized convolutions. *Studia Math.* 23, 217–245.

F. A. Valentine, 1964. *Convex Sets*. New York: McGraw-Hill.

F. Wagner, 1965. Statistical topological spaces. *Notices Amer. Math. Soc.* 12, 588.

A. Wald, 1943. On a statistical generalization of metric spaces. *Proc. Nat. Acad. Sci. USA* 29, 196–197. [Note: A slightly extended version appears in *Rep. Math. Colloq.* (2nd ser.), 5–6 (1944), 76–79; reprinted in Abraham Wald, *Selected Papers in Statistics and Probability*, pp. 413–416. New York: Wiley, 1955.]

H. Weyl, 1952. *Symmetry*. Princeton, NJ: Princeton Univ. Press.

D. V. Widder, 1946. *The Laplace Transform*. Princeton, NJ: Princeton Univ. Press.

S. S. Wilks, 1962. *Mathematical Statistics*. New York: Wiley.

W. A. Wilson, 1935. On certain types of continuous transformations of metric spaces. *Amer. J. Math.* 57, 62–68.

A. Wintner, 1938. *Asymptotic Distributions and Infinite Convolutions*. Ann Arbor, Michigan: Edwards Brothers, Inc.

E. F. Wolff, 1981. N-dimensional measures of dependence. *Stochastica* 4, 175–188.

A. F. S. Xavier, 1968. On the product of probabilistic metric spaces. *Portugal. Math.* 27, 137–147.

Y. Yakimovsky, 1976. Boundary and object detection in real world maps. *J. Assoc. Comput. Mach.* 23, 599–618.

L. A. Zadeh, 1965. Fuzzy sets. *Inform. and Control* 8, 338–353.

E. C. Zeeman, 1962. The topology of the brain and visual perception, in *Topology of 3-Manifolds* (M. K. Fort, Jr., ed.), pp. 240–256. Englewood Cliffs, NJ: Prentice-Hall.

E. Zermelo, 1896a. Über einen Satz der Dynamik und die mechanische Wärmetheorie. *Ann. Physik* 57, 485–494.

———, 1896b. Über mechanische Erklärungen irreversibler Vorgänge: Eine Antwort auf Hrn. Boltzmann's "Entgegnung." *Ann. Physik* 59, 791–801.

Index

L

M

T

Errata

PAGE	LINE	IN PLACE OF	READ
4	8	(1.2.2)–(1.2.4)	(1.2.2)–(1.2.4) and Menger's (and our) tacit assumption that $T(0,1) = 0$.
37	-8	the Cartesian plane,	the Cartesian plane and $z = (0,0)$,
39	6	$\mathcal{V}$-space	V-space
42	-7	In contrast, the c^θ-closure	In contrast, the topology determined by the c^θ-closure
45	-1	$G(y) \leq F((y+h)+)$	$G(y) \leq \ell^+ F(y+h) + h$
49	11	$\{\varepsilon_a + 1/n\}$	$\{\varepsilon_a - 1/n\}$
55	9	commutative	associative and commutative
55	10	a familiar induction yields	familiar inductions yield
59	-3	(5.3.5) and (5.3.11),	(5.3.5), (5.3.10) and (5.3.11),
64	-3	$f_2(x) = f_1(\alpha x)$	$f_1(x) = f_2(\alpha x)$
68	-2	$k_2(x) = (k_1(x))^\alpha$	$k_1(x) = (k_2(x))^\alpha$

PAGE	LINE	IN PLACE OF	READ
69	-10	$g_1 \circ f_2$ is linear.	$g_1 \circ f_2$ is linear on $[0, g_2(0)]$.
75	10	$\Psi^{-1}(L(\Psi(x), \Psi(y))$	$\Psi^{-1}(L(\Psi(x), \Psi(y)))$
75	13	$\Psi\left(T\left(\Psi^{-1}(x), \Psi^{-1}(y)\right),\right.$	$\Psi\left(T\left(\Psi^{-1}(x), \Psi^{-1}(y)\right)\right),$
81		See the notes on Chapter 6 for a clearer exposition of Lemma 6.1.8.	
83	17	n-copula	n-subcopula
87	7	But by (5.4.4) and (5.3.13),	But by Theorem 5.5.2,
108		The hypotheses on L in Theorem 7.4.3 and Corollary 7.4.4 are too broad. See the comments on Chapter 7.	
108	-8	$f_\alpha^{-1}(f(x)\, f(y))$	$f_\alpha^{-1}(f_\alpha(x)\, f_\alpha(y))$
109	17	if C and L are associative	if $\overline{C}$ and L are associative
109	19	and associative.	and associative, and let $\overline{C}$ be associative.
109	-8	**C**	$\overline{\mathbf{C}}$
110	5	(7.4.6).	(7.4.7).
133	6	(7.2.24)	(7.2.26)
139	-8	$(d(p,q) + d(p,r))^\alpha$	$(d(p,q) + d(q,r))^\alpha$

PAGE	LINE	IN PLACE OF	READ
142	5	Delete the entire sentence beginning "As usual"	
146	6	metrics	pseudometrics
142	-4	PM	PPM
145	1, 2	P-measurable	random variables
145	9	a	an
145	9	Delete the word "canonical"	
145	10	Menger space	Menger PPM space
145	13	PM space	PPM space
167	-4	$\int_{\lvert\sigma\mathbf{u}-\mathbf{s}\rvert<x}$	$\int_{\lvert\sigma\mathbf{v}-\mathbf{s}\rvert<x}$
182	16	$P\left(D\left(x\right)\right)$	$P^{(2)}\left(D\left(x\right)\right)$
195	7	8.2.8	8.2.9
198	-9	(S, d, G)	(S, d, G) with G in $\mathcal{D}^{+}$
199	-1	12.1.5	12.1.6
200	2	(8.6.5)	(8.6.6)
204	15	$F_{pq}^{\wedge}(c) \leq \alpha F_{pq}^{\wedge}(c)$	$F_{\varphi p \varphi q}^{\wedge}(c) \leq \alpha F_{pq}^{\wedge}(c)$
206	-10	nondecreasing	nonincreasing

PAGE	LINE	IN PLACE OF	READ
207	-13	$\mathcal{D}$ is closed	$\mathcal{D}^+$ is closed
210	1	f is stronger than g	$f \geq g$
221	8	(3.5.6)	(3.5.3)
212	1	Theorem 12.7.10 is nonsense. See the comments on Chapter 12.	
239	-5	from V into K	from S into R
239	-4	on V.	on S.
246	8	$J(A) = \frac{1}{c} \log P(A)$	$J(A) = -\frac{1}{c} \log P(A)$

Notes

Chapter 1. Introduction and Historical Survey

For general updates on the historical development of the subject, see [Schweizer 2003; Sempi 2004]. There have been particularly noteworthy developments in the following areas:

1. Associative functions, particularly t-norms, for which see the notes to Chapter 5 and [Schweizer 2005].

2. Copulas and quasi-copulas, for which see the notes to Chapter 6 and [Dall'Aglio 1991; Schweizer 1991; Sklar 1996a; Nelsen 2005].

3. Binary operations and other functions on Δ and Δ^+, for which see the notes to Chapter 7 and [Schweizer 2005].

4. Distributional chaos, for which see the notes to Chapter 11 and [Schweizer 1996].

5. New definitions of probabilistic normed and probabilistic inner product spaces, plus applications to cluster analysis and symbolic data analysis, for which see the notes to Chapter 15.

Chapter 4. Distribution Functions

1. Besides the modified Lévy metric of Definition 4.2.1, a number of other metrics for weak convergence of distribution functions have been introduced over the years: see e.g., [Sempi 1982; Radu 1984, 1986; Taylor 1985; Pascali and Sempi 1997].

2. If f is left-continuous on $(a,b]$, then the function $f^{\vee}$ given in (4.4.5) is right-continuous on $[f(a), f(b))$, with $f^{\vee}(f(b)) = b$. In this case

$$f\left(f^{\vee}(y)\right) \leq y \text{ for all } y \text{ in } \left[f(a), f(b)\right]$$

and

$$f^{\vee}\left(f(x)\right) \geq x \text{ for all } x \text{ in } [a,b].$$

Moreover

$$f^{\vee} \circ f \circ f^{\vee} = f^{\vee} \text{ and } f \circ f^{\vee} \circ f = f.$$

In the language of lattice theory, f is a *residuated mapping*, $f^{\vee}$ a *residual mapping* and $(f, f^{\vee})$ a *residuated-residual* pair [Blyth and Janowitz 1972; Janowitz and Schweizer 1989].

3. The model for cluster analysis developed in [Janowitz and Schweizer 1989] is order-theoretic and has led to the study of Δ and Δ^{+} from this point of view. If the (pointwise) supremum of any subset of $(\Delta, \leq)$ is defined in the usual way and if the infimum is defined as the supremum of the set of all lower bounds, then $(\Delta, \leq)$ is a complete lattice. As shown in [Powers 1988], the join irreducible elements of $(\Delta, \leq)$ are the functions $\delta_{a,b}$ defined for any a in $\mathbf{R}$ and b in I via $\delta_{a,b} = b\varepsilon_a + (1-b)\varepsilon_{\infty}$. Using this and the fact that any F in Δ may be represented as $F = \sup_{a \in \mathbf{R}} \delta_{a,F(a)}$, Powers completely characterized the order automorphisms of $(\Delta, \leq)$ and, in addition, showed that they are continuous in (Δ, d_L). (They are automatically sup-continuous.)

Simple examples show that weak continuity neither implies nor is implied by sup-continuity.

Chapter 5. Associativity

1. Since the early 1980s, there has been an explosion of interest in t-norms and t-conorms, fueled primarily by the central role that these functions play in multivalued logic, fuzzy set theory and, of course, the theory of PM spaces. The literature is vast. There are now two books detailing the developments—[Klement, Mesiar,

and Pap 2000] and [Alsina, Frank, and Schweizer, to appear]: here it should be noted that, apart from the necessary preliminaries, there is remarkably little overlap between them. A summary of the central ideas of [Klement, Mesiar and Pap 2000], together with some more recent results and additional references, is given in the papers [Klement, Mesiar, and Pap 2004 I, II, III]. There is also a recent conference proceedings [Conf 2005a] where some of the history of the subject is sketched in [Schweizer 2005].

2. Since the 1981 thesis of Krause (see also [Krause 1983]), it has become apparent that in dealing with questions of existence and uniqueness of representations, one begins, not with the assumption of full associativity, but with the weaker assumption of *power-associativity*, defined as follows:

A binary operation T on a set S is *power-associative* if for every x in S and for all positive integers m, n

$$\delta_{m+n}(x) = T\left(\delta_m(x), \delta_n(x)\right) = T\left(\delta_n(x), \delta_m(x)\right),$$

where the *diagonals* δ_n are the functions recursively defined on S via

$$\delta_1(x) = x \text{ and } \delta_{n+1}(x) = T\left(x, \delta_n(x)\right) \text{ for all integers } n \geq 1.$$

Various assumptions of continuity or monotonicity then generally enable one to pass from power-associativity to full-associativity: see e.g., [Krause 1981; Bacchelli 1986].

The diagonal δ_2 is particularly important in such investigations: see, e.g., [Darsow and Frank 1983; Section 9.3 in Kuczma, Choczewski, and Ger 1990; Bézivin and Tomás 1993; Frank 1996; Sklar 1996b].

Results in this area are intimately connected with the second part of Hilbert's Fifth Problem [Aczél 1989].

3. It is useful to add the following paragraph to the end of Section 5.3:

Note that in Part iii of Theorem 5.3.8 each function T_α in $\mathcal{O}(S_\alpha)$ is isomorphic to a function T'_α in $\mathcal{O}[0,1]$ via:

$$T_\alpha(x,y) = a_\alpha + (e_\alpha - a_\alpha)\, T'_\alpha\left(\frac{x - a_\alpha}{e_\alpha - a_\alpha}, \frac{y - a_\alpha}{e_\alpha - a_\alpha}\right). \tag{5.3.15}$$

Since the functions in $\mathcal{O}[0,1]$ are of special importance to us, we make the convention that instead of saying "the semigroup $([a,e], T)$ is an ordinal sum of the semigroups (S_α, T_α)", we say "T is an ordinal sum of the functions T'_α".

4. Problem 5.8.1 has been settled, negatively, in [Krause, to appear]. A thorough exposition of Krause's result is contained in Appendix B.1 of [Klement, Mesiar, and Pap 2000]. Problems 5.8.2 and 5.8.3 remain open. There are much longer lists of problems on associative function in [Alsina, Frank, and Schweizer 2003], [Klement, Mesiar, and Pap 2004] and Appendix B of [Alsina, Frank, and Schweizer, to appear].

Chapter 6. Copulas

1. Since the original publication of this book, there has been a phenomenal growth, not only of interest in copulas *per se*, but of their actual use in statistics and related areas. In these notes we can only directly touch upon a few aspects of this growth. We begin with a fairly comprehensive catalog of the most important references:

For the prehistory of the subject, see [Dall'Aglio 1991; Sklar 1996a]. The period from 1959 through 1989 is nicely summarized in [Schweizer 1991]. The following decade was marked by conferences largely devoted to copulas in Rome [Conf 1991], Seattle [Conf 1996], and Prague [Conf 1997], and capped by the publication of [Nelsen 1999] which has become the standard overview of the subject.

Since then, there have been conferences in Barcelona [Conf 2002] and Québec [Conf 2005 b, c]. The Québec conference highlighted the use of copulas in finance and risk management. And apart from the articles in the various conference proceedings, other published works in which copulas are prominent include [Cherubini, Luciano, and Vecchiato 2004; Embrechts, Höing, and Juri 2003; Fisher 1997; Genest and MacKay 1986a, 1986b; Hutchinson and Lai 1990; Joe 1997; Drouet Mari and Kotz 2001; Scarsini 1989; Singh and Zhang 2004; Singh 2005]. Further references will appear below.

2. While the proof of the fundamental Lemma 6.1.8 is correct, its exposition deserves some clarification, which we trust the following rewrite accomplishes:

PROOF. If $n = 2$, then, since H is 2-increasing, (6.1.3) yields

$$H(y, e_2) - H(x, e_2) - H(y, x_2) + H(x, x_2) \geq 0$$

if x, y are in A_1 and

$$H(e_1, y) - H(e_1, x) - H(x_1, y) + H(x_1, x) \geq 0$$

if x, y are in A_2. The two inequalities together yield (6.1.6).

If $n > 2$ and x, y are in A_m, then we define $n-1$ distinct n-boxes B_k for $1 \leq k \leq n$, $k \neq m$, as follows: Let

$$B_k = [(u_1, \ldots, u_n), (v_1, \ldots, v_n)]$$

where $u_m = x$, $u_k = x_k$ and $u_i = a_i$ for each $i \neq m, k$ while $v_m = y$, $v_k = e_k$, and $v_i = e_i$ for any $i < k$, $i \neq m$, $v_i = x_i$ for any $i > k$, $i \neq m$. (For example, if $n = 3$, $m = 2$, then $B_1 = [(x_1, x, a_3), (e_1, y, x_3)]$ and $B_3 = [(a_1, x, x_3)(e_1, y, e_3)]$.) Since H is n-increasing, $V_H(B_k) \geq 0$ for each k, whence

$$\sum_{1 \leq k \leq n,\ k \neq m} V_H(B_k) \geq 0. \tag{6.1.7}$$

Now, since H is grounded, for each $V_H(B_k)$ the expression (6.1.2) reduces to a 4-term sum involving the 4 vertices of B_k that contain no a_i. When these $n-1$ sums are substituted into the left-hand side of (6.1.7), that side telescopes to:

$$\begin{aligned} &H(e_1, \ldots, e_{m-1}, y, e_{m+1}, \ldots, e_n) \\ &\quad - H(x_1, \ldots, x_{m-1}, y, x_{m+1}, \ldots, x_n) \\ &\quad - H(e_1, \ldots, e_{m-1}, x, e_{m+1}, \ldots, e_n) \\ &\quad + H(x_1, \ldots, x_{m-1}, x, x_{m+1}, \ldots, x_n), \end{aligned}$$

whence (6.1.7) reduces to (6.1.6).

3. In [Deheuvels 1978] there is a direct probabilistic proof of Theorem 6.2.5. A combinatorial (non-probabilistic) proof of Theorem 6.2.6, from which Theorem 6.2.5 follows immediately, appears as Appendix 1 in [Sklar 1996a]. A proof of Theorem 6.2.5 by "checkerboard approximation" is given in [Carley and Taylor 2002].

4. In the statement of Theorem 6.3.6, replace the expression "real-analytic" by "infinitely differentiable." Real-analyticity is a consequence of complete monotonicity.

The first part of Theorem 6.3.6 can be strengthened as follows: $(-1)^n f^{(n)} \geq 0$ for $n = 1, \ldots, N$ if and only if T^{n-1} is a copula for $n = 1, \ldots, N$, (see Section 4.4 of [Alsina, Frank, and Schweizer, to appear]). M. J. Frank [2002] has considered the often difficult problem of determining the greatest possible value of N.

5. The problems at the end of the chapter remain open, though relevant later work does exist. Thus, both [Frank 2002] and Appendix 2 in [Sklar 1996a] are related to "the most important open question" which is "to find necessary and/or sufficient conditions for the compatibility of any given finite set of copulas." And Problem 6.7.2, and even the more general problem of giving a probabilistic characterization of *any* copula, has a partial answer in [Mikusiński, Sherwood, and Taylor 1991] and an unexpected answer in an approximate sense in [Li, Mikusiński, Sherwood, and Taylor 1997]. Quoting the latter: "...it is [now] well known (see, for example [Mikusiński, Sherwood, and Taylor 1992] or [Vitale 1990]) that any copula can be approximated arbitrarily closely in the uniform sense by copulas which correspond to deterministic dependence between a pair of random variables [the so-called shuffles of Min]. This has the somewhat unexpected implication that any sort of dependence between a pair of random variables (even independence) can be approximated as closely as one might wish by a completely deterministic relation."

6. In [Li, Mikusiński, Sherwood, and Taylor 1997], the work of this "Orlando group" of authors (who have also treated the intimate connections between copulas, doubly stochastic measures, and Markov operators, see, e.g., [Sherwood and Taylor 1988]) connects

with the work of the "Chicago group" who have shown, in [Darsow, Nguyen, and Olsen 1992; Darsow and Olsen 1995; Olsen, Darsow, and Nguyen 1996] that Markov processes and Markov operators can be usefully defined and studied in terms of copulas. To do so, they introduced a remarkable associative binary operation $*$ on the set of all 2-copulas, defined as follows:

For any two 2-copulas C_1 and C_2,

$$(C_1 * C_2)(x, y) = \int_0^1 (D_2 C_1)(x, z) \cdot (D_1 C_2)(z, y)\, dz$$

where D_1 and D_2 are the partial derivatives (a.e.) with respect to the first and second place, respectively. Under certain natural conditions it follows that if $\{X_t | t \in T\}$ is a real stochastic process with parameter set T and if, for any s, t in T, C_{st} is the copula of X_s and X_t, then the transition probabilities of the process satisfy the Chapman-Kolmogorov equation if and only if $C_{st} = C_{su} * C_{ut}$ for all s, u, t in T such that $s < u < t$.

7. Copulas have been used to define and study measures of dependence, first in [Schweizer and Wolff 1981] and then by many others (see [Nelsen 1999]). In [Scarsini 1984] it was shown that they can also be used to define and study measures of concordance [Nelsen 2002; Taylor, to appear].

8. An *n-quasi-copula*, $n \geq 2$, is an n-ary operation on $[0, 1]$ that is non-decreasing in each place and satisfies the same boundary conditions (6.2.1) as well as the same Lipschitz condition (6.1.9) as an n-copula. Thus a 2-quasi-copula is a binary operation on $[0, 1]$ satisfying

$$Q(x, 0) = Q(0, x) = 0 \text{ and } Q(x, 1) = Q(1, x) = x, \text{ for all } x \text{ in } I$$

and

$$|Q(x_1, y_1) - Q(x_2, y_2)| \leq |x_1 - x_2| + |y_1 - y_2| .$$

The origins of this notion will be discussed in the notes on Chapter 7. Here we note the following: For any $n \geq 2$,

a. Any n-copula is an n-quasi-copula.

b. Any n-quasi-copula is continuous.

c. W^{n-1} is a n-quasi-copula and, for any n-quasi-copula Q_n,

$$W^{n-1} \leq Q_n \leq M^{n-1}.$$

d. Under the usual pointwise ordering, the set of n-quasi-copulas is a complete lattice.

e. The set of n-quasi-copulas is compact in the topology of uniform convergence.

For more details and references, see [Nelsen 2005].

Chapter 7. Triangle Functions

1. Add the following to Definition 7.1.1: "A triangle function is *continuous* if it is continuous at every pair in $\Delta^+ \times \Delta^+$ in the metric space (Δ^+, d_L) (see (3.1.14))."

2. As noted in the errata, Theorem 7.4.3 and Corollary 7.4.4 are incorrect as stated. Thus the text from the sentence beginning "This is most emphatically not the case..." through the statement of Corollary 7.4.4 should be replaced by the following:

"This is most emphatically not the case, for M. J. Frank [1991] has proved the following remarkable

Theorem 7.4.3 *Let C be a copula and let L in $\mathcal{L}$ have the form*

$$L(u,v) = h^{-1}\left(h\left(u\right) + h\left(v\right)\right), \tag{7.4.14}$$

where h is a continuous, strictly increasing function from $\mathbf{R}^+$ onto $\mathbf{R}^+$. Then $\sigma_{C,L}$ is associative (and in fact is a triangle function) if and only if either $C = \Pi$ or $C = M$ or C is an ordinal sum of Π's and M's.

Note that (7.4.14) includes the particular case $L = Sum$, for which case Theorem 7.4.3 was proved in [Frank 1975]."

3. Problems 7.9.1–7.9.3 remain open, except that the following simple argument shows that the set $\mathcal{D}^+$ is closed under τ_W and therefore under any triangle function $\tau \geq \tau_W$:

For any F, G in $\mathcal{D}^+$ and any positive $\eta < 1$, there is a positive x such that $F(x) > 1 - \eta/2$ and $G(x) > 1 - \eta/2$. Hence

$$\tau_W (F, G) (2x) \geq W (F (x), G (x)) \geq 1 - \eta.$$

Since η can be arbitrarily small, this means that $\tau_W(F, G)$ is in $\mathcal{D}^+$.

Problem 7.9.4 remains open, though in his thesis, Y.-H. Shyu [1989] showed, by quite intricate arguments, that under very strong conditions on T and L, and conditions stronger than absolute continuity on F and G, one obtains a stronger result than absolute continuity for $\tau_{T,L} (F, G)$.

4. In [Ying 1992] it is shown that $\tau_{T,L}$ is a continuous triangle function if and only if T is continuous and L is commutative, associative and has 0 as identity. This settles the second part of Problem 7.9.6. In addition, some partial answers to the first part are also given.

Aside from the above, Problems 7.9.4–7.9.12 remain open.

5. The answer to the question in Problem 7.9.13 is "No". The first example of a non-causal triangle function is due to T. B. M. McMaster [1987]. Subsequently M. S. Ying [1989] exhibited a one-parameter family of simpler examples. However, as pointed out by these authors, their triangle functions are not continuous on Δ^+. More recently, R. M. Tardiff (private communication) showed that if T is a continuous t-norm and L^* a continuous composition law, then the function τ_{T,L^*} defined on $\Delta^+ \times \Delta^+$ via

$$\tau_{T,L^*} (F, G) (x) = \ell^- \inf \{T (F (u), G (v)) \mid L^* (u, v) = x\}$$

is a continuous triangle function and that if, furthermore, both T and L^* are strictly increasing in each place, then τ_{T,L^*} is not causal.

The family of triangle functions τ_{T,L^*} is interesting in its own right and merits further study.

6. For any c in (0,1) and any pair of distribution functions F, G in Δ, the *mixture* of F and G is the distribution function φ defined by

$$\varphi(F, G) = cF + (1 - c)G.$$

In [Alsina and Schweizer 1988] it is shown that φ is not derivable from any binary operation on random variables; and, similarly, that the geometric mean $\sqrt{FG}$ is not derivable. In [Alsina, Nelsen, and Schweizer 1993] these results are extended. There a binary operation φ on Δ is said to be *induced pointwise* by a binary operation Φ on [0,1] if, for any F, G in Δ and any t in $\mathbf{R}$,

$$\varphi(F, G)(t) = \Phi(F(t), G(t)),$$

and the class of binary operations on Δ which are both induced pointwise and derivable from binary operations on random variables is completely characterized. In [Nelsen, Quesada-Molina, Schweizer, and Sempi 1996] this characterization is extended from binary to n-ary operations. The key to these characterizations is the notion of a quasi-copula. The original definition, as given in the abovementioned papers, is rather cumbersome. In [Genest, Quesada-Molina, Rodriguez-Lallena, and Sempi 1999] it is shown that for 2-quasi-copulas this definition is equivalent to the one given in the notes on Chapter 6; and in [Cuculescu and Theodorescu 2001/2002] this equivalence is shown to hold for n-quasi copulas, $n > 2$, as well. It is generally much easier to work with the newer definition: this has led to a rapid development of the theory of quasi-copulas. The results obtained shed light on the structure of the set of copulas: e.g., in [Nelsen and Úbeda-Flores, to appear] it is shown that the lattice of 2-quasi-copulas is the Dedekind-MacNeille completion of the poset of 2-copulas. For $n > 2$ the question is open [Nelsen 2005].

7. Using results of [Powers 1988] (see the notes on Chapter 4), T. Riedel [1991] considered Cauchy's equation for triangle functions τ, viz. $\varphi(\tau(F, G)) = \tau(\varphi(F), \varphi(G))$ on Δ^+. He obtained all the order automorphism solutions when $\tau = \tau_T$ and T is a continuous Archimedean t-norm. He also obtained a representation for all sup-continuous (equivalently, residuated) solutions when T is strict. In

[Riedel 1992] he extended these results to certain classes of triangle functions of the form $\tau_{T,L}$; and in [Riedel 1994], in connection with Problem 7.9.6, he used this lattice-theoretic approach to characterize a class of such triangle functions. Lastly, [Powers and Riedel 1996] showed that weakly continuous isotone bijections of $(\Delta, \leq, d_L)$ need not be order isomorphisms and obtained a necessary and sufficient condition which guarantees that they are.

8. Since probability theory in general deals with the whole set Δ rather than just the subset Δ^+, it is natural to consider binary operations on Δ that have the same nice properties as the triangle functions on Δ^+. Accordingly, following [Alsina, Schweizer, Sempi, and Sklar 1997] we define a *multiplication* on Δ as a binary operation on Δ that is associative, commutative, non-decreasing in each place, and whose restriction to $\Delta^+ \times \Delta^+$ is a triangle function.

Convolution is the best-known multiplication on Δ. Other multiplications can be derived by simply extending the definition of a triangle function from $\Delta^+ \times \Delta^+$ to $\Delta \times \Delta$: this is immediate for $\mathbf{T}$, where T is any t-norm, and as pointed out in [Schweizer 1975a], if T is left-continuous, also works for τ_T. Furthermore, as stated in [Alsina, Schweizer, Sempi, and Sklar 1997], minor modifications of the arguments given in [Schweizer 1975a] suffice to show that, for any continuous t-conorm T^*, the triangle function τ_{T^*} can also be extended to a multiplication on Δ. It would seem that this procedure should work as well for the more general functions $\tau_{T,L}$ and $\tau_{T^*,L}$, but this has yet to be shown explicitly.

Chapter 8. Probabilistic Metric Spaces

1. Since (Δ^+, d_L) is a metric space, it follows that (Δ^+, d_L, G) is a simple space for any G in Δ^+ distinct from ε_0 and ε_∞. But there are more interesting ways of imposing a PM-structure on Δ^+, such as those presented in [Sherwood and Taylor 1974] and [Höhle 1978] (papers referred to in the text in an inappropriate connection with the incorrect Theorem 12.7.10). Sherwood and Taylor define

a binary operation α_p on Δ^+ for each $p \geq 1$ by: $\alpha_p(F,G)$ is the function in Δ^+ whose quasi-inverse $(\alpha_p(F,G))^\wedge$ is given by:

$$(\alpha_p(F,G))^\wedge(x) = \left(\int_0^x |(F(t))^p - (G(t))^p|\,dt\right)^{1/p}$$

for x in [0,1]. It then follows from the Minkowski inequality, (7.7.2) and Corollary 4.4.3 that $(\Delta^+, \alpha_p, \tau_M)$ is a PM-space.

Höhle, starting with a continuous triangle function τ, defines a binary operation $\mathcal{F}_\tau$ on Δ^+ by:

$$\mathcal{F}_\tau(F,G) = \sup\{H \text{ in } \Delta^+ \mid \tau(F,H) \leq G,\ \tau(G,H) \leq F\}.$$

Then $(\Delta^+, \mathcal{F}_\tau, \tau)$ is a PM-space tailor-made to fit the triangle function τ.

2. If T is an Archimedean t-norm with additive generator f, then the Menger triangle inequality (1.2.1) may be expressed in the form

$$f\,F_{pr}(x+y) \leq f\,F_{pq}(x) + f\,F_{qr}(y)$$

[Schweizer and Sklar 1961a]. Using this fact, V. Radu [1986] showed that if $(S, \mathcal{F}, T)$ is a Menger space with T as above, then the function $d_f : S \times S \to \mathbf{R}^+$ defined by

$$d_f(p,q) = \sup\{t \mid t \leq f\,F_{pq}(t)\}$$

is a metric on S, which induces the ε, λ-topology. For $T = W$, so that $f = 1 - j$,

$$d_{1-j}(p,q) = \sup\{t \mid F_{pq}(t) \leq 1 - t\}.$$

Comparison with Lemma 4.3.3 yields that if $(S, \mathcal{F}, \tau)$ is a PM space with $\tau \geq \tau_W$ then the function β defined on $S \times S$ via $\beta(p,q) = d_L(F_{pq}, \varepsilon_0)$ is a metric. When the PM space in question is Höhle's space $(\Delta^+, \mathcal{F}_\tau, \tau)$ one obtains a class of metrics on Δ^+ that includes the classical Lévy metric [Radu 1987].

Lastly, as shown in [Radu 1984], expressing the conjugate transform $C_T F$ in (8.2.14) in terms of an additive generator f, rather than a multiplicative generator k, of T yields

$$d_{T,x}(p,q) = \inf_{t>0}\{xt + f\,F_{pq}(t)\}$$

and, as above, $d_{W,1}$ leads to a metric which is metrically equivalent to the Lévy metric.

3. The discussion of hysteresis in Section 5 can be supplemented by the material in several later papers. In [Weinstock, Erber, and Nisenoff 1988] there is an analysis of measurement by a SQUID (superconducting quantum interference device) of magnetic hysteresis in extremely pure iron through a millionfold range in intensity of an applied magnetic field. The analysis confirms linear dependence in weak fields and quadratic dependence in stronger ones.

Computer simulations of mechanical hysteresis in steel frameworks [Erber and Guralnick 1988] and actual observed hysteresis in steel specimens [Erber, Guralnick, and Michels 1993] yield similar results.

Chapter 9. Random Metric Spaces

1. In the last fifteen years, Tie-Xin Guo (along with various colleagues, e.g., Zhao-Yong You, Xi Lin, Fu-Zhou Gong, and Lin-Fu Zhu) has been engaged in a profound, in-depth study of random metric, random normed, and random inner product spaces. He has approached these constructs from the point of view of a functional analyst—his aim being to formulate and establish results of classical functional analysis in the random space setting, i.e., to provide a new approach to random functional analysis. His studies have led him, when necessary, to modify and extend the straightforward definition given on page 240 in various directions. Thus he frequently replaces the probability space (Ω, A, P) by a measure space, often works with equivalence classes rather than individual elements and assumes, as in Definition 9.3.1, that each of the defining conditions only holds almost everywhere. The latter assumption leads to delicate questions of measurability.

The attempt to formulate a proper and fruitful definition of a conjugate space has let Guo to the notion of a random normed module, i.e., a random normed space satisfying an additional condition. Armed with this notion, he has been able to establish the random normed module equivalents of the Hahn-Banach Theorem, the Riesz Representation Theorem, and James' Theorem and has established key relations between random normed modules and normed spaces as well as between random conjugate spaces and classical conjugate spaces. In this setting he has solved a number of difficult measurability problems as well as various problems of best-approximation, e.g., in Lebesgue-Bochner function spaces. For details see [Guo 2001, 2004; Guo and Ma 2004; Guo and Li, to appear].

The achievements outlined above are only a beginning. Much of the work may be viewed as a new and fruitful approach to standard stochastic analysis and promises to enrich that subject in many significant ways. How much of this will carry over to general probabilistic metric, normed, and inner product spaces is an open and challenging question.

2. Chapter 7 of [Höhle 2001] is devoted to the study of Boolean-valued topological properties of random metric spaces (in the general sense, as given in Definition 9.3.1). In particular, convergence of sequences, Boolean-valued completions and a principle of uniformly continuous extension are discussed. These ideas are summarized and carried further in [Höhle 2005].

3. We note that a version of the three-way equivalence among E-spaces, metrically-generated spaces and random metric spaces plays a central role in [Regenwetter and Marley 2001].

Chapter 10. Distribution-Generated Spaces

The subject matter of this chapter—in particular, the concept of a C-space—has found its way into the work of several additional physicists, e.g., [Guz 1984; Prugovečki 1984; Roy 1998; Requardt and Roy 2001] (see Section 10.6). Indeed, the papers [Rosen 1947,1962] and the quote from [Menger 1949] exhibited in our Preface are the leitmotifs of Roy's book.

Chapter 11. Transformation-Generated Spaces

1. The consideration of transformation-generated spaces has led to the development of a far-reaching theory of distributional chaos. For the early history of this development see [Schweizer 1996], and for a general survey, with examples and a comparison with other notions of chaos, see [Schweizer, Sklar, and Smítal 2001/2002].

Given a metric space (S, d) and a transformation Ψ on S, we can define, parallel to the definition of F_{pq} via (11.1.5) and (11.1.6), a function F^*_{pq} for each pair (p, q) in $S \times S$, by

$$F^*_{pq} = \ell^{-1}\varphi^*_{pq},$$

where

$$\varphi^*_{pq}(x) = \limsup_{n\to\infty} F^{(n)}_{pq}(x) \text{ for each real } x.$$

In Section 11.2 we considered situations in which the limit inferior in (11.1.5) can be replaced by a bona-fide limit, i.e., situations in which $F_{pq} = F^*_{pq}$. Now we consider the opposite situation, i.e., $F_{pq} \neq F^*_{pq}$, which means, since $F_{pq} \leq F^*_{pq}$, that $F_{pq}(x) < F^*_{pq}(x)$ for all x in some non-empty open interval. Accordingly, if there are pairs (p, q) in $S \times S$ such that $F_{pq} \neq F^*_{pq}$, then we say that Ψ is *distributionally chaotic* or exhibits *distributional chaos*.

In [Schweizer and Smítal 1994], the original paper on distributional chaos, this notion was investigated in the case $C(J, J)$, i.e., S a compact real interval J under the usual metric, and Ψ a continuous function from J into J. They showed that Ψ is distributionally chaotic if and only if Ψ has positive topological entropy (for the important but complicated notion of topological entropy, see Chapter 14 in [Denker, Grillenberger, and Sigmund 1976]). They also showed that for any distributionally chaotic Ψ in $C(J, J)$ there are points p, q in J such that

$$F^*_{pq} = \varepsilon_0 \text{ and } F_{pq} \leq \varepsilon_a \text{ for some } a > 0.$$

This is the strongest possible version of distributional chaos, and as such is labeled DCI in [Balibrea, Smítal, and Stefánková, to appear]. Among the very many distributionally chaotic functions in

$C(J, J)$ are the familiar "tent map" and the "trapezoid functions" (see Section 5 in [Schweizer, Sklar, and Smítal 2001/2002]).

In [Sklar and Smítal 2000] it is shown that any continuous transformation on a compact metric space that has either the so-called specification property (see Chapter 21 in [Denker, Grillenberger, and Sigmund 1976]) or the generalized specification property introduced in [Balibrea, Schweizer, Sklar, and Smítal 2003] is distributionally chaotic. Moreover, as in the case of chaotic functions in $C(J, J)$, such transformations are DCI and have positive topological entropy. The set of such transformations includes the so-called Anosov diffeomorphisms, so in particular those favorite functions of ergodic theorists, the so-called hyperbolic automorphisms of the n-torus (see Chapters 22–24 in [Denker, Grillenberger, and Sigmund 1976] and, among the many later papers on this topic, see [Gebremeskel and Sklar 1999; Martins and Sousa Ramos 1999]).

What is perhaps the most remarkable feature of these examples is that Anosov diffeomorphisms are all mixing transformations, which means, by Theorem 11.3.4, that for each such diffeomorphism there is a function F in Δ^+ such that $F^*_{pq} = F_{pq} = F$ for almost all pairs (p, q). In other words, almost all pairs are not only "non-chaotic," but are non-chaotic in exactly the same way. In a statistical sense, this is extremely *orderly* behavior. On the other hand, the "chaotic" pairs (p, q), i.e., those for which $F^*_{pq} \neq F_{pq}$, though only forming a set of measure 0, are dense in $S \times S$, so that "order" and "chaos" are inextricably intermixed. The tent map and the functions C_n defined in (11.3.7) also exhibit this behavior.

Now, it is clear that DCI implies the much weaker result that there are pairs (p, q) such that

$$\liminf_{n\to\infty} \left(d\left(\Psi^n\left(p\right), \Psi^n\left(q\right)\right)\right) = 0$$

and

$$\limsup_{n\to\infty} \left(d\left(\Psi^n\left(p\right), \Psi^n\left(q\right)\right)\right) > 0,$$

which is the criterion for chaos in the sense of Li and Yorke (see Section 2 of [Schweizer, Sklar, and Smítal 2001/2002]). Some, though not all, weaker forms of distributional chaos also imply Li-Yorke chaos: a 2-dimensional transformation that exhibits distributional chaos but not Li-Yorke chaos appears in [Babilonová 1997]. And, as might be expected from the respective definitions, there are examples, even very simple ones, that show that Li-Yorke chaos does not imply distributional chaos.

While distributional chaos is equivalent to positive topological entropy (PTE) for continuous functions mapping compact real intervals into themselves, this is not the case in general: [Forti, Paganoni, and Smítal 1995] and [Liao and Fan 1998] show that distributional chaos does not necessarily imply PTE, while [Forti and Paganoni 1998] demonstrates the stronger result that distributional chaos does not imply positive sequence topological entropy for any admissible sequence. For DCI in particular, the non-implications go both ways: [Forti, Paganoni, and Smítal 1999] shows that DCI does not imply PTE, and [Smítal and Stefánková 2004] that PTE does not imply DCI. Whether PTE implies *any* form of distributional chaos remains an open problem.

On comparing PTE with distributional chaos as indicators of chaotic behavior (for PTE has been used as such), one finds that for PTE the numerical value of the entropy has no relation to the actual extent of chaos [Schweizer, Sklar, and Smítal 2001/2002, Section 3] while distributional chaos enables one to define natural measures of such actual extent: e.g., for compact metric spaces (S, d) one such measure is

$$sup\left\{\int_0^{diam\, S}\left(F^*_{pq}(x) - F_{pq}(x)\right)dx \mid p, q \text{ in } S\right\}\Big/ diam\, S;$$

there are others [ibid, Sections 4 and 6].

2. Theorem 11.4.2 has been extended from mixing transformations to weakly mixing transformations [Sempi 1985; Fatkić 1992]; and, as Fatkić has shown, from the ordinary diameter to the diameter of any finite order (see 11.6.1). The latter result is a first step toward the resolution of Problem 11.6.5.

Chapter 12. The Strong Topology

1. The literature on fixed point theorems in PM spaces seems to be endless. A presentation of the results obtained through 1989 is given in [Constantin and Istrăţescu 1989]; a recent, comprehensive treatment is given in [Hadžić and Pap 2001]; and an up-to-date survey may be found in [Hadžić and Pap 2005].

Most of the work in this area is devoted to straightforward generalizations (of generalizations) of known metric space results. Here the hypotheses are often such that the contractions on the PM spaces in question are contractions on associated, topologically equivalent, metric spaces (see p. 204 of the text and [Hadžić and Pap 2001; Radu 1987; Schweizer, Sherwood, and Tardiff 1988]).

2. An interesting result, complementing the work of H. Sherwood, is given in [Tardiff 1992] where it is shown that if all the distance distribution functions F_{pq} of a complete PM space $(S, \mathcal{F}, \tau)$ with $\tau \geq \tau_W$ satisfy the growth condition

$$\int_1^\infty ln\, u\, dF_{pq}(u) < \infty,$$

e.g., if their first moments exist, then any contraction of the form (12.6.1) has a unique fixed point. When applied to a complete E-space, this result gives a partial answer to Problem 12.11.1.

3. Problem 12.11.3 has been solved—negatively—by H. Sherwood who constructed a simple counterexample in which $S = \{0, 1, 2\}$ (see [Alsina, Frank, and Schweizer, to appear]).

4. Problem 12.11.4 has been partially solved. In [Tardiff 1980] it is shown that if T_1 and T_2 are strict t-norms with additive generators t_1 and t_2, respectively, then $T_1 \gg T_2$ if and only if the function $h = t_1 \circ t_2^{-1}$ satisfies the inequality

$$\begin{aligned} h^{-1}[h(a+b) + h(c+d)] &\leq h^{-1}[h(a) + h(c)] \\ &+ h^{-1}[h(b) + h(d)] \end{aligned}$$

for all a, b, c, d in $\mathbf{R}^+$. A sufficient condition for this (and a more general) inequality, due to H. P. Mulholland [1950], is that h be continuous, strictly increasing with $h(0) = 0$ and that both h and $\log \circ h \circ \exp$

be convex. From this it follows at once that the dominance relation is transitive on any collection D_M of strict t-norms having the property that, for any T_1, T_2 in D_M, the function $h = t_1 \circ t_2^{-1}$ satisfies Mulholland's condition. Another sufficient condition [Tardiff 1984] is that h be differentiable, strictly increasing with $h(0) = 0$ and that both h and $\log \circ h' \circ \exp$ be convex. Using this, Tardiff established dominance on any collection D_T of strict t-norms satisfying his conditions as well as the condition that, for any t_1 and t_2, the function $t_1 \circ t_2^{-1}$ has a third derivative.

In [Jarczyk and Matkowski 2002] it is shown that Mulholland's condition implies Tardiff's, but not conversely; and furthermore, that a continuous function $f : \mathbf{R}^+ \to \mathbf{R}^+$ is such that $\log \circ f \circ \exp$ is convex if and only if $f\left(\sqrt{xy}\right) \leq \sqrt{f(x)f(y)}$, i.e., if and only if f is convex with respect to the geometric mean.

Lastly, in [Sherwood 1984], using a direct argument, it is shown that the dominance relation is transitive on the family of t-norms $\{T_p\}$ given by (5.6.9). For the subfamily of strict t-norms, i.e., $\{T_p | p \leq 0\}$, transitivity follows from the argument given above.

5. Expanding on Lemma 12.7.4, for any continuous t-norms, T_1 and T_2, the following are equivalent [Tardiff 1975]: $T_1 \gg T_2$, $\mathbf{T}_1 \gg \mathbf{T}_2$, $\tau_{T_1} \gg \tau_{T_2}$, $\tau_{T_2^*} \gg \tau_{T_1^*}$.

6. Theorem 12.7.10 as it stands is nonsense: The function $\mathcal{F}$ cannot be a triangle function; for if it were, by (8.1.1) and the fact that a triangle function is non-decreasing, for every F in Δ^+, we would have $\varepsilon_0 = \mathcal{F}(F, F) \leq \mathcal{F}(F, \varepsilon_0) = F$. For a discussion of the papers [Sherwood and Taylor 1974] and [Höhle 1978] see the notes to Chapter 8.

7. The paper mentioned in the footnote on page 214 has appeared as [Alsina and Schweizer 1983].

Chapter 13. Profile Functions

Other interesting examples of the Poincaré Paradox, using the idea of many-valued equalities, are given in Section 3.2 of [Höhle 2005]; and an insightful discussion of the question of indistinguishability is given in [Trillas and Valverde 1984].

Chapter 15. Supplements

1. The definition of a probabilistic normed (PN) space given in Section 15.1, which goes back to the work of A. N. Šerstnev, can now be replaced by the following more comprehensive one introduced in [Alsina, Schweizer, and Sklar 1993]:

A *probabilistic normed space* (briefly, a *PN space*) is a quadruple (S, η, τ, τ^*) where S is a real linear space, τ and τ^* are continuous triangle functions, and η is a function from S into Δ^+ such that (writing N_p for $\eta(p)$) the following conditions hold for all p, q in S:

(N1) $N_p = \varepsilon_0$ if and only if $p = \theta$, where θ is the null-vector in S;

(N2) $N_{-p} = N_p$;

(N3) $N_{p+q} \geq \tau(N_p, N_q)$;

(N4) $N_p \leq \tau^*\left(N_{\alpha p}, N_{(1-\alpha)p}\right)$ for all α in $[0, 1]$.

In the 1993 paper the authors show that if (and only if) $\tau^* = \tau_M$ and equality holds in (N4), then we have a PN space in the original sense of Šerstnev, so such a space may aptly be called a *Šerstnev PN space*. They also define a *Menger PN space* to be one in which $\tau = \tau_T$ and $\tau^* = \tau_{T^*}$, for some continuous t-norm T and its t-conorm T^*, and denote such a space by (S, η, T). Thus, since $M \leq T^*$ for any t-conorm T^*, any Šerstnev PN space in which $\tau = \tau_T$ for some continuous t-norm T is also a Menger PN space (S, η, T) for that T.

In a later paper [Alsina, Schweizer, and Sklar 1997] the authors, using the above definition of a PN space, endow the space S with the strong topology induced by η. They then show that the function

ν from $S \times S$ into Δ^+ given by $\nu(p,q) = N_{p+q}$ is uniformly continuous, and the function M from $R \times S$ into S given by $M(\alpha, p) = \alpha p$ is uniformly continuous in its second place, but need not be continuous in its first place. Thus, in general a PN space need not be a topological vector space. However, if τ^* is Archimedean in the sense of Section 14.1, i.e., if τ^* admits no idempotent elements other than ε_0 and ε_∞, then M is continuous in its first place (but not uniformly continuous, as erroneously stated in the 1997 paper) and so is jointly continuous.

The study of probabilistic normed spaces as defined above has been undertaken, e.g., in [Lafuerza-Guillén, Rodriguez-Lallena, and Sempi 1997] and [Lafuerza-Guillén and Sempi 2003].

2. A new definition of a probabilistic inner product space was introduced in [Alsina, Schweizer, Sempi, and Sklar 1997]. It is based on the new definition of a PN space given above, uses the notion of a multiplication on Δ (see the notes on Chapter 7), and goes as follows:

Let S be a real linear space, and $\mathcal{G}$ a function from $S \times S$ into Δ whose value at the pair (p,q) in $S \times S$ is denoted by $G_{p,q}$. Define a function η from S into Δ^+ by

$$N_p(x) = \begin{cases} G_{p,p}(x^2), & x > 0, \\ 0, & x \leq 0, \end{cases}$$

where N_p is the value of η at the vector p in S. Then a *probabilistic inner product space* (briefly, a *PIP space*) is a quadruple $(S, \mathcal{G}, \tau, \tau^*)$ where τ and τ^* are multiplications on Δ such that $\tau \leq \tau^*$ and the following conditions hold for all p, q, r in S:

(P1) $G_{p,p}$ is in Δ^+, and $G_{p,p} = \theta$ if and only if $p = \theta$, where θ is the null vector in S;

(P2) $G_{\theta,p} = \varepsilon_0$;

(P3) $G_{p,q} = G_{q,p}$;

(P4) $G_{-p,q} = \overline{G}_{p,q}$ where, for any F in Δ, $\overline{F}(x) = \ell^-(1-F)(-x)$;

(P5) $N_{p+q} \geq \tau\left(N_p, N_q\right)$;

(P6) $N_p \leq \tau^*\left(N_{\alpha p}, N_{(1-\alpha)p}\right)$ for every α in [0,1];

(P7) $\tau\left(G_{p,r}, G_{q,r}\right) \leq G_{p+q,r} \leq \tau^*\left(G_{p,r}, G_{q,r}\right)$.

Note that (S, η, τ, τ^*) is a PN space. If $\tau^* = \tau_M$ and equality holds in (P6), then we have a *Šerstnev PIP space*.

If $\tau = \tau_T$ and $\tau^* = \tau_{T^*}$ for some continuous t-norm T and its t-conorm T^*, then $(S, \mathcal{G}, \tau_T, \tau_{T^*})$ is a *Menger PIP space* denoted by $(S, \mathcal{G}, T)$.

In their 1997 paper, the authors, after discussing several examples, show that if $T = M$ in a Menger PIP space, then the space is actually a real inner product space, i.e., that there is a real inner product $\langle\cdot,\cdot\rangle$ such that $G_{p,q} = \varepsilon_{\langle p,q\rangle}$ for all p, q in S. They then remark that "anyone tempted to 'enrich' the theory of PIP spaces [by elaborating the properties of spaces of the form $(S, \mathcal{G}, M)$] is strongly advised to take careful note of [this result]."

3. Probabilistic topological spaces and fuzzy topological spaces are closely related. The latter have received much attention and will not be discussed here. A good overview of those spaces and related structures is given in [Lowen 1997].

Modelling his approach on that of [Čech 1966], L. C. Florescu [1989] introduced probabilistic convergence structures in two ways: via the notion of a probabilistic convergence function $p(\lambda, x)$ which models the probability that the net λ converges to the point x; and via the notion of a probabilistic convergence class, which is a family C_t of related convergence classes (in the sense of Čech) in which (λ, x) belonging to C_t is interpreted as the probability that the net λ converges to the point x is not less than t. These notions are equivalent and yield a probabilistic convergence structure which, in general, is bijectively related to a probabilistic topological space. These convergence structures are also considered, but now from the view of filters rather than nets, in [Richardson and Kent 1996] and [Minkler, Minkler, and Richardson 1998].

4. The idea of using distribution functions as numbers (see p. 123) has been put into practice in three areas:

(i) *Chaos Theory*: Specifically, distributional chaos as against Li-Yorke chaos—see the notes on Chapter 11.

(ii) *Cluster Analysis* is one of many techniques of organizing and extracting information from large data sets. A large class of cluster methods start with a set S of elements to be clustered (classified) and a *dissimilarity coefficient*, i.e., a mapping $d : S \times S \to \mathbf{R}^+$ such that $d(x, x) = 0$ and $d(x, y) = d(y, x)$, which is derived from the attributes on which the classification is based. Often these attributes are distributed or stochastic and the dissimilarity coefficient in question is a mean, median or other typical value of the data. Condensing the data in this fashion leads to a loss of information before the clustering even begins. It is therefore desirable to have cluster methods which, instead of summarizing and then classifying, classify and then summarize, i.e., cluster methods that begin by working directly with the distributed data. In short, the numerical dissimilarity coefficients should be replaced by distribution-valued ones. Following up on this idea has led to the theory of *percentile clustering*. The basics of this theory are worked out in [Janowitz and Schweizer 1989], where, in addition, various specific algorithms are developed and applied to bona-fide data sets, with good results. But here, while the first stage of the clustering algorithms is probabilistic, subsequent stages employ older deterministic algorithms. For while one can easily define a probabilistic distance between sets (see Section 12.9), a "probabilistic theory of hazy lumps" [Menger 1968] that can be implemented algorithmically still needs to be worked out.

(iii) *Symbolic Data Analysis* is another technique of data mining. Initiated and developed by E. Diday and his colleagues [Bock and Diday 2000], it usually begins with a rectangular array whose entries are often numerical. Some years ago these entries were replaced by distribution functions, leading to the following construct: Let $(\Omega, \mathcal{A}, P)$ be a probability space and F a mapping from $\mathbf{R} \times \Omega$ to I. Then (ignoring measurability questions) for each fixed ω in Ω, $F(t, \omega)$ is in Δ, and for each fixed t in $\mathbf{R}$, $F(t, \omega)$ is a random variable with values in I (i.e., F is a stochastic process). For any x in I and t in $\mathbf{R}$, the two-place function

$$G(t, x) = P\{\omega \in \Omega \mid F(t, \omega) \leq x\}$$

is called a *distribution of distributions*. These objects form the basis from which the desired information is to be extracted. Since, for $t_1 \neq t_2$, $F(t_1, \omega)$ and $F(t_2, \omega)$ are generally dependent, copulas come into play. The details are given in [Diday and Vrac, to appear]. This theory has been applied with remarkable success to a problem in meteorology [Vrac, Diday, and Chédin 2004].

Supplementary References

These supplementary references follow the same format as the references in the original text—alphabetically by author(s) and under each author, by date—except that it has proved convenient to head the list by references to six conference proceedings, listed under the heading Conf. References to individual papers in such proceedings are then referred to the appropriate Conf listing. Any reference in the notes dated before 1983 should be first looked for in the original bibliography. Further references to papers dealing with probabilistic metric and related spaces may be found in Section 54E70 of Mathematical Reviews.

Conf 1991. G. Dall'Aglio, S. Kotz, and G. Salinetti, eds. *Advances in Probability Distribution Functions with Given Marginals: Beyond the Copulas.* Mathematics and its Applications 67. Kluwer, Dordrecht.

Conf 1996. L. Rüschendorf, B. Schweizer, and M. D. Taylor, eds. *Distributions with Fixed Marginals and Related Topics.* IMS Lecture Notes-Monograph Series 28. Institute of Mathematical Statistics, Hayward.

Conf 1997. V. Beneš and J. Štěpán, eds. *Distributions with Given Marginals and Moment Problems.* Kluwer, Dordrecht .

Conf 2002. C. M. Cuadras, J. Fortiana, and J. A. Rodriguez-Lallena, eds. *Distributions with Given Marginals and Statistical Modelling.* Kluwer, Dordrecht.

Conf 2005a. E. P. Klement and R. Mesiar, eds. *Logical, Algebraic, Analytic and Probabilistic Aspects of Triangular Norms.* Elsevier, Amsterdam.

Conf 2005b. C. Genest et al., eds. *Proceedings of the Conference on Dependence Modeling: Statistical Theory and its Applications in Finance and Insurance (DeMoSTAFI).* Theory: Special Issue of the *Canadian J. of Statistics.*

Conf 2005c.————. Applications: Special Issue of *Insurance: Mathematics and Economics.*

J. Aczél, 1989. The state of the second part of Hilbert's fifth problem. *Bull. Amer. Math. Soc.* 20, 153–163.

C. Alsina, M. J. Frank, and B. Schweizer, 2003. Problems on associative functions. *Aequationes Math.* 66, 128–140.

————, *Associative Functions on Intervals*, to appear.

C. Alsina, R. B. Nelsen, and B. Schweizer, 1993. On the characterization of a class of a binary operations on distribution functions. *Statist. Prob. Lett.* 17, 85–89.

C. Alsina and B. Schweizer, 1983. The countable product of probabilistic metric spaces. *Houston J. Math.* 9, 303–310.

————, 1988. Mixtures are not derivable. *Found. Phys. Lett.* 1, 171–174.

C. Alsina, B. Schweizer, C. Sempi, and A. Sklar, 1997. On the definition of a probabilistic inner product space. *Rend. Mat.* (7) 17, 115–127.

C. Alsina, B. Schweizer, and A. Sklar, 1993. On the definition of a probabilistic normed space. *Aequationes Math.* 46, 91–98.

————, 1997. Continuity properties of probabilistic norms. *J. Math. Anal. Appl.* 208, 446–452.

B. Bacchelli, 1986. Representation of continuous associative functions. *Stochastica* 10, 13–28.

M. Babilonová, 1997. Distributional chaos for triangular maps. *Annales Math. Silesiana* 13, 33–38.

F. Balibrea, B. Schweizer, A. Sklar, and J. Smítal, 2003. Generalized specification property and distributional chaos. *Internat. J. Bifurcation and Chaos* 13, 1683–1694.

F. Balibrea, J. Smítal, and M. Stefánková: The three versions of distributional chaos, to appear.

J. P. Bézivin and M. S. Tomás, 1993. On the determination of strict t-norms on some diagonal segments. *Aequationes Math.* 45, 239–245.

T. S. Blyth and M. F. Janowitz, 1972. *Residuation Theory*. Pergamon, London.

H. H. Bock and E. Diday, eds. 2000. *Analysis of Symbolic Data. Exploratory Methods for Extracting Statistical Information from Complex Data.* Springer, Heidelberg .

H. Carley and M. D. Taylor, 2002. A new proof of Sklar's theorem. [Conf 2002] 29–34.

U. Cherubini, E. Luciano, and W. Vecchiato, 2004. *Copula Methods in Finance* : Wiley, Hoboken, NJ.

G. Constantin and I. Istrăţescu, 1989. *Elements of Probabilistic Analysis with Applications.* Editura Academiei Romania. Kluwer, Bucharest and Dordrecht.

I. Cuculescu and R. Theodorescu, 2001/2002. Copulas: diagonals, tracks. *Rev. Roumaine Math. Pures Appl.* 46, 731–742.

G. Dall'Aglio, 1991. Fréchet classes: the beginnings. [Conf 1991] 1–12.

W. F. Darsow and M. J. Frank, 1983. Associative functions and Abel-Schröder systems. *Publ. Math. Debrecen* 30, 253–272.

W. F. Darsow, B. Nguyen, and E. T. Olsen, 1992. Copulas and Markov processes. *Illinois J. Math.* 36, 600–642.

W. F. Darsow and E. T. Olsen, 1995. Norms for copulas. *Internat. J. Math. Math. Sci.* 18, 417–436.

P. Deheuvels, 1978. Caractérisation complète des lois extrêmes multivariées et de la convergence des types extrêmes. *Publ. Inst. Statist. Univ. Paris* 23, 1-37.

M. Denker, C. Grillenberger, and K. Sigmund, 1976. *Ergodic Theory on Compact Spaces.* Springer Lecture Notes in Mathematics 527.

E. Diday and M. Vrac. Mixture decomposition of distributions by copulas in the symbolic data analysis framework, to appear.

D. Drouet Mari and S. Kotz, 2001. *Correlation and Dependence.* Imperial College Press, London.

P. Embrechts, A. Höing, and A. Juri, 2003. Using copulas to bound the Value-at-Risk for functions of dependent risks. *Finance and Stochastics* 7, 145–167.

T. Erber and S. A. Guralnick, 1988. Hysteresis and incremental collapse: the iterative evolution of a complex system. *Annals of Physics* 181, 25–53.

T. Erber, S. A. Guralnick, and S. C. Michels, 1993. Hysteresis and fatigue. *Annals of Physics* 224, 157–192.

H. Fatkić, 1992. Note on weakly mixing transformations. *Aequationes Math.* 43, 38–44.

N. I. Fisher, 1997. Copulas, in S. Kotz, B. Read, and D. L. Banks, eds. *Encyclopedia of Statistical Sciences*, Update Volume 1, 159–163.

L. C. Florescu, 1989. Probabilistic convergence structures. *Aequationes Math.* 38, 123–145.

G.-L. Forti and L. Paganoni, 1998. A distributionally chaotic triangular map with zero sequence topological entropy. *Math. Pannon.* 9, 147–152.

G.-L. Forti, L. Paganoni, and J. Smítal, 1995. Strange triangular maps of the interval. *Bull. Austral. Math. Soc.* 51, 395–415.

———, 1999. Dynamics of homeomorphisms on minimal sets generated by triangular mappings. *Bull. Austral. Math. Soc.* 59, 1–20.

M. J. Frank, 1991. Convolutions for dependent random variables. [Conf 1991] 75–93.

———, 1996. Diagonals of copulas and Schröder's equation. *Aequationes Math.* 51, 150.

———, 2002. Multivariate Archimedean copulas. *Aequationes Math.* 64, 175.

K. Gebremeskel and A. Sklar, 1999. Ergodic automorphisms of the n-torus are not embeddable. *Gräzer Math. Ber.* 339, 189–194.

C. Genest and J. MacKay, 1986a. Copules archimédiennes et familles de lois bidimensionnelles dont les marges sont données. *Canadian J. Statist.* 14, 145–159.

———, 1986b. The joy of copulas: bivariate distributions with uniform marginals. *Amer. Statist.* 40, 280–285.

C. Genest, J. J. Quesada-Molina, J. A. Rodriguez-Lallena, and C. Sempi, 1999. A characterization of quasi-copulas. *J. Multivariate Anal.* 69, 193–205.

Tie-Xin Guo, 2001. Survey of recent developments of random metric theory and its applications in China (I), (II). *Acta Analysis Functionalise Applicatae.* 3, 129–158, 208–230.

———, 2004. The theory of random normed modules and its applications, in Peide Liu, ed. *Functional Space Theory and its Applications: Proceedings of International Conference & 13th Academic Symposium in China.* U. K. Research Information Ltd., Hemel Hempstead, 57–66.

Tie-Xin Guo and Shao-Bin Li. The James Theorem in complete random normed modules and its applications. *J. Math. Anal. Appl.*, to appear.

Tie-Xin Guo and Rai-Ping Ma, 2004. Some reviews of various definitions of a random conjugate space together with various kinds of boundedness of a random linear functional. *Acta Analysis Functionalise Applicatae* 6, 16–38.

W. Guz, 1984. Statistical phase spaces, fuzzy sets, and statistical metric spaces. *Found. Phys.* 14, 821–848.

O. Hadžić and E. Pap, 2001. *Fixed Point Theory in Probabilistic Metric Spaces.* Kluwer, Dordrecht.

———, 2005. Triangular norms in probabilistic metric spaces and fixed point theorems. [Conf 2005a] 443–472.

U. Höhle, 2001. *Many Valued Topology and its Applications.* Kluwer, Dordrecht.

———, 2005. Many-valued equalities and their representations. [Conf 2005a] 301–320.

T. P. Hutchinson and C. D. Lai, 1990. *Continuous Bivariate Distributions, Emphasizing Applications.* Rumsby Scientific Publishing, Adelaide.

M. F. Janowitz and B. Schweizer, 1989. Ordinal and percentile clustering. *Math. Social Sciences* 18, 135–186.

W. Jarczyk and J. Matkowski, 2002. On Mulholland's inequality. *Proc. Amer. Math. Soc.* 130, 3243-3247.

H. Joe, 1997. *Multivariate Models and Dependence Concepts.* Chapman & Hall, London.

E. P. Klement, R. Mesiar, and E. Pap, 2000. *Triangular Norms.* Kluwer, Dordrecht.

———, 2004 I. Triangular norms I: Basic analytical and algebraic properties. *Fuzzy Sets and Systems* 143, 5–26.

———, 2004 II. Triangular norms II: General constructions and parameterized families. *Fuzzy Sets and Systems* 145, 411–438.

———, 2004 III. Continuous t-norms. *Fuzzy Sets and Systems* 145, 439–454.

———, 2004. Problems on triangular norms and related operators. *Fuzzy Sets and Systems.* 145, 471–479.

G. M. Krause, 1983. Interior idempotents and non-representability of groupoids. *Stochastica* 7, 5–10.

———. The devil's terraces, a discontinuous associative function, to appear.

M. Kuczma, B. Choczewski, and R. Ger, 1990. *Iterative Functional Equations.* Cambridge University Press.

B. Lafuerza-Guillén, J. A. Rodriguez-Lallena, and C. Sempi, 1997. Some classes of probabilistic normed spaces. *Rend. Mat.* (7) 17, 237–252.

B. Lafuerza-Guillén and C. Sempi, 2003. Probabilistic norms and convergence of random variables. *J. Math. Anal. Appl.* 280, 9–16.

X. Li, P. Mikusiński, H. Sherwood, and M. D. Taylor, 1997. On approximation of copulas. [Conf 1997] 107–116.

G. Liao and Q. Fan, 1998. Minimal subshifts which display Schweizer-Smítal chaos and have zero topological entropy. *Sci. China Ser.* A41, 33–38.

R. Lowen, 1997. *Approach Spaces. The Missing Link in the Topology-Uniformity-Metric Triad.* Clarendon Press, Oxford.

T. B. M. McMaster, 1987. A non-causal triangle function. *Math. Proc. Camb. Phil. Soc.* 101, 287–290.

P. Martins and J. Sousa Ramos, 1999. Topological and geometrical properties of isentropic torus automorphisms. *Gräzer Math. Ber.* 339, 251–260.

P. Mikusiński, H. Sherwood, and M. D. Taylor, 1991. Probabilistic interpretations of copulas and their convex sums. [Conf 1991] 95–112.

————, 1992. Shuffles of Min. *Stochastica* 13, 61–74.

J. Minkler, G. Minkler, and G. D. Richardson, 1998. T-regular probabilistic convergence spaces. *J. Austral. Math. Soc., Ser. A.* 64, 210–221.

H. P. Mulholland, 1950. On generalizations of Minkowski's inequality in the form of a triangle inequality. *Proc. London Math. Soc.* (2) 51, 294–307.

R. B. Nelsen, 1999. *An Introduction to Copulas.* Springer Lecture Notes in Statistics 139. Springer, New York.

————, 2002. Concordance and copulas: a survey. [Conf 2002] 169–177.

————, 2005. Copulas and quasi-copulas: An introduction to their properties and applications. [Conf 2005a] 391–414.

R. B. Nelsen, J. J. Quesada-Molina, B. Schweizer, and C. Sempi, 1996. Derivability of some operations on distribution functions. [Conf 1996] 233–243.

R. B. Nelsen and M. Úbeda-Flores. The lattice-theoretic structure of sets of bivariate copulas and quasi-copulas, to appear.

E. T. Olsen, W. F. Darsow, and B. Nguyen, 1996. Copulas and Markov operators. [Conf 1996] 244–259.

E. Pascali and C. Sempi, 1997. Two Lévy-type metrics for distribution functions. *Richerche Mat.* 46, 49–60.

R. C. Powers, 1988. Order automorphisms of spaces of nondecreasing functions. *J. Math. Anal. Appl.* 136, 112-123.

R. C. Powers and T. Riedel, 1996. When are continuous isotone bijections order automorphisms? *Real Analysis Exchange* 21, 154–164.

E. Prugovečki, 1984. *Stochastic Quantum Mechanics and Quantum Spacetime.* Reidel, Dordrecht.

V. Radu, 1984. On a family of metrics for the distribution functions, in M. Iosifescu, ed. *Proceedings of the Seventh Conference on Probability Theory*. Editura Acad. R. S. România, Bucharest, 487–492.

————, 1986. On obtaining Lévy's metric by probabilistic metrics. *An. Univ. Timisoara, seria st. matematice* 24, 60–65.

————, 1987. Some fixed point theorems in probabilistic metric spaces. *Springer Lecture Notes in Math.* 1233, 125–133.

M. Regenwetter and A. A. J. Marley, 2001. Random relations, random utilities and random functions. *J. Math. Psych* 45, 864–912.

M. Requardt and S. Roy, 2001. Quantum spacetimes as a statistical geometry of fuzzy lumps and the connection with random metric spaces. *Classical Quantum Gravity* 18, 3039–3057.

G. D. Richardson and D. C. Kent, 1996. Probabilistic convergence spaces. *J. Austral. Math. Soc., Ser. A.* 61, 400–420.

T. Riedel, 1991. Cauchy's equation on Δ^+. *Aequationes Math.* 41, 192–211.

———, 1992. Cauchy's equation on Δ^+: further results. *Aequationes Math.* 44, 236–248.

———, 1994. On sup-continuous triangle functions. *J. Math. Anal. Appl.* 184, 382–388.

S. Roy, 1998. *Statistical Geometry and Applications to Microphysics and Cosmology.* Kluwer, Dordrecht.

M. Scarsini, 1984. On measures of concordance. *Stochastica* 8, 201–218.

———, 1989. Copulae of probability measures on product spaces. *J. Multivariate Anal.* 31, 201–219.

B. Schweizer, 1991. Thirty years of copulas. [Conf 1991], 13–50.

———, 1996 The genesis of the notion of distributional chaos. *Rendiconti del Seminario Matematico e Fisico di Milano* 66, 159–167.

———, 2003. Commentary on probabilistic geometry, in B. Schweizer, K. Sigmund, and A. Sklar, eds., *Karl Menger, Selecta Mathematica* V. 2, 409–432. Springer, Vienna.

———, 2005. Triangular norms, looking back—triangle functions, looking ahead. [Conf 2005a] 3-15.

B. Schweizer, H. Sherwood, and R. M. Tardiff, 1988. Contractions on probabilistic metric spaces: examples and counterexamples. *Stochastica* 12, 5–17.

B. Schweizer, A. Sklar, and J. Smítal, 2001/2002. Distributional (and other) chaos and its measurement. *Real Analysis Exchange* 27, 495–524.

B. Schweizer and J. Smítal, 1994. Measures of chaos and a spectral decomposition of dynamical systems in the interval. *Trans. Amer. Math. Soc.* 344, 737–754.

C. Sempi, 1982. On the space of distribution functions. *Riv. Mat. Univ. Parma* (4) 8, 243–250.

———, 1985. Weakly mixing transformations on metric spaces. *Radovi Mat.* 1, 3–7.

———, 2004. Probabilistic metric spaces, in K. P. Hart, J. Nagata, and J. E. Vaughn, eds. *Encyclopedia of General Topology.* Kluwer, Dordrecht.

H. Sherwood, 1984. Characterizing dominates on a family of triangular norms. *Aequationes Math.* 27, 255–273.

H. Sherwood and M. D. Taylor, 1988. Doubly stochastic measures with hairpin support. *Prob. Theory and Rel. Fields* 78, 617–626.

Y.-H. Shyu, 1989. Absolute continuity in the τ_T-operations. Ph. D. Thesis, Illinois Institute of Technology, Chicago.

V. P. Singh, ed., 2005. *Copulas in Hydrology.* Special Issue of the *Journal of Hydrologic Engineering.*

V. P. Singh and L. Zhang, 2004. Stochastic dependence modeling in environmental hydrology. *Proc. Int. Conf. on Hydraulic Engineering; Research and Practice*, V. 1, 46–59. Indian Inst. of Technology, Roorkee, India.

A. Sklar, 1996a. Random variables, distribution functions, and copulas — a personal look backward and forward. [Conf 1996] 1–14.

———, 1996b. Representation of associative functions via simultaneous Schröder equations. *Aequationes Math.* 51, 157–158.

A. Sklar and J. Smítal, 2000. Distributional chaos on compact metric spaces via specification properties. *J. Math. Anal. Appl.* 241, 181–188.

J. Smítal and M. Stefánková, 2004. Distributional chaos for triangular maps. *Chaos, Solitons, and Fractals* 21, 1125–1128.

R. M. Tardiff, 1984. On a generalized Minkowski inequality and its relation to dominates for t-norms. *Aequationes Math.* 27, 308–316.

———,1992. Contraction maps on probabilistic metric spaces. *J. Math. Anal. Appl.* 165, 517–523.

M. D. Taylor, 1985. New metrics for weak convergence of distribution functions. *Stochastica* 9, 5–17.

———, Multivariate measures of concordance, to appear.

E. Trillas and E. Valverde, 1984. An inquiry into indistinguishability operators, in H. J. Skala, S. Termini, and E. Trillas, eds. *Aspects of Vagueness*. Reidel, Dordrecht.

R. A. Vitale, 1990. Stochastic dependence on an interesting class of degenerate distributions, in H. Block, A. R. Sampson, and T. Savits, eds. *Topics in Statistical Dependence.* IMS Lecture Notes-Monograph Series 16. Institute of Mathematical Statistics, Hayward.

M. Vrac, E. Diday, and A. Chédin, 2004. Décomposition de mélange de distributions et application à des données climatiques. *Revue de Statistiques Appliquées*, 52, 67–96.

H. Weinstock, T. Erber, and M. Nisenoff, 1988. Threshold of Barkhausen emission for onset of hysteresis in iron. *Physical Review B*, 31, 1535–1553.

M. Ying, 1989. On a class of non-causal triangle functions. *Math. Proc. Camb. Phil. Soc.* 106, 467–469.

———, 1992. On an open problem by Schweizer and Sklar. *J. Jiangxi Normal University* 16, 112–118.

A CATALOG OF SELECTED

DOVER BOOKS

IN SCIENCE AND MATHEMATICS

Astronomy

CHARIOTS FOR APOLLO: The NASA History of Manned Lunar Spacecraft to 1969, Courtney G. Brooks, James M. Grimwood, and Loyd S. Swenson, Jr. This illustrated history by a trio of experts is the definitive reference on the Apollo spacecraft and lunar modules. It traces the vehicles' design, development, and operation in space. More than 100 photographs and illustrations. 576pp. 6 3/4 x 9 1/4. 0-486-46756-2

EXPLORING THE MOON THROUGH BINOCULARS AND SMALL TELESCOPES, Ernest H. Cherrington, Jr. Informative, profusely illustrated guide to locating and identifying craters, rills, seas, mountains, other lunar features. Newly revised and updated with special section of new photos. Over 100 photos and diagrams. 240pp. 8 1/4 x 11. 0-486-24491-1

WHERE NO MAN HAS GONE BEFORE: A History of NASA's Apollo Lunar Expeditions, William David Compton. Introduction by Paul Dickson. This official NASA history traces behind-the-scenes conflicts and cooperation between scientists and engineers. The first half concerns preparations for the Moon landings, and the second half documents the flights that followed Apollo 11. 1989 edition. 432pp. 7 x 10. 0-486-47888-2

APOLLO EXPEDITIONS TO THE MOON: The NASA History, Edited by Edgar M. Cortright. Official NASA publication marks the 40th anniversary of the first lunar landing and features essays by project participants recalling engineering and administrative challenges. Accessible, jargon-free accounts, highlighted by numerous illustrations. 336pp. 8 3/8 x 10 7/8. 0-486-47175-6

ON MARS: Exploration of the Red Planet, 1958-1978--The NASA History, Edward Clinton Ezell and Linda Neuman Ezell. NASA's official history chronicles the start of our explorations of our planetary neighbor. It recounts cooperation among government, industry, and academia, and it features dozens of photos from Viking cameras. 560pp. 6 3/4 x 9 1/4. 0-486-46757-0

ARISTARCHUS OF SAMOS: The Ancient Copernicus, Sir Thomas Heath. Heath's history of astronomy ranges from Homer and Hesiod to Aristarchus and includes quotes from numerous thinkers, compilers, and scholasticists from Thales and Anaximander through Pythagoras, Plato, Aristotle, and Heraclides. 34 figures. 448pp. 5 3/8 x 8 1/2. 0-486-43886-4

AN INTRODUCTION TO CELESTIAL MECHANICS, Forest Ray Moulton. Classic text still unsurpassed in presentation of fundamental principles. Covers rectilinear motion, central forces, problems of two and three bodies, much more. Includes over 200 problems, some with answers. 437pp. 5 3/8 x 8 1/2. 0-486-64687-4

BEYOND THE ATMOSPHERE: Early Years of Space Science, Homer E. Newell. This exciting survey is the work of a top NASA administrator who chronicles technological advances, the relationship of space science to general science, and the space program's social, political, and economic contexts. 528pp. 6 3/4 x 9 1/4. 0-486-47464-X

STAR LORE: Myths, Legends, and Facts, William Tyler Olcott. Captivating retellings of the origins and histories of ancient star groups include Pegasus, Ursa Major, Pleiades, signs of the zodiac, and other constellations. "Classic." – *Sky & Telescope.* 58 illustrations. 544pp. 5 3/8 x 8 1/2. 0-486-43581-4

A COMPLETE MANUAL OF AMATEUR ASTRONOMY: Tools and Techniques for Astronomical Observations, P. Clay Sherrod with Thomas L. Koed. Concise, highly readable book discusses the selection, set-up, and maintenance of a telescope; amateur studies of the sun; lunar topography and occultations; and more. 124 figures. 26 halftones. 37 tables. 335pp. 6 1/2 x 9 1/4. 0-486-42820-6

Chemistry

Engineering

FUNDAMENTALS OF ASTRODYNAMICS, Roger R. Bate, Donald D. Mueller, and Jerry E. White. Teaching text developed by U.S. Air Force Academy develops the basic two-body and n-body equations of motion; orbit determination; classical orbital elements, coordinate transformations; differential correction; more. 1971 edition. 455pp. 5 3/8 x 8 1/2. 0-486-60061-0

INTRODUCTION TO CONTINUUM MECHANICS FOR ENGINEERS: Revised Edition, Ray M. Bowen. This self-contained text introduces classical continuum models within a modern framework. Its numerous exercises illustrate the governing principles, linearizations, and other approximations that constitute classical continuum models. 2007 edition. 320pp. 6 1/8 x 9 1/4. 0-486-47460-7

ENGINEERING MECHANICS FOR STRUCTURES, Louis L. Bucciarelli. This text explores the mechanics of solids and statics as well as the strength of materials and elasticity theory. Its many design exercises encourage creative initiative and systems thinking. 2009 edition. 320pp. 6 1/8 x 9 1/4. 0-486-46855-0

FEEDBACK CONTROL THEORY, John C. Doyle, Bruce A. Francis and Allen R. Tannenbaum. This excellent introduction to feedback control system design offers a theoretical approach that captures the essential issues and can be applied to a wide range of practical problems. 1992 edition. 224pp. 6 1/2 x 9 1/4. 0-486-46933-6

THE FORCES OF MATTER, Michael Faraday. These lectures by a famous inventor offer an easy-to-understand introduction to the interactions of the universe's physical forces. Six essays explore gravitation, cohesion, chemical affinity, heat, magnetism, and electricity. 1993 edition. 96pp. 5 3/8 x 8 1/2. 0-486-47482-8

DYNAMICS, Lawrence E. Goodman and William H. Warner. Beginning engineering text introduces calculus of vectors, particle motion, dynamics of particle systems and plane rigid bodies, technical applications in plane motions, and more. Exercises and answers in every chapter. 619pp. 5 3/8 x 8 1/2. 0-486-42006-X

ADAPTIVE FILTERING PREDICTION AND CONTROL, Graham C. Goodwin and Kwai Sang Sin. This unified survey focuses on linear discrete-time systems and explores natural extensions to nonlinear systems. It emphasizes discrete-time systems, summarizing theoretical and practical aspects of a large class of adaptive algorithms. 1984 edition. 560pp. 6 1/2 x 9 1/4. 0-486-46932-8

INDUCTANCE CALCULATIONS, Frederick W. Grover. This authoritative reference enables the design of virtually every type of inductor. It features a single simple formula for each type of inductor, together with tables containing essential numerical factors. 1946 edition. 304pp. 5 3/8 x 8 1/2. 0-486-47440-2

THERMODYNAMICS: Foundations and Applications, Elias P. Gyftopoulos and Gian Paolo Beretta. Designed by two MIT professors, this authoritative text discusses basic concepts and applications in detail, emphasizing generality, definitions, and logical consistency. More than 300 solved problems cover realistic energy systems and processes. 800pp. 6 1/8 x 9 1/4. 0-486-43932-1

THE FINITE ELEMENT METHOD: Linear Static and Dynamic Finite Element Analysis, Thomas J. R. Hughes. Text for students without in-depth mathematical training, this text includes a comprehensive presentation and analysis of algorithms of time-dependent phenomena plus beam, plate, and shell theories. Solution guide available upon request. 672pp. 6 1/2 x 9 1/4. 0-486-41181-8

HELICOPTER THEORY, Wayne Johnson. Monumental engineering text covers vertical flight, forward flight, performance, mathematics of rotating systems, rotary wing dynamics and aerodynamics, aeroelasticity, stability and control, stall, noise, and more. 189 illustrations. 1980 edition. 1089pp. 5 5/8 x 8 1/4. 0-486-68230-7

MATHEMATICAL HANDBOOK FOR SCIENTISTS AND ENGINEERS: Definitions, Theorems, and Formulas for Reference and Review, Granino A. Korn and Theresa M. Korn. Convenient access to information from every area of mathematics: Fourier transforms, Z transforms, linear and nonlinear programming, calculus of variations, random-process theory, special functions, combinatorial analysis, game theory, much more. 1152pp. 5 3/8 x 8 1/2. 0-486-41147-8

A HEAT TRANSFER TEXTBOOK: Fourth Edition, John H. Lienhard V and John H. Lienhard IV. This introduction to heat and mass transfer for engineering students features worked examples and end-of-chapter exercises. Worked examples and end-of-chapter exercises appear throughout the book, along with well-drawn, illuminating figures. 768pp. 7 x 9 1/4. 0-486-47931-5

BASIC ELECTRICITY, U.S. Bureau of Naval Personnel. Originally a training course; best nontechnical coverage. Topics include batteries, circuits, conductors, AC and DC, inductance and capacitance, generators, motors, transformers, amplifiers, etc. Many questions with answers. 349 illustrations. 1969 edition. 448pp. 6 1/2 x 9 1/4. 0-486-20973-3

BASIC ELECTRONICS, U.S. Bureau of Naval Personnel. Clear, well-illustrated introduction to electronic equipment covers numerous essential topics: electron tubes, semiconductors, electronic power supplies, tuned circuits, amplifiers, receivers, ranging and navigation systems, computers, antennas, more. 560 illustrations. 567pp. 6 1/2 x 9 1/4. 0-486-21076-6

BASIC WING AND AIRFOIL THEORY, Alan Pope. This self-contained treatment by a pioneer in the study of wind effects covers flow functions, airfoil construction and pressure distribution, finite and monoplane wings, and many other subjects. 1951 edition. 320pp. 5 3/8 x 8 1/2. 0-486-47188-8

SYNTHETIC FUELS, Ronald F. Probstein and R. Edwin Hicks. This unified presentation examines the methods and processes for converting coal, oil, shale, tar sands, and various forms of biomass into liquid, gaseous, and clean solid fuels. 1982 edition. 512pp. 6 1/8 x 9 1/4. 0-486-44977-7

THEORY OF ELASTIC STABILITY, Stephen P. Timoshenko and James M. Gere. Written by world-renowned authorities on mechanics, this classic ranges from theoretical explanations of 2- and 3-D stress and strain to practical applications such as torsion, bending, and thermal stress. 1961 edition. 560pp. 5 3/8 x 8 1/2. 0-486-47207-8

PRINCIPLES OF DIGITAL COMMUNICATION AND CODING, Andrew J. Viterbi and Jim K. Omura. This classic by two digital communications experts is geared toward students of communications theory and to designers of channels, links, terminals, modems, or networks used to transmit and receive digital messages. 1979 edition. 576pp. 6 1/8 x 9 1/4. 0-486-46901-8

LINEAR SYSTEM THEORY: The State Space Approach, Lotfi A. Zadeh and Charles A. Desoer. Written by two pioneers in the field, this exploration of the state space approach focuses on problems of stability and control, plus connections between this approach and classical techniques. 1963 edition. 656pp. 6 1/8 x 9 1/4. 0-486-46663-9

Mathematics–Logic and Problem Solving

PERPLEXING PUZZLES AND TANTALIZING TEASERS, Martin Gardner. Ninety-three riddles, mazes, illusions, tricky questions, word and picture puzzles, and other challenges offer hours of entertainment for youngsters. Filled with rib-tickling drawings. Solutions. 224pp. 5 3/8 x 8 1/2. 0-486-25637-5

MY BEST MATHEMATICAL AND LOGIC PUZZLES, Martin Gardner. The noted expert selects 70 of his favorite "short" puzzles. Includes The Returning Explorer, The Mutilated Chessboard, Scrambled Box Tops, and dozens more. Complete solutions included. 96pp. 5 3/8 x 8 1/2. 0-486-28152-3

THE LADY OR THE TIGER?: and Other Logic Puzzles, Raymond M. Smullyan. Created by a renowned puzzle master, these whimsically themed challenges involve paradoxes about probability, time, and change; metapuzzles; and self-referentiality. Nineteen chapters advance in difficulty from relatively simple to highly complex. 1982 edition. 240pp. 5 3/8 x 8 1/2. 0-486-47027-X

SATAN, CANTOR AND INFINITY: Mind-Boggling Puzzles, Raymond M. Smullyan. A renowned mathematician tells stories of knights and knaves in an entertaining look at the logical precepts behind infinity, probability, time, and change. Requires a strong background in mathematics. Complete solutions. 288pp. 5 3/8 x 8 1/2. 0-486-47036-9

THE RED BOOK OF MATHEMATICAL PROBLEMS, Kenneth S. Williams and Kenneth Hardy. Handy compilation of 100 practice problems, hints and solutions indispensable for students preparing for the William Lowell Putnam and other mathematical competitions. Preface to the First Edition. Sources. 1988 edition. 192pp. 5 3/8 x 8 1/2. 0-486-69415-1

KING ARTHUR IN SEARCH OF HIS DOG AND OTHER CURIOUS PUZZLES, Raymond M. Smullyan. This fanciful, original collection for readers of all ages features arithmetic puzzles, logic problems related to crime detection, and logic and arithmetic puzzles involving King Arthur and his Dogs of the Round Table. 160pp. 5 3/8 x 8 1/2. 0-486-47435-6

UNDECIDABLE THEORIES: Studies in Logic and the Foundation of Mathematics, Alfred Tarski in collaboration with Andrzej Mostowski and Raphael M. Robinson. This well-known book by the famed logician consists of three treatises: "A General Method in Proofs of Undecidability," "Undecidability and Essential Undecidability in Mathematics," and "Undecidability of the Elementary Theory of Groups." 1953 edition. 112pp. 5 3/8 x 8 1/2. 0-486-47703-7

LOGIC FOR MATHEMATICIANS, J. Barkley Rosser. Examination of essential topics and theorems assumes no background in logic. "Undoubtedly a major addition to the literature of mathematical logic." – *Bulletin of the American Mathematical Society*. 1978 edition. 592pp. 6 1/8 x 9 1/4. 0-486-46898-4

INTRODUCTION TO PROOF IN ABSTRACT MATHEMATICS, Andrew Wohlgemuth. This undergraduate text teaches students what constitutes an acceptable proof, and it develops their ability to do proofs of routine problems as well as those requiring creative insights. 1990 edition. 384pp. 6 1/2 x 9 1/4. 0-486-47854-8

FIRST COURSE IN MATHEMATICAL LOGIC, Patrick Suppes and Shirley Hill. Rigorous introduction is simple enough in presentation and context for wide range of students. Symbolizing sentences; logical inference; truth and validity; truth tables; terms, predicates, universal quantifiers; universal specification and laws of identity; more. 288pp. 5 3/8 x 8 1/2. 0-486-42259-3

Mathematics–Algebra and Calculus

VECTOR CALCULUS, Peter Baxandall and Hans Liebeck. This introductory text offers a rigorous, comprehensive treatment. Classical theorems of vector calculus are amply illustrated with figures, worked examples, physical applications, and exercises with hints and answers. 1986 edition. 560pp. 5 3/8 x 8 1/2. 0-486-46620-5

ADVANCED CALCULUS: An Introduction to Classical Analysis, Louis Brand. A course in analysis that focuses on the functions of a real variable, this text introduces the basic concepts in their simplest setting and illustrates its teachings with numerous examples, theorems, and proofs. 1955 edition. 592pp. 5 3/8 x 8 1/2. 0-486-44548-8

ADVANCED CALCULUS, Avner Friedman. Intended for students who have already completed a one-year course in elementary calculus, this two-part treatment advances from functions of one variable to those of several variables. Solutions. 1971 edition. 432pp. 5 3/8 x 8 1/2. 0-486-45795-8

METHODS OF MATHEMATICS APPLIED TO CALCULUS, PROBABILITY, AND STATISTICS, Richard W. Hamming. This 4-part treatment begins with algebra and analytic geometry and proceeds to an exploration of the calculus of algebraic functions and transcendental functions and applications. 1985 edition. Includes 310 figures and 18 tables. 880pp. 6 1/2 x 9 1/4. 0-486-43945-3

BASIC ALGEBRA I: Second Edition, Nathan Jacobson. A classic text and standard reference for a generation, this volume covers all undergraduate algebra topics, including groups, rings, modules, Galois theory, polynomials, linear algebra, and associative algebra. 1985 edition. 528pp. 6 1/8 x 9 1/4. 0-486-47189-6

BASIC ALGEBRA II: Second Edition, Nathan Jacobson. This classic text and standard reference comprises all subjects of a first-year graduate-level course, including in-depth coverage of groups and polynomials and extensive use of categories and functors. 1989 edition. 704pp. 6 1/8 x 9 1/4. 0-486-47187-X

CALCULUS: An Intuitive and Physical Approach (Second Edition), Morris Kline. Application-oriented introduction relates the subject as closely as possible to science with explorations of the derivative; differentiation and integration of the powers of x; theorems on differentiation, antidifferentiation; the chain rule; trigonometric functions; more. Examples. 1967 edition. 960pp. 6 1/2 x 9 1/4. 0-486-40453-6

ABSTRACT ALGEBRA AND SOLUTION BY RADICALS, John E. Maxfield and Margaret W. Maxfield. Accessible advanced undergraduate-level text starts with groups, rings, fields, and polynomials and advances to Galois theory, radicals and roots of unity, and solution by radicals. Numerous examples, illustrations, exercises, appendixes. 1971 edition. 224pp. 6 1/8 x 9 1/4. 0-486-47723-1

AN INTRODUCTION TO THE THEORY OF LINEAR SPACES, Georgi E. Shilov. Translated by Richard A. Silverman. Introductory treatment offers a clear exposition of algebra, geometry, and analysis as parts of an integrated whole rather than separate subjects. Numerous examples illustrate many different fields, and problems include hints or answers. 1961 edition. 320pp. 5 3/8 x 8 1/2. 0-486-63070-6

LINEAR ALGEBRA, Georgi E. Shilov. Covers determinants, linear spaces, systems of linear equations, linear functions of a vector argument, coordinate transformations, the canonical form of the matrix of a linear operator, bilinear and quadratic forms, and more. 387pp. 5 3/8 x 8 1/2. 0-486-63518-X

Mathematics–Geometry and Topology

PROBLEMS AND SOLUTIONS IN EUCLIDEAN GEOMETRY, M. N. Aref and William Wernick. Based on classical principles, this book is intended for a second course in Euclidean geometry and can be used as a refresher. More than 200 problems include hints and solutions. 1968 edition. 272pp. 5 3/8 x 8 1/2. 0-486-47720-7

TOPOLOGY OF 3-MANIFOLDS AND RELATED TOPICS, Edited by M. K. Fort, Jr. With a New Introduction by Daniel Silver. Summaries and full reports from a 1961 conference discuss decompositions and subsets of 3-space; n-manifolds; knot theory; the Poincaré conjecture; and periodic maps and isotopies. Familiarity with algebraic topology required. 1962 edition. 272pp. 6 1/8 x 9 1/4. 0-486-47753-3

POINT SET TOPOLOGY, Steven A. Gaal. Suitable for a complete course in topology, this text also functions as a self-contained treatment for independent study. Additional enrichment materials make it equally valuable as a reference. 1964 edition. 336pp. 5 3/8 x 8 1/2. 0-486-47222-1

INVITATION TO GEOMETRY, Z. A. Melzak. Intended for students of many different backgrounds with only a modest knowledge of mathematics, this text features self-contained chapters that can be adapted to several types of geometry courses. 1983 edition. 240pp. 5 3/8 x 8 1/2. 0-486-46626-4

TOPOLOGY AND GEOMETRY FOR PHYSICISTS, Charles Nash and Siddhartha Sen. Written by physicists for physics students, this text assumes no detailed background in topology or geometry. Topics include differential forms, homotopy, homology, cohomology, fiber bundles, connection and covariant derivatives, and Morse theory. 1983 edition. 320pp. 5 3/8 x 8 1/2. 0-486-47852-1

BEYOND GEOMETRY: Classic Papers from Riemann to Einstein, Edited with an Introduction and Notes by Peter Pesic. This is the only English-language collection of these 8 accessible essays. They trace seminal ideas about the foundations of geometry that led to Einstein's general theory of relativity. 224pp. 6 1/8 x 9 1/4. 0-486-45350-2

GEOMETRY FROM EUCLID TO KNOTS, Saul Stahl. This text provides a historical perspective on plane geometry and covers non-neutral Euclidean geometry, circles and regular polygons, projective geometry, symmetries, inversions, informal topology, and more. Includes 1,000 practice problems. Solutions available. 2003 edition. 480pp. 6 1/8 x 9 1/4. 0-486-47459-3

TOPOLOGICAL VECTOR SPACES, DISTRIBUTIONS AND KERNELS, François Trèves. Extending beyond the boundaries of Hilbert and Banach space theory, this text focuses on key aspects of functional analysis, particularly in regard to solving partial differential equations. 1967 edition. 592pp. 5 3/8 x 8 1/2. 0-486-45352-9

INTRODUCTION TO PROJECTIVE GEOMETRY, C. R. Wylie, Jr. This introductory volume offers strong reinforcement for its teachings, with detailed examples and numerous theorems, proofs, and exercises, plus complete answers to all odd-numbered end-of-chapter problems. 1970 edition. 576pp. 6 1/8 x 9 1/4. 0-486-46895-X

FOUNDATIONS OF GEOMETRY, C. R. Wylie, Jr. Geared toward students preparing to teach high school mathematics, this text explores the principles of Euclidean and non-Euclidean geometry and covers both generalities and specifics of the axiomatic method. 1964 edition. 352pp. 6 x 9. 0-486-47214-0

Mathematics-History

THE WORKS OF ARCHIMEDES, Archimedes. Translated by Sir Thomas Heath. Complete works of ancient geometer feature such topics as the famous problems of the ratio of the areas of a cylinder and an inscribed sphere; the properties of conoids, spheroids, and spirals; more. 326pp. 5 3/8 x 8 1/2. 0-486-42084-1

THE HISTORICAL ROOTS OF ELEMENTARY MATHEMATICS, Lucas N. H. Bunt, Phillip S. Jones, and Jack D. Bedient. Exciting, hands-on approach to understanding fundamental underpinnings of modern arithmetic, algebra, geometry and number systems examines their origins in early Egyptian, Babylonian, and Greek sources. 336pp. 5 3/8 x 8 1/2. 0-486-25563-8

THE THIRTEEN BOOKS OF EUCLID'S ELEMENTS, Euclid. Contains complete English text of all 13 books of the Elements plus critical apparatus analyzing each definition, postulate, and proposition in great detail. Covers textual and linguistic matters; mathematical analyses of Euclid's ideas; classical, medieval, Renaissance and modern commentators; refutations, supports, extrapolations, reinterpretations and historical notes. 995 figures. Total of 1,425pp. All books 5 3/8 x 8 1/2.

Vol. I: 443pp. 0-486-60088-2
Vol. II: 464pp. 0-486-60089-0
Vol. III: 546pp. 0-486-60090-4

A HISTORY OF GREEK MATHEMATICS, Sir Thomas Heath. This authoritative two-volume set that covers the essentials of mathematics and features every landmark innovation and every important figure, including Euclid, Apollonius, and others. 5 3/8 x 8 1/2. Vol. I: 461pp. 0-486-24073-8
Vol. II: 597pp. 0-486-24074-6

A MANUAL OF GREEK MATHEMATICS, Sir Thomas L. Heath. This concise but thorough history encompasses the enduring contributions of the ancient Greek mathematicians whose works form the basis of most modern mathematics. Discusses Pythagorean arithmetic, Plato, Euclid, more. 1931 edition. 576pp. 5 3/8 x 8 1/2.
0-486-43231-9

CHINESE MATHEMATICS IN THE THIRTEENTH CENTURY, Ulrich Libbrecht. An exploration of the 13th-century mathematician Ch'in, this fascinating book combines what is known of the mathematician's life with a history of his only extant work, the Shu-shu chiu-chang. 1973 edition. 592pp. 5 3/8 x 8 1/2.
0-486-44619-0

PHILOSOPHY OF MATHEMATICS AND DEDUCTIVE STRUCTURE IN EUCLID'S ELEMENTS, Ian Mueller. This text provides an understanding of the classical Greek conception of mathematics as expressed in Euclid's Elements. It focuses on philosophical, foundational, and logical questions and features helpful appendixes. 400pp. 6 1/2 x 9 1/4. 0-486-45300-6

BEYOND GEOMETRY: Classic Papers from Riemann to Einstein, Edited with an Introduction and Notes by Peter Pesic. This is the only English-language collection of these 8 accessible essays. They trace seminal ideas about the foundations of geometry that led to Einstein's general theory of relativity. 224pp. 6 1/8 x 9 1/4. 0-486-45350-2

HISTORY OF MATHEMATICS, David E. Smith. Two-volume history – from Egyptian papyri and medieval maps to modern graphs and diagrams. Non-technical chronological survey with thousands of biographical notes, critical evaluations, and contemporary opinions on over 1,100 mathematicians. 5 3/8 x 8 1/2.

Vol. I: 618pp. 0-486-20429-4
Vol. II: 736pp. 0-486-20430-8

Physics

THEORETICAL NUCLEAR PHYSICS, John M. Blatt and Victor F. Weisskopf. An uncommonly clear and cogent investigation and correlation of key aspects of theoretical nuclear physics by leading experts: the nucleus, nuclear forces, nuclear spectroscopy, two-, three- and four-body problems, nuclear reactions, beta-decay and nuclear shell structure. 896pp. 5 3/8 x 8 1/2. 0-486-66827-4

QUANTUM THEORY, David Bohm. This advanced undergraduate-level text presents the quantum theory in terms of qualitative and imaginative concepts, followed by specific applications worked out in mathematical detail. 655pp. 5 3/8 x 8 1/2. 0-486-65969-0

ATOMIC PHYSICS AND HUMAN KNOWLEDGE, Niels Bohr. Articles and speeches by the Nobel Prize–winning physicist, dating from 1934 to 1958, offer philosophical explorations of the relevance of atomic physics to many areas of human endeavor. 1961 edition. 112pp. 5 3/8 x 8 1/2. 0-486-47928-5

COSMOLOGY, Hermann Bondi. A co-developer of the steady-state theory explores his conception of the expanding universe. This historic book was among the first to present cosmology as a separate branch of physics. 1961 edition. 192pp. 5 3/8 x 8 1/2. 0-486-47483-6

LECTURES ON QUANTUM MECHANICS, Paul A. M. Dirac. Four concise, brilliant lectures on mathematical methods in quantum mechanics from Nobel Prize-winning quantum pioneer build on idea of visualizing quantum theory through the use of classical mechanics. 96pp. 5 3/8 x 8 1/2. 0-486-41713-1

THE PRINCIPLE OF RELATIVITY, Albert Einstein and Frances A. Davis. Eleven papers that forged the general and special theories of relativity include seven papers by Einstein, two by Lorentz, and one each by Minkowski and Weyl. 1923 edition. 240pp. 5 3/8 x 8 1/2. 0-486-60081-5

PHYSICS OF WAVES, William C. Elmore and Mark A. Heald. Ideal as a classroom text or for individual study, this unique one-volume overview of classical wave theory covers wave phenomena of acoustics, optics, electromagnetic radiations, and more. 477pp. 5 3/8 x 8 1/2. 0-486-64926-1

THERMODYNAMICS, Enrico Fermi. In this classic of modern science, the Nobel Laureate presents a clear treatment of systems, the First and Second Laws of Thermodynamics, entropy, thermodynamic potentials, and much more. Calculus required. 160pp. 5 3/8 x 8 1/2. 0-486-60361-X

QUANTUM THEORY OF MANY-PARTICLE SYSTEMS, Alexander L. Fetter and John Dirk Walecka. Self-contained treatment of nonrelativistic many-particle systems discusses both formalism and applications in terms of ground-state (zero-temperature) formalism, finite-temperature formalism, canonical transformations, and applications to physical systems. 1971 edition. 640pp. 5 3/8 x 8 1/2. 0-486-42827-3

QUANTUM MECHANICS AND PATH INTEGRALS: Emended Edition, Richard P. Feynman and Albert R. Hibbs. Emended by Daniel F. Styer. The Nobel Prize–winning physicist presents unique insights into his theory and its applications. Feynman starts with fundamentals and advances to the perturbation method, quantum electrodynamics, and statistical mechanics. 1965 edition, emended in 2005. 384pp. 6 1/8 x 9 1/4. 0-486-47722-3

Physics

INTRODUCTION TO MODERN OPTICS, Grant R. Fowles. A complete basic undergraduate course in modern optics for students in physics, technology, and engineering. The first half deals with classical physical optics; the second, quantum nature of light. Solutions. 336pp. 5 3/8 x 8 1/2. 0-486-65957-7

THE QUANTUM THEORY OF RADIATION: Third Edition, W. Heitler. The first comprehensive treatment of quantum physics in any language, this classic introduction to basic theory remains highly recommended and widely used, both as a text and as a reference. 1954 edition. 464pp. 5 3/8 x 8 1/2. 0-486-64558-4

QUANTUM FIELD THEORY, Claude Itzykson and Jean-Bernard Zuber. This comprehensive text begins with the standard quantization of electrodynamics and perturbative renormalization, advancing to functional methods, relativistic bound states, broken symmetries, nonabelian gauge fields, and asymptotic behavior. 1980 edition. 752pp. 6 1/2 x 9 1/4. 0-486-44568-2

FOUNDATIONS OF POTENTIAL THERY, Oliver D. Kellogg. Introduction to fundamentals of potential functions covers the force of gravity, fields of force, potentials, harmonic functions, electric images and Green's function, sequences of harmonic functions, fundamental existence theorems, and much more. 400pp. 5 3/8 x 8 1/2. 0-486-60144-7

FUNDAMENTALS OF MATHEMATICAL PHYSICS, Edgar A. Kraut. Indispensable for students of modern physics, this text provides the necessary background in mathematics to study the concepts of electromagnetic theory and quantum mechanics. 1967 edition. 480pp. 6 1/2 x 9 1/4. 0-486-45809-1

GEOMETRY AND LIGHT: The Science of Invisibility, Ulf Leonhardt and Thomas Philbin. Suitable for advanced undergraduate and graduate students of engineering, physics, and mathematics and scientific researchers of all types, this is the first authoritative text on invisibility and the science behind it. More than 100 full-color illustrations, plus exercises with solutions. 2010 edition. 288pp. 7 x 9 1/4. 0-486-47693-6

QUANTUM MECHANICS: New Approaches to Selected Topics, Harry J. Lipkin. Acclaimed as "excellent" (*Nature*) and "very original and refreshing" (*Physics Today*), these studies examine the Mössbauer effect, many-body quantum mechanics, scattering theory, Feynman diagrams, and relativistic quantum mechanics. 1973 edition. 480pp. 5 3/8 x 8 1/2. 0-486-45893-8

THEORY OF HEAT, James Clerk Maxwell. This classic sets forth the fundamentals of thermodynamics and kinetic theory simply enough to be understood by beginners, yet with enough subtlety to appeal to more advanced readers, too. 352pp. 5 3/8 x 8 1/2. 0-486-41735-2

QUANTUM MECHANICS, Albert Messiah. Subjects include formalism and its interpretation, analysis of simple systems, symmetries and invariance, methods of approximation, elements of relativistic quantum mechanics, much more. "Strongly recommended." – *American Journal of Physics*. 1152pp. 5 3/8 x 8 1/2. 0-486-40924-4

RELATIVISTIC QUANTUM FIELDS, Charles Nash. This graduate-level text contains techniques for performing calculations in quantum field theory. It focuses chiefly on the dimensional method and the renormalization group methods. Additional topics include functional integration and differentiation. 1978 edition. 240pp. 5 3/8 x 8 1/2. 0-486-47752-5

Physics

MATHEMATICAL TOOLS FOR PHYSICS, James Nearing. Encouraging students' development of intuition, this original work begins with a review of basic mathematics and advances to infinite series, complex algebra, differential equations, Fourier series, and more. 2010 edition. 496pp. 6 1/8 x 9 1/4. 0-486-48212-X

TREATISE ON THERMODYNAMICS, Max Planck. Great classic, still one of the best introductions to thermodynamics. Fundamentals, first and second principles of thermodynamics, applications to special states of equilibrium, more. Numerous worked examples. 1917 edition. 297pp. 5 3/8 x 8. 0-486-66371-X

AN INTRODUCTION TO RELATIVISTIC QUANTUM FIELD THEORY, Silvan S. Schweber. Complete, systematic, and self-contained, this text introduces modern quantum field theory. "Combines thorough knowledge with a high degree of didactic ability and a delightful style." – *Mathematical Reviews.* 1961 edition. 928pp. 5 3/8 x 8 1/2. 0-486-44228-4

THE ELECTROMAGNETIC FIELD, Albert Shadowitz. Comprehensive undergraduate text covers basics of electric and magnetic fields, building up to electromagnetic theory. Related topics include relativity theory. Over 900 problems, some with solutions. 1975 edition. 768pp. 5 5/8 x 8 1/4. 0-486-65660-8

THE PRINCIPLES OF STATISTICAL MECHANICS, Richard C. Tolman. Definitive treatise offers a concise exposition of classical statistical mechanics and a thorough elucidation of quantum statistical mechanics, plus applications of statistical mechanics to thermodynamic behavior. 1930 edition. 704pp. 5 5/8 x 8 1/4. 0-486-63896-0

INTRODUCTION TO THE PHYSICS OF FLUIDS AND SOLIDS, James S. Trefil. This interesting, informative survey by a well-known science author ranges from classical physics and geophysical topics, from the rings of Saturn and the rotation of the galaxy to underground nuclear tests. 1975 edition. 320pp. 5 3/8 x 8 1/2. 0-486-47437-2

STATISTICAL PHYSICS, Gregory H. Wannier. Classic text combines thermodynamics, statistical mechanics, and kinetic theory in one unified presentation. Topics include equilibrium statistics of special systems, kinetic theory, transport coefficients, and fluctuations. Problems with solutions. 1966 edition. 532pp. 5 3/8 x 8 1/2. 0-486-65401-X

SPACE, TIME, MATTER, Hermann Weyl. Excellent introduction probes deeply into Euclidean space, Riemann's space, Einstein's general relativity, gravitational waves and energy, and laws of conservation. "A classic of physics." – *British Journal for Philosophy and Science.* 330pp. 5 3/8 x 8 1/2. 0-486-60267-2

RANDOM VIBRATIONS: Theory and Practice, Paul H. Wirsching, Thomas L. Paez and Keith Ortiz. Comprehensive text and reference covers topics in probability, statistics, and random processes, plus methods for analyzing and controlling random vibrations. Suitable for graduate students and mechanical, structural, and aerospace engineers. 1995 edition. 464pp. 5 3/8 x 8 1/2. 0-486-45015-5

PHYSICS OF SHOCK WAVES AND HIGH-TEMPERATURE HYDRO DYNAMIC PHENOMENA, Ya B. Zel'dovich and Yu P. Raizer. Physical, chemical processes in gases at high temperatures are focus of outstanding text, which combines material from gas dynamics, shock-wave theory, thermodynamics and statistical physics, other fields. 284 illustrations. 1966–1967 edition. 944pp. 6 1/8 x 9 1/4. 0-486-42002-7